Technology, Science and Culture - A Global Vision, Volume V

Edited by Israel Cedillo Lazcano and Martín Alejandro Serrano Meneses

Published in London, United Kingdom

Technology, Science and Culture - A Global Vision, Volume V
http://dx.doi.org/10.5772/intechopen.113949
Edited by Israel Cedillo Lazcano and Martín Alejandro Serrano Meneses

Contributors
Gabriela Stefanie Solís Santos, Juana Beatriz Romero Rivera

First published in London, United Kingdom, 2025 by IntechOpen
IntechOpen is the global imprint of INTECHOPEN LIMITED, registered in England and Wales, registration number: 11086078, 167-169 Great Portland Street, London, W1W 5PF, United Kingdom

For EU product safety concerns: IN TECH d.o.o., Prolaz Marije Krucifikse Kozulić 3, 51000 Rijeka, Croatia, info@intechopen.com or visit our website at intechopen.com.

British Library Cataloguing-in-Publication Data
A catalogue record for this book is available from the British Library

Technology, Science and Culture - A Global Vision, Volume V
Edited by Israel Cedillo Lazcano and Martín Alejandro Serrano Meneses
p. cm.
Print ISBN 978-0-85014-219-8
Online ISBN 978-0-85014-220-4
eBook (PDF) ISBN 978-0-85014-221-1

UDLAP®

Universidad de las Américas Puebla

Technology, Science and Culture – A Global Vision, Volume V

IntechOpen

Technology, Science and Culture – A Global Vision, Volume V
2023

Editors
Israel Cedillo Lazcano
Martín Alejandro Serrano Meneses

Knowledge area co-editors
Aura Matilde Jiménez Garduño
Nelly Ramírez Corona
José Luis Sánchez Salas
Enrique Ajuria Ibarra
Zobeida Jezabel Guzmán Zavaleta

Contributors
Gabriela Stefanie Solis Santos, Juana Beatriz Romero Rivera

unesco Chair

UDLAP®

UNESCO Chair on
Hydrometeorogical Risks,
Universidad de las Américas Puebla

The number and impact of water-related natural disasters has increased since the middle of last century. As result of increased climate variability and the effects of global warming, the hydrometeorological risk has increased and spread, while the resilience of societies, in many cases, is not adequate. Consequently, the risk has increased. Floods and droughts, particularly in a changing climate, require greater understanding to generate better forecasts and proper management of these phenomena. Mexico, like other countries in the world, and of course in Latin America and the Caribbean region, suffers from both weather extremes.

The UNESCO Chair on Hydrometeorological Risks, held at the University of Americas Puebla, is devoted to the analysis, measurement, modelling and management of extreme hydrometeorological events in the context of a more urbanized world, climate change and further vulnerable regions. Focused on the development of basic and applied research for the design of adaptation and mitigation measures, dissemination and preparation of decision makers as well as the public. In its activities keeps a gender focus, directed in particular to reduce the vulnerability of women to hydrometeorological disasters.

The Chair acts in the following fields:

1. Hydrometeorological risks and climate change.

2. Modelling and forecasting of hydrometeorological risks.

3. Integrated management of hydrometeorological risks.

4. Gender and hydrometeorological risks.

A detailed description of the UNESCO Chair on Hydrometeorological Risks, members and publications, can be obtained at its Website https://www.udlap.mx/catedraunesco/

The Chair publishes a quarterly Newsletter, in Spanish and English, that can be consulted at https://www.udlap.mx/catedraunesco/newsletters.aspx

Contents

Preface

This new volume of the series "Technology, Science and Culture. A Global Vision" that emerges from the efforts of PhD students and researchers at different stages of their respective careers, continues with the multidisciplinary set of contributions that has define it since its conception. As the reader can verify, following the structure of the previous volumes, the research pieces found within this work are related to the areas of Food Sciences, Smart Systems, Molecular Biomedicine, Creation and Theories of Culture, and Water Sciences, and they offer different views of different challenges that are not isolated among them. If one reads carefully, one can identify common elements that offer a systemic view of the risks and the opportunities that our societies face around the orb. For instance, we can open the book and enjoy with a relevant introduction to our times on evolutionary computation, which will help us to identify the leitmotiv that inspires the development of forecasting systems, pharmaceuticals, and even literary works following the tradition of authors like Lovecraft.

At the same time, we can identify how the technologies that are used to define the Fourth Industrial Revolution are employed to improve the management of basic and valuable resources, such as, water and the development of new techniques to address health issues and, of course, of food alternatives and complements to improve the quality of life.

From these contributions, I am certain that our readership will learn about novel and rather interesting issues, and find inspiration to research and work on the lines presented herein.

Chapter

An Introduction to Evolutionary Computation

Carlos A. Coello Coello

1. Introduction

Evolutionary computation is a research area within computer science that focuses on the study of the properties of a series of stochastic metaheuristics (which are generically called "Evolutionary Algorithms"), which are inspired on the "survival of the fittest" principle from Charles Darwin's theory of evolution [1].

A metaheuristic is a high-level procedure that applies a rule or set of rules which are based on a knowledge source [2]. The main emphasis of metaheuristics is to explore the search space in a relatively efficient manner (i.e., more efficiently than using exhaustive search), although not necessarily optimum. Metaheuristics are often considered as search and optimization strategies for general use in spite of their theoretical limitations [3]. Furthermore, they normally require little information about the domain and are very useful when the search space is very large, little known and/or difficult to navigate. When used for optimization, they cannot guarantee, in general, convergence to the best possible solution (i.e., to the global optimum of the problem), but in practice, they normally generate reasonably good approximations of the global optimum in reasonably short computational times.

Evolutionary Algorithms (EAs) are stochastic metaheuristics because their behavior is based on the use of random numbers and, therefore, they do not necessarily generate the same result each time they are executed. EAs have been successfully applied to a wide variety of problems [4]. Over the years, at least two different types of EAs have been proposed based on the type of problem to be solved [5]: (1) those designed to solve optimization problems (mainly nonlinear and combinatorial) and (2) those designed to solve machine learning problems (i.e., classification).

However, research in evolutionary computation is not limited to the development of new algorithms, since it also includes the design of new operators, new selection mechanisms and new encodings. Additionally, other topics of interest include theoretical aspects of EAs, study of the sources of difficulty of a search space and their correlation with the operators of an EA and hybrids of EAs and other types of techniques, among many others.

The remainder of this chapter is organized as follows. First, we will review some of the history of EAs. Then, we will describe the three main types of EAs that were originally proposed (i.e., evolution strategies, evolutionary programming and genetic algorithms), as well as a very popular variant of the genetic algorithm called "genetic programming". The descriptions provided include some relevant applications as well as some of the related research that has been conducted in Mexico. In the final part of the chapter, other bio-inspired metaheuristics which have been adopted in Mexico are also briefly discussed. Additionally, we provide a short overview of the historical origins of research on EAs in Mexico. The chapter concludes with some future perspectives in the area.

2. Historical highlights of evolutionary algorithms

Evolutionary algorithms have a long and fascinating history. In their origins, several researchers proposed, in a completely independent way, different computational models with specific goals, all of which had in common that they were inspired on Darwin's theory of natural evolution. Over time, the evident similarities among these techniques led to the use of the generic term "Evolutionary Algorithm" in the 1990s [6]. In this section, we will briefly discuss some of the most relevant historical highlights of Evolutionary Algorithms. It is worth indicating that most of the foundations of computer science were established in the 1960s and 1970s, which is also the historical period in which the three most popular Evolutionary Algorithms' paradigms were proposed: (1) evolutionary programming, (2) evolution strategies and (3) genetic algorithms. Although these three techniques were inspired on the same ideas, their details are different, including the use of different encoding schemes and different types of operators.

Additionally, we will discuss other proposals that are considered relevant because of their historical relevance, in spite of the fact that their use has never been popular.

Wright [7] and Cannon [8] were the first researchers who saw natural evolution as a learning process in which the genetic information of the species continuously changes through a trial-and-error process. This way, evolution can be seen as a mechanism whose goal is to maximize the adaptation capacity of species in their environment. Wright went farther and introduced the concept of "fitness landscape", as a way of representing the space of all the possible genotypes together with their fitness values. This concept has been extensively adopted to study the behavior of EAs [9]. Additionally, Wright developed one of the earliest studies that related selection with mutation and concluded that these must have a proper balance in order to have a successful evolutionary process. Wright's ideas were extended by Campbell [10], who claimed that blind variation combined with a proper selection process constitutes the fundamental principle of evolution.

Turing can also be considered as an evolutionary computation pioneer, since he recognized a connection (which seemed obvious to him) between evolution and machine learning [11]. Specifically, Turing claimed that, for developing a computer that could pass his famous (now called Turing) test, so that a computer could be considered to be "intelligent", it could be required to develop a learning scheme based on natural evolution. Turing had suggested before that methods based on evolution could be used to train a set of networks analogous to what we call today "neural networks" [12]. In fact, Turing even used the term "genetic search" to refer to this sort of scheme. However, since Turing's supervisor considered that this technical report (from 1948) looked more like a college essay than a scientific proposal, this work was published until 1968 [13] when evolutionary algorithms had been already proposed. Consequently, Turing's ideas were, unfortunately, disregarded.

Toward the end of the 1950s and the beginning of the 1960s, there were several relevant breakthroughs that occurred during the same period in which computers became more accessible. During this period, there were several analyses of the population dynamics that appeared during evolution. A survey of the state-of-the-art in this topic was presented by Crosby in 1967 [13], who identified three types of schemes:

1. Methods based on the study of mathematical formulations that emerge in the deterministic analysis of the population dynamics. These schemes usually assume an infinite population size and analyze all the gene distributions[1] under different circumstances.

2. Approaches that simulate evolution but without explicitly representing a population [14]. In these methods, the frequency of the different potential genotypes[2] is stored for each generation, resulting in a non-scalable scheme in terms of the number of genes. Additionally, it is necessary to have a mathematical representation of the evolutionary system.

3. Schemes that simulated evolution using an explicit representation of the population. Fraser [15] pioneered these methods, which are described in a series of papers published over a decade [16]. The main conclusions derived from these studies were gathered in a seminal book on this topic that was published in 1970 [17]. Since its inception, this model was widely accepted and it was adopted by several researchers. In spite of the fact that there exist several implementation details in which Fraser's simulation differ from modern evolutionary algorithms, some authors consider the work of Fraser as the first actual proposal of what we call today a "genetic algorithm" [6]. It is worth mentioning that, although at the implementation level, the main difference between Fraser's proposal and the modern genetic algorithms lies on the fact that the former used diploid chromosomes (i.e., two chromosomes for each individual) and the latter uses haploid chromosomes (i.e., a single chromosome for each individual); conceptually, Fraser's work focused on analyzing population dynamics and not on solving problems, which is the typical application of modern evolutionary algorithms.

In this same period, other researchers proposed schemes similar to modern evolutionary algorithms. From them, it is worth mentioning Friedman, Box and Friedberg, because of the further impact that their work had. Friedman [18] proposed the automated design of control circuits using a scheme inspired on natural evolution that he called "selective feedback". His proposal was very different from modern evolutionary algorithms. For example, he did not adopt the notions of "population[3]" or "generation[4]". Furthermore, he did not implement his method and stated some

[1] A gene is a sequence of DNA that occupies a specific position in a chromosome. A chromosome is a structure inside the nucleus of a cell that contains the genetic material that conforms an individual.

[2] A genotype is the genetic information of an organism. This information (contained in the chromosomes of the organism) can or cannot be observed in the individual.

[3] Most modern evolutionary algorithms operate on a set of potential solutions to the problem at hand. Such a set of solutions is called "population".

[4] Evolutionary algorithms are run for a certain number of iterations, which are called "generations".

assumptions that turned out to be too optimistic. Nevertheless, his work is considered as pioneering within an area that is known today as "evolvable hardware", in which the aim is to design circuits using evolutionary algorithms (in this area, hardware implementations are commonly adopted).

Box [19] proposed a technique called "Evolutionary Operation" (EVOP) to optimize the handling of processes in chemical engineering. EVOP uses a baseline solution (the parent) to generate several additional solutions (the offspring), modifying a few production parameters at a time. Then, an individual is selected to pass on to the next generation based on certain statistical evidence. This was not an autonomous system, since it required human assistance in the variation and selection processes. Thus, EVOP can be considered as the first interactive evolutionary algorithm.

The most important contribution made by Friedberg [20] was his attempt to automatically generate computer programs using an evolutionary scheme. In this earliest work [21], he did not identify any specific relationship between his proposal and natural evolution. However, this link was identified in further publications of his co-authors [22]. In his first attempts, the goal was to automatically generate small programs with some simple functionality based on a set of tailored instructions. In order to direct the search toward promising regions, specific domain information was incorporated through the definition of a variation scheme specifically designed for the problem to be solved. In spite of the use of specialized operators, the success of this technique was rather limited. Nevertheless, some of the analysis undertaken by Friedberg and his co-authors were particularly important for the further development of evolutionary computation. Among his most important contributions, we can enumerate the following:

1. The idea of dividing candidate solutions into classes is related to the notions of implicit parallelism[5] and schema[6] that John Holland would propose some years later [23].

2. Several sets of instructions were tested, showing, for the first time, the influence of the mapping from the genotype to the phenotype.

3. A credit assignment algorithm was adopted to measure the influence of the instructions, in an isolated manner. This idea is closely related to the work that Holland would undertake some years later when developing the genetic algorithm and the classifier systems.

There are several other papers that did not attract much attention when they were originally published but that proposed techniques that were re-invented years later. For example, Reed et al. [24] provided several contributions that are remarkably

[5] Implicit parallelism is a property attributed to genetic algorithms that allows them to explore in a more efficient way the search space through the statistical estimation of the quality of the solutions.

[6] A *schema* is an abstraction used by Holland to represent the solutions processed by a genetic algorithm. Holland used this concept to provide a mathematical analysis of the way in which a genetic algorithm operates.

similar to the concepts of self-adaptation,[7] crossover[8] and coevolution[9] that are currently used in evolutionary computation.

2.1 Evolutionary programming

This type of evolutionary algorithm was originally proposed by Lawrence Fogel in the early 1960s, when he was conducting basic research on artificial intelligence for the US National Science Foundation [25]. At that time, most of the existing attempts to generate intelligent behavior used man as a model. However, Fogel considered that, since natural evolution was able to create humans and other intelligent creatures, its simulation could lead to intelligent behaviors. The basic ideas of Fogel were similar to those of the works previously described: he aimed to use selection and variation mechanisms for evolving candidate solutions that were the best adapted to the desired goal. Nevertheless, since Fogel was not aware of these works, his proposal can be seen as a re-invention of the schemes based on natural evolution. Philosophically, the encoding structures used in Evolutionary Programming are an abstraction of the phenotype[10] of different species [26]. Consequently, the encoding of candidate solutions may freely adapt to satisfy the problem requirements (i.e., there is no need to encode the candidate solutions in a particular alphabet as it happens, for example, with the genetic algorithm in which solutions are normally encoded in binary). Since recombination between different species is not possible, Evolutionary Programming does not incorporate a crossover or recombination operator. The lack of a crossover operator, the freedom to adapt the encoding and the use of a probabilistic selection operator[11] (which was not incorporated in the early versions of the algorithm) are the main characteristics that distinguish Evolutionary Programming from the other types of evolutionary algorithms.

In his early versions of Evolutionary Programming, Fogel realized that one of the main features of human behavior is the capability to predict the environment, together with a mechanism that allows to translate predictions to a proper response in such a way that the desired goal is achieved. On the other hand, a fine states transductor is a machine that allows to transform a sequence of input symbols into a sequence of output symbols, such as it is shown in **Figure 1**.

Depending on its states and transitions, this machine can be used to model different situations and can produce different transformations. Therefore, Fogel considered this machine to be an appropriate mechanism to generate intelligent behavior. In his first

---

[7] Self-adaptation is a mechanism that allows an evolutionary algorithm to adapt (or tune) by itself the values of its parameters.

[8] Crossover is an operator used in evolutionary algorithms to allow the exchange of genetic material between two individuals (the parents) by generating offpsring (or children) that combine such material.

[9] Coevolution is a model used in evolutionary computation in which two populations interact with each other in such a way that the fitness of the members of one of them depend on the other population and viceversa. In other words, in this model, the evolution of a certain species is conditioned to the evolution of another species.

[10] Phenotype is a term adopted in genetics to denote the visible physical features of an individual. In evolutionary algorithms, the phenotype refers to the decoded value of a candidate solution.

[11] The use of a probabilistic selection operator implies that the least fit individuals in the population will have a probability greater than zero of being selected.

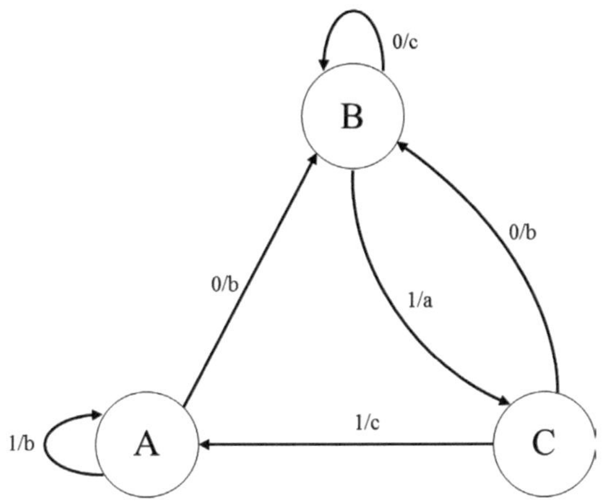

Figure 1.
A finite state machine like the ones used by Lawrence Fogel in his early experiments with evolutionary programming. The symbols to the left and to the right of the "/" denote input and output, respectively. The initial state in this case is C.

experiments, Fogel's goal was to develop a machine capable of predicting the behavior of an environment that was represented as a sequence of symbols from an input alphabet. In other words, given the symbols that had been previously generated by the environment, he aimed to predict the next symbol that would emerge from the environment. For instance, if we want to obtain a finite state transductor such as the one shown in **Figure 1**, we need to create a transitions table such as the one shown in **Table 1**.

Table 1 shows, for each state, the next state according to the input, as well as the output symbol. Fogel tried to generate these finite state machines in an automated way (by simulating evolution), using as his input the output symbols from the transitions table. The goal was that Evolutionary Programming could generate a finite state machine that generated the output symbols that we provided without human intervention.

The original Evolutionary Programming algorithm worked as follows: First, an initial population having *N* randomly generated finite state machines was produced. Then, each member of the population was mutated, giving rise to a population of offspring of the same size as the population of parents. Fogel adopted five types of mutation operators: (1) add a state, (2) delete a state, (3) change the next state of a transition, (4) change the output symbol of a transition, or (5) change the initial state. The mutation operators were randomly selected[12] and, in some cases, the

Current state	C	B	C	A	A	B
Input symbol	0	1	1	1	0	0
Next state	B	C	A	A	B	B
Output symbol	b	a	c	b	b	c

Table 1.
This table shows the transitions of the finite state machine from **Figure 1**.

[12] Fogel experimented with several schemes to choose the most appropriate mutation operator, but at the end, he decided to adopt a random selection method.

offspring were mutated more than once. Finally, the offspring were evaluated, and the *N* best machines (from the union of parents and offspring) were selected. Each finite state machine was evaluated in terms of its ability to predict, in a correct way, the next symbols in the known sequences. Initially, fitness[13] was computed by considering a small number of symbols, but as evolution progressed, more symbols were added to the training set. In his first experiments, Fogel performed tests with different types of prediction tasks: periodic sequences of numbers, sequences with noise, non-stationary environments, etc. Then, he considered more complex tasks such as pattern recognition, classification and the design of control systems. All these works originated from what is considered as the first book on evolutionary computation in history [27]. This work was very important because it presents, among other things, one of the first applications of coevolution. It is also important to emphasize that in most of the early Evolutionary Programming studies, the computational results that were presented were not too extended due to the limited computational power available at that time. Most of these initial experiments were retaken and extended several years later, providing a greater understanding of the true capabilities of Evolutionary Programming [28].

In the 1970s, most of the research conducted on Evolutionary Programming was undertaken under the supervision of Dearholt. One of the main contributions of his work was the application of Evolutionary Programming to practical problems. For example, pattern recognition in regular expressions [29] and in handwritten characters [30], as well as for classifying different types of electrocardiograms [31]. These works incorporated several algorithmic novelties. From them, perhaps the most important had been the simultaneous use of several mutation operators and the dynamic adaptation of the probabilities associated with different mutation operators.

At the beginning of the 1980s, Evolutionary Programming diversified using other arbitrary encodings for its solutions with the aim of solving different types of problems. Eventually, it was used, for example, to solve routing problems using an encoding based on the use of integer permutations [28]. It was also used to generate computer programs in an automated manner, adopting a tree encoding [32]. Furthermore, vectors of real numbers were used to solve continuous optimization problems [33] as well as to train neural networks [34]. During this period, it was commonly accepted that the design of specialized encodings for a particular problem, together with an associated set of operators, allowed to perform a more efficient search [35]. In fact, several researchers defended the position that the design of smart mutation schemes could avoid the use of recombination [36].

In the 1990s, the use of Evolutionary Programming in continuous optimization problem was considerably extended. During this period, great efforts were made to develop self-adaptive schemes. When using self-adaptation, some parameters of the algorithm are encoded together with the decision variables of the problem, so that the evolutionary operators act directly over them to define their most suitable values. Under this sort of scheme, the decision variables of a problem are called object parameters and the parameters of the evolutionary algorithm are called strategy parameters. In the first attempts to incorporate self-adaptation in Evolutionary Programming, the scaling factor of the Gaussian mutation was the parameter to be adapted [37]. In other, more advanced versions, several mutation

[13] The "fitness" of an individual is a measure of merit that allows to compare a solution with respect to the others. If we are solving an optimization problem, fitness is normally defined in terms of the objective function that we aim to optimize.

operators were simultaneously adapted [33]. Since these self-adaptive schemes turned out to provide a superior performance with respect to the traditional variants of Evolutionary Programming, this mechanism became very popular in many of the implementations used to solve real-world problems, particularly when adopting real numbers encoding [4]. It is worth noting that the self-adaptation mechanism adopted in Evolutionary Programming for continuous optimization problems is very similar to the one adopted in Evolution Strategies. The most significant difference is that Evolutionary Programming does not incorporate a recombination operator. Additionally, some implementation details were also different in the first self-adaptive variants of Evolutionary Programming [1]. For example, the order in which the object parameters and the strategy parameters were mutated was different in Evolutionary Programming and in Evolution Strategies, although these differences vanished over the years. Additionally, there is evidence indicating that the exclusive use of the mutation operator can produce better results than when incorporating recombination and *vice versa* [38]. There is also evidence indicating that it is beneficial to adapt the parameters associated with the crossover operator adopted by Evolution Strategies [39], which decreases even more the differences between Evolutionary Programming and Evolution Strategies. It is worth noting, however, that the first mechanism to self-adapt parameters in continuous optimizaton was produced in Evolution Strategies, as we will see later on. Evolutionary Programming is the least popular evolutionary algorithm in current use. Although its original use was for prediction problems, today is more frequently used for continuous optimization. This trend is reflected also in Mexico, where we could find very little evidence of the use of Evolutionary Programming for both single-objective [40] and multi-objective [41] optimization in continuous search spaces.

2.2 Evolution strategies

During the mid-1960s, three graduate students at the Technical University of Berlin (Bienert, Schwefel and Rechenberg) were studying practical problems in fluid mechanics and other related areas with the aim of building robots that could solve engineering problems in an automated manner [6]. In order to achieve this goal, they formulated these tasks as optimization problems and designed autonomous machines that could be automatically modified when applying certain rules. They also tried to apply classical optimization algorithms. However, since the problems that they aimed to solve had noise and were multimodal, classical optimization algorithms were unable to find acceptable solutions. So, to deal with this situation, they decided to apply mutation and selection methods that were similar to those adopted in natural evolution. Rechenberg published the first report on this new technique that was called "evolution strategy" applied to the minimization of the total drag of a body inside a wind tunnel [42]. Furthermore, they solved other complex problems including the design of pipes and of hydrodynamic nozzles [6].

From a philosophical perspective, the encoding adopted by Evolution Strategies is an abstraction of the phenotype of individuals [26]. Because of that, it is possible to freely adopt any sort of encoding of solutions that provides the best match with respect to the problem requirements. Additionally, evolution strategies allow the recombination of individuals, although this operator was not incorporated in the original version of the algorithm.

The first version of this algorithm is known as the (1 + 1)-ES (ES stands for "Evolution Strategy"). It works as follows: first, an initial solution is randomly created and is considered as "the parent". This solution is mutated (i.e., the value of one of its decision variables is randomly modified), and the mutated individual is called "the offspring". Mutation was applied following a binomial distribution so that small mutations could be applied more often than large mutations, as it happens in nature. The mutated solution would then become the parent if it were better or equal to its parent. Then, the new parent would be mutated again. This process was repeated until reaching a certain stopping criterion. The experiments performed by the creators of this technique showed that it worked better in complex optimization problems than the classical optimization techniques available at that time.

In 1965, Hans-Paul Schwefel implemented for the first time an Evolution Strategy in a general use computer and applied it to continuous optimization problems [43]. The most important contribution made by Schwefel was that he incorporated Gaussian mutation with zero mean, which is an operator that is still being used today. This mutation operator could be controlled through the adjustment of the standard deviations (i.e., the step size) of the Gaussian distribution. The studies conducted by Schwefel on the step size originated the first theoretical analyses of the Evolution Strategy. Toward the end of the 1960s, a new version of the Evolution Strategy that used a population was introduced. This initial population-based version was called $(\mu + 1)$-ES, and it adopted μ individuals which were used to create a new individual through the application of recombination and mutation operators. Furthermore, the μ best individuals from the $\mu + 1$ that were available, were selected to survive. This scheme is very similar to the so-called steady-state selection [44] that would become popular with genetic algorithms a few years later. Over time, Evolution Strategies allowed any number of parents and offspring. Additionally, two selection schemes became standard in the area: the $(\mu + \lambda)$-ES and the (μ,λ)-ES. In the first one, μ parents generate λ offspring and the μ best individuals from their union (parents and offspring) survive. In the second scheme, μ parents generate λ offspring and the μ best offspring survive. The first papers on Evolution Strategies were published in German, which limited their access. However, in 1981, Schwefel published the first book on Evolution Strategies written in English [45]. Since then, an important number of researchers around the world have studied and applied them extensively. In Mexico, Evolution Strategies have been used to solve image processing [46] and constrained nonlinear optimization problems [47].

Schwefel's book focuses on the use of Evolution Strategies for continuous optimization, which is the domain in which this technique has been most commonly applied. In his book, Schwefel discusses, among other topics, the use of different types of recombination operators as well as the adaptation of the Gaussian distributions used for the mutation operator.

In the early studies of the recombination operator, four possible schemes were proposed based on the combination of two different properties. First, two options were provided for the number of parents involved in the creation of an offspring: (i) bi-sexual recombination (or local) and (ii) multi-sexual recombination (or global). In the first, two parents are selected, and they are recombined, producing the candidate solution for the offspring. In the second, different parents are selected to generate each variable of the candidate solution of the offspring. There are also two options to combine the values of the parents: the so-called "discrete recombination", in which one of the values is randomly selected and the so-called "intermediate recombination", in which the average of the values of the parents is computed. In these schemes, it is not possible to specify the number of parents who will take part in the recombination. Rechenberg proposed a generalization that changes this constraint. In this case, a new parameter, called ρ, is adopted to indicate the number of parents that will participate in the creation of each individual. These originated the variants called (μ/ρ, λ)-ES and (μ/ρ + λ)-ES. Later on, other versions that assigned weights to the contribution of each of the individuals to the recombination, were introduced [48].

In spite of the use of recombination, the main operator in Evolution Strategies is mutation. As it was previously indicated, this operator is normally based on a Gaussian distribution with zero mean. In most cases, individuals are perturbated through the addition of a random vector that is generated from a multivariate Gaussian distribution. In other words, the following expression is adopted:

$$x_i' = x_i + N(0,C)$$

where x_i' is the new individual generated from x_i, C represents a covariance matrix and $N(0,C)$ returns a random number with a zero mean. Thus, the new individual (i.e., the "offspring") is created from the previous one (the "parent") through a random perturbation to one of its decision variables. From the early days of Evolution Strategies, it was clear that C could have a significant effect on the performance of the algorithm. Consequently, several methods to adapt C during the execution of the algorithm have been proposed. The first schemes of this sort originated from the studies undertaken by Rechenberg [49], who analyzed the convergence properties of Evolution Strategies with respect to the relative frequency of successful mutations (if an offspring is better than its parent in terms of its objective function value, a mutation is considered to be successful). This work led to the first adaptation scheme in which the size of the global step is adjusted through a procedure that is aligned with the goal of producing successful mutations with a rate of 1/5. This procedure is known as the "1/5 success rule". The idea is to adjust the standard deviation when the success rate is not exactly 1/5, under the assumption that this is indeed the ideal success rate. It is important to notice that this rule is not general. This eventually originated self-adaptive versions of Evolution Strategies. In the first of such self-adaptive schemes, the size of the global step was the only parameter subject to adaptation. However, over the years, much more sophisticated self-adaptive schemes were developed. Another interesting aspect related to Evolution Strategies is that if the whole covariance matrix is adapted, correlations among the mutations performed can be induced, with the consequent involvement of different parameters, which makes it easier for the algorithm to deal with non-separable functions. The first scheme in which the whole covariance matrix is adapted was proposed by Schwefel [45]. In that case, the coordinate system in which the control of the step size takes place, as well as the step sizes themselves, are self-adapted. It is worth noting, however, that this scheme has not been too successful, mainly because it requires very large population sizes in order to operate properly. A more efficient alternative is the Evolution Strategy with an adaptation scheme of the covariance matrix proposed by Hansen and Ostermeir [50], which is known as CMA-ES. In this scheme, two different self-adaptation techniques are combined with the aim of building covariance matrices that maximize the creation of mutation vectors that were successful at previous generations. This is a very complex technique and it requires of several parameters, which has led to the development of more simple (and with less parameters) variants [51].

Finally, there are also schemes that favor certain search directions without adapting the covariance matrix. For example, Poland and Zell [52] proposed to use the main descent direction to guide the search. The main advantage of this sort of scheme is that it reduces the computational complexity in terms of time and space. Nevertheless, since the time required to compute the covariance matrices is not significant with respect to the evaluation of the objective function (particularly, if the objective function is computationally expensive), this sort of scheme is not very popular, although it can be useful in problems having a large number of decision variables, since in that case, the cost of computing the covariance matrices will significantly increase.

Evolution Strategies are a very good choice for solving continuous optimization problems in which all the decision variables are real numbers. However, it is worth noting that in recent years, they have been replaced by other evolutionary algorithms such as differential evolution which is easier to implement and use.

2.3 Genetic algorithms

John Holland [53] identified the relationship between the adaptation process that exists in natural evolution and optimization, and conjectured that such a process could play a critical role in other areas such as learning, automatic control, or mathematics. His initial goal was to study adaptation in a more formal manner, and Holland proposed an algorithmic framework inspired by these ideas. Such a framework was called "reproductive and adaptive plans", but a few years later, it was renamed as "Genetic Algorithm". The way in which a Genetic Algorithm works is based on the progressive modification of a set of data structures with the aim of adapting them to a certain environment. The specific way in which such adaptations take place is inspired on genetics and on natural evolution. However, very quickly, Genetic Algorithms became problem solvers in areas such as machine learning[14] and automatic control. Nevertheless, they have been more commonly adopted in optimization both for discrete and continuous search spaces.

From a philosophical perspective, the encoding adopted in Genetic Algorithms is an abstraction of the genotype of individuals. In this case, regardless of the type of decision variables of the problem, they must be encoded in binary. Although nowadays it is possible to use other types of encoding, Holland argued that the binary alphabet is universal and that its use provides several advantages both from a pragmatic and from a theoretical point of view. However, in spite of its advantages, the use of binary encoding has the evident disadvantage of requiring a decoding process to transform a binary string into the decision variable that it encodes. Furthermore, the use of binary encoding may, in some cases, introduce problems due to the discretization of a search space that was originally not discrete (e.g., when the decision variables are real numbers). Such problems go from using an inappropriate accuracy up to the use of very long binary strings, which makes inefficient the application of the operators of the Genetic Algorithm.

Each binary string that encodes a decision variable is called *gene*, and each bit within a gene is called an *allele*. The concatenation of all genes (i.e., the set of all of decision variables of a problem) is called *chromosome*, as shown in **Figure 2**. An individual is formed by a chromosome (i.e., we are considering haploid chromosomes, although diploid chromosomes are the most common in Nature).

The way in which a Genetic Algorithm operates is very simple. It starts with a set of randomly generated solutions which is called "initial population". Then, we have to

[14] Genetic Algorithms gave rise to the so-called *Learning Classifier Systems*, in which sets of rules of the type IF-THEN-ELSE are evolved in order to solve classification problems.

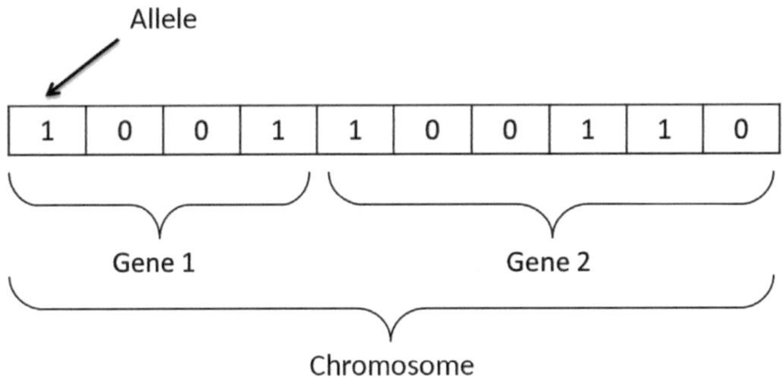

Figure 2.
Structure of a chromosomic string for a problem having two decision variables ("gene 1" and "gene 2" correspond to these two decision variables).

compute the fitness of each of these solutions (or "individuals"). The fitness value is a normalized version of the objective function that we aim to optimize so that we avoid zero and negative values. Parents are selected to recombine based on their fitness values (i.e., individuals with a larger fitness value have a higher probability of being selected). The original version of the Genetic Algorithm uses proportional selection, which means that the fittest individuals have a higher survival probability. However, the least fit individuals have a probability different from zero of being selected and could also become parents. Selected individuals are recombined using a crossover operator to generate the population of offspring to which the mutation operator is applied. The mutated offspring form the new population, which completely replaces the previous population. This process is repeated until a certain stopping criterion is reached, which is normally a maximum number of generations (or iterations). **Table 2** shows a hypothetical population of a genetic algorithm consisting of five individuals for which their chromosomes and fitness values are shown.

Figure 3 shows the way in which a proportional selection method known as "roulette wheel" works. The black triangle is a pivot. When using this selection method, we simulate a random movement of the wheel that contains all the individuals in the population. Each time that the wheel turns, we select the individual indicated by the pivot when the wheel stops to become a parent.

The portion that each individual occupies in the wheel corresponds to its fitness contribution so that the fittest individuals will have a higher probability of being selected so that they can become parents and can recombine using one of the several

Individual #	Chromosome	Fitness	% from the total
1	11,010,110	254	24.5
2	10,100,111	47	4.5
3	00110110	457	44.1
4	01110010	194	18.7
5	11,110,010	85	8.2
TOTAL		1037	100.0

Table 2.
A hypothetical example of a population with five individuals having chromosomes of eight bits in length. The third column shows the fitness values of each individual.

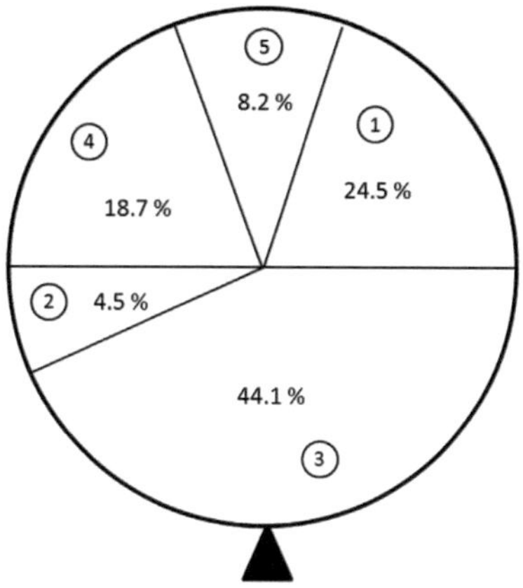

Figure 3.
Illustration of roulette wheel selection, which is one of the most popular proportional selection mechanisms used with genetic algorithms. The five individuals from **Table 2** *are placed in a wheel in which each subdivision corresponds to their fitness contribution. In this selection technique, we simulate the turn of a roulette wheel such that we select as a parent to the individual to which the pivot points when the wheel stops. Clearly, the fittest individuals will have a higher probability of being selected, but the least fit individuals may also be selected, although their probability for that to happen will be very low.*

crossover operators that are available. **Figures 4** and **5** illustrate the way in which one-point crossover and two-point crossover work, respectively. Finally, each of the offspring that are generated using crossover is mutated, as shown in **Figure 6**.

Over the years, different variants of the genetic algorithm have been developed, some of which include deterministic (rather than probabilistic) and steady-state (in which one individual is replaced at a time) selection schemes, non-binary encodings, more sophisticated or highly specialized crossover and mutation operators, diploid and multiploid chromosomes, as well as additional operators such as inversion, translocation and segregation [5].

A distinctive aspect of genetic algorithms is that it has been traditionally considered that the crossover operator is its main power source and that the mutation operator is only a secondary operator which is used to maintain diversity in the population and to keep the search space fully connected. This philosophy is exactly the opposite

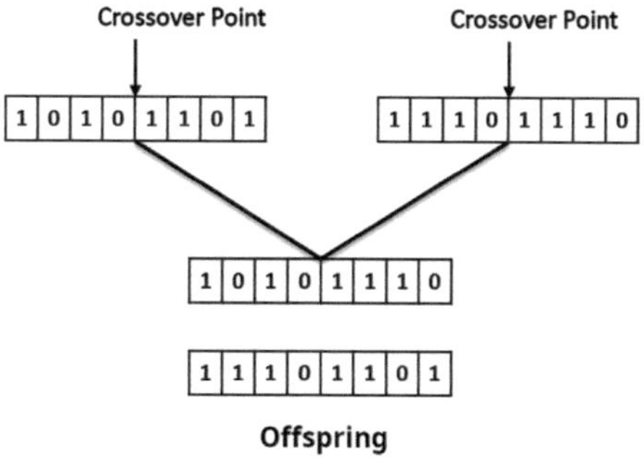

Figure 4.
One-point crossover. It is worth noting that the recombination of two parents produces two offsprings. The first child has the first four bits from parent 1 (left) and the other four from parent 2 (right). The second child has the first four bits from parent 2 and the other four from parent 1.

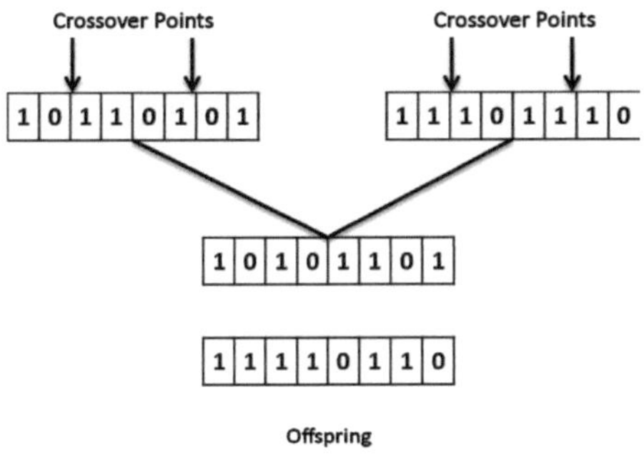

Figure 5.
Two-points crossover. The first child has the first two bits and the last two bits from parent 1 (left) and the rest from parent 2 (right). The second child has the two first bits and the last two bits from parent 2 and the rest from parent 1.

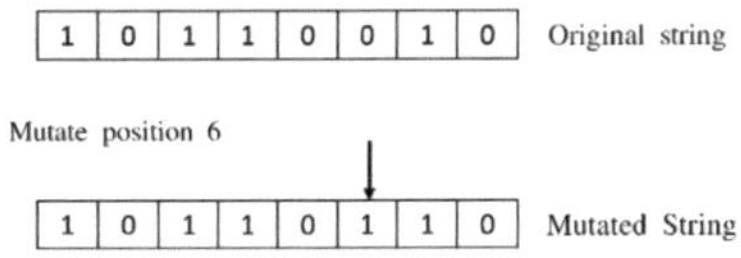

Figure 6.
An example of the use of the mutation operator of a genetic algorithm. In this case, position 6 of the original string is mutated. The operator applies a "NOT" to the original position, which changes the original zero in that position by a one.

from the one adopted in Evolution Strategies in which mutation is the main operator and crossover only acts as a secondary operator.

In his book from 1975, Holland tried to model the behavior of a genetic algorithm through his "schema theorem", in which he derived a formula that provides the probability that a certain pattern of bits (called "schema") survives the effects of selection, crossover and mutation [23]. Over the years, much more sophisticated mathematical models have been developed to try to explain the way in which a genetic algorithm works, including those based on Markov chains and on statistical mechanics [1]. The first proof of convergence of an ordinary genetic algorithm [54] (called "canonical") showed that an operator called "elitism" is required to guarantee convergence. This operator consists in retaining the best individual at each generation and copying it intact to the next generation, without applying crossover or mutation to it.

In Mexico, very relevant research on genetic algorithms has been conducted both related to machine learning[15] as well as to single- and multi-objective optimization.[16] There is also research on the use of genetic algorithms and their variants in electronics (an area known as "evolvable hardware"),[17] computer vision,[18] bioinformatics,[19]

[15] For example, at Tec de Monterrey campus Monterrey and at the Instituto Nacional de Astrofísica, Óptica y Electrónica (INAOE).

[16] There are research groups working in these areas at the Centro de Investigación y de Estudios Avanzados del Instituto Politécnico Nacional (CINVESTAV-IPN), the Universidad Veracruzana, the Universidad Autónoma de Querétaro, the Centro de Investigación Científica y de Educación Superior de Ensenada (CICESE), Centro de Investigación en Matemáticas (CIMAT) and the Universidad Nacional Autónoma de México (UNAM).

[17] At INAOE and CIMAT.

[18] At CICESE.

[19] At CICESE and Tec de Monterrey campus Estado de México.

finance,[20] cryptography,[21] image processing[22] and design.[23] There are also research groups in México that have done basic research on evolutionary operators,[24] evolutionary multi-objective optimization[25] and on theoretical aspects of genetic algorithms.[26] Additionally, there has been also research on genetic programming,[27] which is a genetic algorithm variant proposed by John Koza [55] in which trees are used to evolve computer programs.

3. Other bio-inspired metaheuristics

There are several other bio-inspired metaheuristics that have also been studied in Mexico. From them, we can highlight the following:

- Particle swarm optimization
- Artificial immune systems
- Differential evolution
- Ant colony optimization

Each of them will be briefly described next.

3.1 Particle swarm optimization

It was proposed by Kennedy and Eberhard in the mid-1990s [56]. This metaheuristic simulates the movement of a flock of birds or a bank of fish. From a mathematical perspective, it simulates the movement of a particle in a fluid (water or air). In this approach, the trajectories are defined from a very simple formula that involves the velocity of each particle, an inertia factor and the influence of the best solution that has been obtained so far for a particular particle and for the entire swarm. Particle swarm optimization has been studied by several research groups in Mexico, particularly for single- and multi-objective optimization.[28]

3.2 Artificial immune systems

This metaheuristic aims to emulate the behavior of our natural immune system, which defends our organism from the attack of diseases (called "antigens"). For that sake, our immune system is able to produce cells called "antibodies" which have the

[20] At Banco de México, CINVESTAV-IPN and Tec de Monterrey campus Monterrey.
[21] At the Centro de Investigación en Computación (CIC-IPN) and CINVESTAV-IPN.
[22] At CIMAT, CINVESTAV-IPN, CICESE, Tec de Monterrey campus Guadalajara and the Universidad de Guadalajara.
[23] At the Instituto Tecnológico Autónomo de México (ITAM).
[24] At the Universidad de Guadalajara and at CINVESTAV-IPN.
[25] At CINVESTAV-IPN, CIMAT, UNAM, CICESE and the Universidad Veracruzana.
[26] At UNAM and CINVESTAV-IPN.
[27] At UNAM, CINVESTAV-IPN, INFOTEC and CICESE.
[28] At CINVESTAV-IPN and CIMAT.

capability to couple to the geometrical shape of an antigen so that it can be nullified. The earliest computational models of an immune system were developed in the 1980s and were popularized by Stephanie Forrest, who also made key contributions to this area [57]. Nowadays, there are several computational models of our immune system, including some based on negative selection, positive selection, clonal selection and immune networks. From them, clonal selection has been the most commonly adopted in the specialized literature (particularly, for solving optimization problems). Artificial immune systems have been adopted by some research groups in Mexico[29] to solve single- and multi-objective optimization problems and for some applications (e.g., cryptography).

3.3 Differential evolution

It was originally proposed by Kenneth Price and Rainer Storn in the mid-1990s [58]. This approach can be seen as a direct search method in which the algorithm tries to estimate the gradient in a region (rather than in a point) through the use of an operator that recombines information from three individuals. Although conceptually simple, this algorithm is a very powerful optimizer for continuous search spaces, and it produces results that are superior to those generated by other evolutionary algorithms, except for evolution strategies. However, in many cases, it is able to match and even outperform evolution strategies. Additionally, differential evolution requires less parameters and has an easier implementation than state-of-the-art evolution strategies. There are research groups in Mexico that have used differential evolution in a variety of single- and multi-objective optimization problems.[30]

3.4 Ant Colony optimization

It was proposed by Marco Dorigo in the early 1990s [59] as a metaheuristic that simulates the behavior of a group of ants that leave their nest looking for food. While moving from their nest to the food source, ants segregate a substance called "feromone", which the other ants can smell. When the different pheromone trails are combined, the resulting path turns out to be the shortest from the nest to the food source. In the computational model developed by Dorigo, a pheromone evaporation factor is considered and the problem to be solved should allow the evaluation of partial solutions. In its original version, this metaheuristic could only be used to solve the traveling salesperson problem (TSP) or any problem that could be transformed into the TSP. However, over the years, more sophisticated versions of the algorithm were developed to tackle other combinatorial optimization problems and even continuous optimization problems [60].

4. Evolutionary computation in Mexico

Research on evolutionary computation originated in Mexico in the late 1980s at the Instituto Tecnológico y de Estudios Superiores de Monterrey (ITESM) campus in Monterrey and at the Universidad Nacional Autónoma de México (UNAM). At ITESM, an Artificial Intelligence Center (later renamed as Intelligent Systems Center)

[29] At Tec de Monterrey Campus Estado de México and at CINVESTAV-IPN.

[30] At the Universidad Veracruzana and CINVESTAV-IPN.

was created in 1989. Almost since its creation, this center had bio-inspired systems as a research line. Manuel Valenzuela Rendón (who obtained a PhD in 1989 under the supervision of David E. Goldgerg, who was then at the University of Alabama) was probably the first researcher who specialized in evolutionary computation in Mexico. In fact, Goldberg mentions Manuel in the acknowledgments of his famous book on genetic algorithms [5].

At UNAM, there are undergraduate theses (see, e.g., [61]) and even PhD theses (see, e.g., [62, 63]) from the late 1980s and the early 1990s, supervised by Dr. Germinal Cocho Gil, which adopt the Monte Carlo method and genetic algorithms. However, the first thesis in which the term "genetic algorithm" appears in the title was produced in 1993.[31]

Toward the second half of the 1990s, there were already courses on evolutionary computation at institutions such as the Centro de Investigación en Computación del Instituto Politécnico Nacional (CIC-IPN), the Masters in Artificial Intelligence of the Universidad Veracruzana and at UNAM.

Currently, there are several graduate programs in Mexico in which it is possible to specialize in evolutionary computation (both at a Master and at a PhD level). Examples of such institutions are the Centro de Investigación y de Estudios Avanzados del Instituto Politécnico Nacional (CINVESTAV-IPN) in its units in Zacatenco (Mexico City) and Tamaulipas (Ciudad Victoria), the Instituto de Investigación en Matemáticas Aplicadas y Sistemas de la UNAM (IIMAS-UNAM), the CIC-IPN, the Universidad de Guadalajara, the Instituto Nacional de Astrofísica, Óptica y Electrónica (INAOE), the Centro de Investigación Científica y de Educación Superior de Ensenada (CICESE), the Centro de Investigación e Innovación en Tecnología de la Información y Comunicación (INFOTEC) and the Tecnológico de Monterrey.

5. Future perspectives

Research on evolutionary computation has experienced an important growth in Mexico in recent years. Today, there is an important number of researchers working with different types of evolutionary algorithms and other bio-inspired metaheuristics, not only focusing on applications but also on basic research (e.g., the design of new algorithms, theoretical aspects of evolutionary algorithms, diversity and population dynamics). Also, as indicated before, there are several graduate programs recognized by CONAHCyT (the main Mexican government funding agency) in which it is possible to specialize in evolutionary computation. Additonally, there are several evolutionary computation research groups in Mexico that have collaborations with researchers from other countries[32] and that have good international visibility.

In recent years, there has also been increased participation of these research groups in the main evolutionary computation conferences: the Genetic and Evolutionary Computation Conference (GECCO), the IEEE Congress on Evolutionary Computation (CEC) and Parallel Problem Solving from Nature (PPSN). However, there are still very few researchers working in Mexico who publish in the top evolutionary computation journals: *IEEE Transactions on Evolutionary Computation*,

[31] Personal communication from Katya Rodríguez Vázquez, who performed several queries to the database of PhD theses of UNAM.

[32] For example, with researchers working in Spain, Australia, China, Chile, Germany, France, UK and Japan, among many others.

Evolutionary Computation, *Genetic Programming and Evolvable Machines* and *ACM Transactions on Learning and Optimization*. Remarkably, several evolutionary computation researchers working in Mexico publish an important number of papers in peer-reviewed journals, some of which have a very good impact factor. Nevertheless, the lack of publications in the top evolutionary computation journals and conferences may limit the visibility of these research groups.

A consolidation of these Mexican research groups (which today consist of an important number of young researchers) is expected to happen in the next few years.

Author details

Carlos A. Coello Coello
Department of Computer Science, CINVESTAV-IPN, Mexico City, Mexico

*Address all correspondence to: carlos.coellocoello@cinvestav.mx

References

[1] Eiben AE, Smith JE. Introduction to Evolutionary Computing. 2nd ed. Berlin: Springer; 2015. 287 p. DOI: 10.1007/978-3-662-44874-8

[2] Blum C, Roli A. Metaheuristics in combinatorial optimization: Overview and conceptual comparison. ACM Computing Surveys. 2003;**35**:268-308. DOI: 10.1145/937503.937505

[3] Wolpert DH, Macready WG. No free lunch theorems for optimization. IEEE Transactions on Evolutionary Computation. 1997;**1**:67-82. DOI: 10.1109/4235.585893

[4] Fogel DB. Evolutionary Computation: Toward a New Philosophy of Machine Intelligence. 3rd ed. USA: IEEE Press; 2006. 296 p. ISBN 978-0471669517

[5] Goldberg DE. Genetic Algorithms in Search, Optimization, and Machine Learning. USA: Addison-Wesley; 1989. 372p. ISBN 978-0-201-15767-3

[6] Fogel DB. Evolutionary Computation: The Fossil Record, Wiley-IEEE Press. USA; 1998. 656p. ISBN 978-078033481-6

[7] Wright S. The roles of mutation, inbreeding, crossbreeding and selection in evolution. VI International Congress of Genetics. 1932;**1**:356-366

[8] Cannon WB. The Wisdom of the Body. W. W. Norton & Company, Inc.; 1932

[9] Richter H. Fitness landscapes: From evolutionary biology to evolutionary computation. In: Richter H, Engelbrecht A, editors. Recent Advances in the Theory and Application of Fitness Landscapes, Emergence, Complexity and Computation. Vol. 6. Berlin: Springer; 2014. pp. 3-31. DOI: 10.1007/978-3-642-41888-4_1

[10] Campbell DT. Blind variation and selective survival as a general strategy in knowledge-processes. Psychological Review. 1960;**67**(6):380-400. DOI: 10.1037/h0040373

[11] Turing AM. Computing machinery and intelligence. Mind. 1950;**59**:433-460. DOI: 10.1093/mind/LIX.236.433

[12] Turing AM. Intelligent Machinery, Technical Report. Teddington, UK: National Physical Laboratory; 1948

[13] Crosby JL. Computers in the study of evolution. Science Progress Oxford. 1967;**55**:279-292

[14] Crosby JL. The use of electronic computation in the study of random fluctuations in rapidly evolving populations. Philosophical Transactions of the Royal Society of London, Series B, Biological Sciences. 1960;**242**(697):550-572. DOI: 10.1098/rstb.1960.0002

[15] Fraser AS. Simulation of genetic systems by automatic digital computers. I. Introduction. Australian Journal of Biological Science. 1957;**10**:484-491. DOI: 10.1071/BI9570484

[16] Fraser AS, Burnell D. Simulation of genetic systems XII. Models of Inversion Polymorphism, Genetics. 1967;**57**:267-282. DOI: 10.1093/genetics/57.2.267

[17] Fraser AS, Burnell D. Computer Models in Genetics. New York, USA: McGraw-Hill; 1970

[18] Friedman GJ. Selective feedback computers for engineering synthesis and nervous system analogy [MSc

dissertation]. Los Angeles, USA: University of California; 1956

[19] Box GEP. Evolutionary operation: A method for increasing industrial productivity. Applied Statistics. 1957;**6**(2):81-101. DOI: 10.2307/2985505

[20] Friedberg RM. A learning machine: Part I. IBM Journal of Research and Development. 1958;**2**(1):2-13. DOI: 10.1147/rd.21.0002

[21] Friedberg RM, Dunham B, North JH. A learning machine: Part II. IBM Journal of Research and Development. 1959;**3**(3):282-287. DOI: 10.1147/rd.33.0282

[22] Dunham B, Fridshal D, Fridshal R, North JH. Design by natural selection. Synthese. 1963;**15**(1):254-259

[23] Holland JH. Adaptation in Natural and Artificial Systems. Cambridge, Massachusetts, USA: The MIT Press; 1975. 232 p. ISBN 978-026258111-0

[24] Reed J, Toombs R, Barricelli NA. Simulation of biological evolution and machine learning: I. Selection of self-reproducing numeric patterns by data processing machines, effects of hereditary control, mutation type and crossing. Journal of Theoretical Biology. 1967;**17**(3):319-342. DOI: 10.1016/0022-5193(67)90097-5

[25] Fogel LJ. Autonomous automata. Industrial Research Magazine. 1962;**4**(2):14-19

[26] Fogel DB. An introduction to simulated evolutionary optimization. IEEE Transactions on Neural Networks. 1994;**5**(1):3-14. DOI: 10.1109/72.265956

[27] Fogel LJ, Owens AJ, Walsh MJ. Artificial Intelligence through Simulated Evolution. USA: John Wiley & Sons; 1966. ISBN 978-0471265160

[28] Fogel LJ, Fogel DB. Artificial Intelligence through Evolutionary Programming, Technical Report PO-9-X56-1102C-1. San Diego, California, USA: U.S. Army Research Institute; 1986

[29] Lyle MR. An investigation into scoring techniques in evolutionary programming [MSc thesis]. Las Cruces, USA; 1972

[30] Cornett FN. An application of evolutionary programming to pattern recognition [MSc Thesis]. Las Cruces, USA; 1972

[31] Dearholt DW. Some experiments on generalization using evolving automata. In: 9th Hawaii International Conference on System Sciences. Honolulu, Hawaii, USA: Western Periodicals; 1976. pp. 131-133

[32] Chellapilla K. Evolving computer programs without subtree crossover. IEEE Transactions on Evolutionary Computation. 1997;**1**(3):209-216. DOI: 10.1109/4235.661552

[33] Yao X, Liu Y, Lin G. Evolutionary programming made faster. IEEE Transactions on Evolutionary Computation. 1999;**3**(2):82-102. DOI: 10.1109/4235.771163

[34] Porto VW, Fogel DB. Alternative neural network training methods [active sonar processing]. IEEE Expert. 1995;**10**(3):16-22. DOI: 10.1109/64.393138

[35] Fogel LJ. Intelligence through Simulated Evolution: Forty Years of Evolutionary Programming. Wiley, USA: Wiley Series on Intelligent Systems; 1999. ISBN 978-0471332503

[36] Fogel DB, Atmar JW. Comparing genetic operators with gaussian mutations in simulated evolutionary

processes using linear systems. Biological Cybernetics. 1990;**63**(2):111-114. DOI: 10.1007/BF00203032

[37] Fogel DB, Fogel LJ, Atmar JW. Meta-evolutionary Programming. In: Twenty-Fifth Asilomar Conference on Signals, Systems & Computers, Pacific Grove, California, USA, IEEE, 4-6 November. 1991. pp. 540-545. DOI: 10.1109/ACSSC.1991.186507

[38] Richter JN, Wright A, Paxton J. Ignoble trails - where crossover is provably harmful. In: Rudolph G, Jansen T, Lucas S, Poloni C, Beume N, editors. Parallel Problem Solving from Nature PPSN X. Vol. 5199. (Lecture Notes in Computer Science). Berlin, Germany: Springer; 2008. pp. 92-101. DOI: 10.1007/978-3-540-87700-4_10

[39] Jain A, Fogel DB. Case studies in applying fitness distributions in evolutionary algorithms II: Comparing the improvements from crossover and Gaussian mutation on simple neural networks. In: 2000 IEEE Symposium on Combinations of Evolutionary Computation and Neural Networks, IEEE, 11-13 May. 2000. pp. 91-97. DOI: 10.1109/ECNN.2000.886224

[40] Coello CA, Landa BR. Adding knowledge and efficient data structures to evolutionary Programming: A cultural algorithm for constrained optimization. In: Langdon WB et al., editors. Proceedings of the Genetic and EvolutionaryComputation Conference, GECCO 2002. San Francisco, California, USA: Morgan Kaufmann Publishers; 2002. pp. 201-209

[41] Coello Coello CA, Landa BR. Evolutionary multiobjective optimization using a cultural algorithm. In: 2003 IEEE Swarm Intelligence Symposium, IEEE Service Center, Indianapolis, Indiana, USA. 2003. pp. 6-13. DOI: 10.1109/SIS.2003.1202240

[42] Rechenberg I. Cybernetic Solution Path of an Experimental Problem, Technical Report, Library Translation 1122. Royal Air Force Establishment; 1965

[43] Schwefel HP. Kybernetische Evolution als Strategie der experimentellen Forschung in der Strömungstechnik [Dipl.-Ing. thesis]. Technical University of Berlin, Hermann Föttinger, Institute for Hydrodynamics; 1965

[44] Whitley D, Kauth J. GENITOR: A Different Genetic Algorithm, Technical Report. USA: Colorado State University, Department of Computer Science; 1988

[45] Schwefel HP. Numerical Optimization of Computer Models. New York, USA: John Wiley & Sons, Inc; 1981

[46] Gómez García HF, González VA, Hernández AA, Marroquín Zaleta JL, Coello Coello CA. Robust multiscale affine 2D-image registration through evolutionary strategies. In: Guervós JJM, Adamidis P, Beyer H-G, Fernández-Villacañas J-L, Schwefel H-P, editors. Parallel Problem Solving from Nature VII, Lecture Notes in Computer Science. Vol. 2439. Granada, Spain: Springer-Verlag; 2002. pp. 740-748. DOI: 10.1007/3-540-45712-7_71

[47] Mezura ME, Coello Coello CA. A simple multi-membered evolution strategy to solve constrained optimization problems. IEEE Transactions on Evolutionary Computation. 2005;**9**(1):1-17. DOI: 10.1109/TEVC.2004.836819

[48] Bäck T, Schwefel HP. An overview of evolutionary algorithms for parameter optimization. Evolutionary Computation. 1993;**1**(1):1-23. DOI: 10.1162/evco.1993.1.1.1

[49] Rechenberg I. Evolutionsstrategie: Optimierung technischer Systeme nach Prinzipien der biologischen Evolution. Stutgart, Germany: Frommann-Holzboog; 1973

[50] Hansen N, Ostermeier A. Completely Derandomized self-adaptation in evolution strategies. Evolutionary Computation. 2001;**9**(2):159-195. DOI: 10.1162/106365601750190398

[51] Beyer HG, Sendhoff B. Covariance matrix adaptation revisited – The CMSA evolution strategy. In: Rudolph G, Jansen T, Lucas S, Poloni C, Beume N, editors. Parallel Problem Solving from Nature - PPSN X. Vol. 5199. Berlin, Germany: Springer; 2008. pp. 123-132. Lecture Notes in Computer Science. DOI: 10.1007/978-3-540-87700-4_13

[52] Poland J, Zell A. Main vector adaptation: A CMA variant with linear time and space complexity. In: Spector L et al., editors. Proceedings of the Genetic and Evolutionary Computation Conference (GECCO 2001). San Francisco, California, USA: Morgan Kaufmann; 2001. pp. 1050-1055

[53] Holland JH. Adaptation in Natural and Artificial Systems. Ann Arbor, Michigan, USA: University of Michigan Press; 1975

[54] Rudolph G. Convergence analysis of canonical genetic algorithms. IEEE Transactions on Neural Networks. 1994;**5**(1):96-101. DOI: 10.1109/72.265964

[55] Koza JR, Programming G. On the Programming of Computers by Means of Natural Selection. Cambridge, Massachusetts, USA: The MIT Press; 1992. ISBN 978-026211170-6

[56] Kennedy J, Eberhart RC. Swarm intelligence. San Francisco, California, USA: Morgan Kaufmann; 2001. ISBN 978-155860595-4

[57] de Castro L, Timmis J. Artificial Immune Systems: A New Computational Intelligence Approach. Springer; 2002. ISBN 978-1-85233-594-6

[58] Storn R, Price K. Differential evolution—A simple and efficient heuristic for global optimization over continuous spaces. Journal of Global Optimization. 1997;**11**(4):341-359. DOI: 10.1023/A:1008202821328

[59] Dorigo M, Di Caro G. The ant colony optimization meta-heuristic. In: Corne D, Dorigo M, Glover F, editors. New Ideas in Optimization. McGraw-Hill, UK; 1999. pp. 11-32

[60] Dorigo M, Stützle T. Ant Colony Optimization. Cambridge, Massachusetts, USA: The MIT Press; 2004. ISBN 978-026204219-2

[61] Miramontes Vidal OR. Algunos Aspectos de la Teoría de Autómatas Celulares y sus Aplicaciones en Biofísica, Tesis de Licenciatura (Física). UNAM: Facultad de Ciencias; 1988

[62] Arce Rincón JH. Estados de Mínima Energía en una Red a Temperatura Cero, Tesis Doctoral (Doctor en Ciencias (Física)). UNAM: Facultad de Ciencias; 1990

[63] Miramontes Vidal PE. Un Esquema de Autómata Celular como Modelo Matemático de la Evolución de los Ácidos Nucleicos, Tesis Doctoral (Doctor en Ciencias (Matemáticas)). UNAM: Facultad de Ciencias; 1992

Chapter

Microbiological Opportunities for the Sustainable Production of Safe Food

Guadalupe Virginia Nevárez-Moorillón, Dayanira Morales-Corral and Daniela Sánchez-Aldana

Abstract

When it comes to sustainably producing safe food, microorganisms can be either a tool or a challenge, considering their role in food deterioration or production. By understanding the behavior of microorganisms in different food production processes, we can identify ways to reduce waste and energy use and improve food safety. For example, using natural fermentation processes to preserve food can increase its shelf life and enhance its nutritional value and flavor. Additionally, probiotics and prebiotics can promote gut health and boost the immune system. Analyzing these microbiological challenges can create a more sustainable and safer food system for everyone. On the other hand, considering the biorefinery concept applied to food production and consumption systems, a sustainable approach would include reducing food loss and waste and revalorizing agroindustrial waste residues to obtain valuable products. In all cases, searching for microorganisms that can help produce safe foods or the reutilization of waste products is possible. Bioprospection of diverse environments can lead to the identification of novel microbial strains that can be used for the abovementioned purposes.

Keywords: biopreservation, fermented foods, food loss and waste, lactic acid bacteria, sustainability

1. Introduction

From ancient times, human societies have been bound to the ability to secure enough food to meet basic survival needs and facilitate its preservation for future use. Food preservation was a key development in the origin of sedentary societies. Agriculture, among the oldest human activities, is connected to the skill of preserving food, initially achieved through techniques developed over centuries based on observation. However, the emergence of food microbiology and food engineering elevated food preservation from an art to a science. This transformation allowed for food processing, distribution, and marketing with confidence in product quality and safety. Besides food production, challenges related to ensuring food security involve affordable healthcare, clean water, clean energy, proper housing, and basic education. Therefore, sustainable food production needs to be studied from an interdisciplinary perspective, much in light of the complexity theory.

Among the numerous challenges of modern global society, the Sustainable Development Goals set by member countries of the United Nations represent a short-term target we all should contribute to their achievement [1]. They serve as a guide to direct scientific and social efforts, are a compass for the required public policies, and are a foundational element in educational strategies at all levels. The goals are linked to the environmental, health, and sustainable use of natural resources. They encourage peace and respect for cultural diversity in all countries across the globe.

One of these objectives is Zero Hunger, which emphasizes the need to promote food security for all. According to the World Health Organization's definition, "Food Security exists when all people, at all times, have physical, social, and economic access to sufficient, safe, and nutritious food that meets their dietary needs and food preferences, allowing them to lead an active and healthy life" [2]. Food security also implies that the food is safe and free from toxins or microorganisms that could compromise its nutritional quality or threaten consumer health. Consequently, there is an ongoing proposal for food preservation strategies, utilizing methods that prolong the shelf life of products or prevent contamination by pathogenic microorganisms. In recent years, efforts in this area have focused on using natural antimicrobials, developing intelligent coatings to maintain the freshness of food, and biocontrol of pathogens using beneficial microorganisms.

On the other hand, the industrial processing of food generates substantial amounts of agro-industrial waste. These residues are an environmental challenge due to their pollution potential, as significant quantities are produced and are not easily degraded or eliminated. The valorization of agroindustrial waste involves deriving valuable products from the waste produced during food processing. It is possible to utilize these residues to obtain other products, such as a substrate for the growth of microorganisms that produce beneficial metabolites or energy production [3].

In this chapter, we will be reviewing the current challenges in producing sufficient and safe food, as well as using microorganisms to reduce food waste. The importance of interdisciplinary studies, such as food security, will also be discussed.

2. The challenge of food for all

The questions we constantly ask about nature enable us to shape and reshape our surroundings. The cumulative scientific knowledge accumulated throughout human history enhances our present quality of life and extends our life expectancy. While there are notable discrepancies in economic development among nations, global populations, as evidenced by statistical trends, have steadily risen in life expectancy. In the early 1800s, the average global life expectancy stood at approximately 35 years. By 1950, it had increased to 49 years; by 2012, it had increased to 70 years [4]. A newborn could potentially surpass the centenary mark due to advancements in medical science. Still, this prolonged life expectancy, while indicative of enhanced capabilities and enjoyment of life globally, poses a predicament due to increasing population growth.

In 1805, the world population was around one billion individuals; by 2023, this figure had ascended to 8 billion. Despite cultural shifts leading to diminished population growth rates, we face the challenge of sustainably feeding an exponentially expanding population [5]. Food production involves social, environmental, economic, and political dimensions. Agriculture currently consumes roughly 70% of global freshwater resources, with nearly half of arable land dedicated to food production. Furthermore, a substantial proportion of greenhouse gas emissions, approximately 26%, stem from food production and waste. The selective breeding of plants and animals for human consumption, subsequently utilized in food production, poses a significant risk to planetary biodiversity [6].

The concept of sustainability entails the concurrent economic, socio-cultural, and technological advancement of the human population, which satisfies the current needs while safeguarding the potential of natural resources for future generations. Sustainability involves balancing economic growth, environmental care, and social benefits. Remarkably, sustainability underpins several Sustainable Development Goals (SDGs) to be achieved by 2030 [1].

As such, some tools have emerged that allow us to measure the impact of human activities on ecosystems. Examples are determining carbon and water footprints, which can be calculated for individuals, communities, or entire nations. Another interesting concept is proposed by the Global Footprint Network, which suggests the concept of Earth Overshoot Day, which is the day when humanity has used up all the resources that can be renewed that particular year [7]. Based on this premise, the organization proposes actions that move toward sustainability and more efficient use of resources. Food production is included among the five areas identified as priorities to reduce the anthropogenic impact on ecosystems. There are examples of how the environmental impact of food production and consumption can be reduced by creating healthy habits for food production, transportation, and consumption.

As mentioned, food security ensures that people always have economic and physical access to sufficient nutritious and safe food to meet their dietary needs [2]. Like other issues previously addressed, food security is a complex problem involving economic, social, and environmental factors. An example is the lack of water for crop production that may not be easily accessible to a particular population. Therefore, food insecurity can be temporary or chronic; as a basic human right, it is our responsibility to reduce it worldwide.

Furthermore, nutritional integrity is an essential consideration within food security discourse, emphasizing the microbiological safety of food products to ensure they are innocuous. Efforts are, thus, directed toward implementing strict quality control measures and employing innovative preservation techniques to mitigate the proliferation of foodborne pathogens and enhance the shelf life of perishable foods.

3. Functional foods and food preservation

After the development of agriculture, the systematic production of plant-based foods, and livestock domestication, an associated challenge occurred: the necessity to maintain foodstuffs over prolonged periods of time while ensuring their safety for consumption (**Figure 1**). There are many examples of how society discovered, through trial and error, ways to preserve food: using natural preservatives such as salt or sugar or food processing techniques like solar drying are effective strategies to extend the shelf life of perishable foods. A well-documented example is how the cuisine of a society or community reflects the use of food additives that aid in food preservation; this is the case with spices and seasoned foods, which are now known to have antimicrobial properties [8].

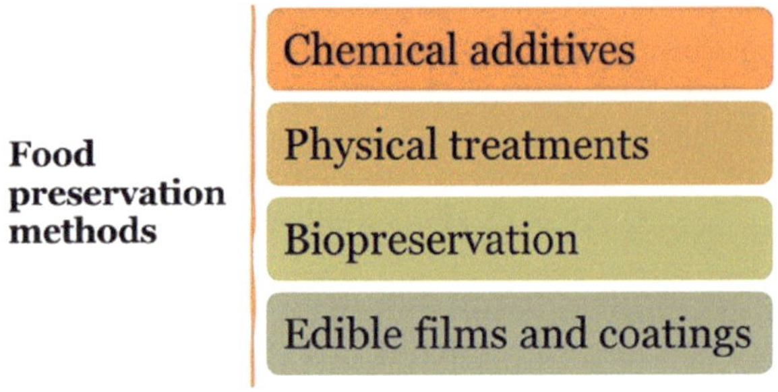

Figure 1.
Food preservation methods.

Another example is the emergence of fermented foods. While they arose spontaneously, foods such as cheese, wine, or fermented milk lasted longer than the original foods. Also, fermented foods have particular flavors and textures that make them appetizable. There are countless fermented products specific to each culture, but the principle is the same: using microorganisms to control pathogenic or food spoilage microorganisms while providing benefits to the consumers [9].

Food preservation methods are based on taking advantage of food product characteristics or components, using additives, or using processing techniques to control unwanted microbial growth. There are standard methods such as drying or adding compounds such as salt or sugars that have been used since ancient times, but there is constant innovation in methods to increase shelf life and food safety.

3.1 Food biopreservation by lactic acid bacteria

Lactic acid bacteria (LAB) are a heterogeneous group of non-sporulating, Gram-positive microorganisms with cocci or bacilli morphology, non-motile, anaerobic aerotolerant, and typically catalase-negative. Their defining characteristic is lactic acid production from lactose, serving as the final or principal product in sugar fermentation. These bacteria thrive in nutrient-rich environments, ranging from plant rhizospheres to various animal microenvironments, including the gastrointestinal tract of mammals [10]. Due to their adaptability to such habitats, LABs find applications in diverse fields such as health, agriculture, livestock, and the food industry. In the latter case, numerous examples of traditional foods result from lactic acid bacteria fermentation [11]. These include fermented vegetables such as sauerkraut or kimchi, bakery products such as San Francisco-style bread, and dairy products such as yogurt or cheese.

Lactic acid bacteria possess various metabolic properties that contribute to biological events such as glycolysis, proteolysis, lipolysis, and diacetyl production, which are important in producing fermented products. Lactic acid is crucial during the production of these products as it helps reduce pH and consequently reduces the risk of contamination by pathogens. Additionally, these bacteria promote the development of desirable sensory properties and ensure the stability and safety of the final product [11].

Recent years have witnessed an emphasis on studying autochthonous bacterial cultures in artisanal products, aiming to understand their diversity and dynamics in the production process to protect the designation of origin [12], add value to functional foods, and explore the preservative, technological, and functional properties of indigenous LAB [10, 11]. Among LAB's functional properties are those related to product preservation, including the production of organic acids, hydrogen peroxide, and, in some strains, peptide compounds known as bacteriocins [13].

Bacteriocins, for instance, hold promise as potential biopreservatives due to their ability to inhibit food pathogens. Moreover, LAB's production of hydrolytic enzymes significantly influences product preservation, taste, and texture. Also, LAB's proteases can generate bioactive peptides from milk proteins, showcasing various health-promoting activities such as antioxidants, immunomodulators, and hypotensive agents. While both bacteriocins and bioactive peptides derive from LAB fermentation, it is crucial to differentiate them; bacteriocins are ribosomally synthesized peptides, whereas bioactive peptides stem from milk protein proteolysis or other sources [14].

One of the fermented foods in which lactic acid bacteria participate is cheese, the result of the acidification process of milk from different mammal species. It is one of the most diverse fermented food products globally, with many variations in its preparation and final use. Processing conditions in different parts of the world are closely linked to historical, cultural, and geographical references [12]. Among the LAB involved in cheese production are two types of microorganisms: starter lactic acid bacteria (LAB) and non-starter lactic acid bacteria (NSLAB) [11]. Flavor is one of the most important criteria determining consumers' choice and acceptance of cheese. Cheese made from raw milk promotes the development of rheological characteristics and contributes mainly to this attribute; however, it has been associated with foodborne illness, promoting the production of cheeses made from pasteurized milk. Nevertheless, it has been demonstrated that pathogenic microorganisms are eliminated during the cheese-ripening process [15].

One of the traditional cheeses in Mexico is Chihuahua cheese, produced in the northern part of the country and originated from a fusion process between the region's merchants' tradition and Mennonite immigrants. Until a few years ago, it was made from unpasteurized milk, posing a sanitation problem and preventing its export [16]. If the ripening time were longer, the number of pathogens present would decrease [15], but the taste would not be pleasant for consumption. To identify the biotechnological capabilities of LAB present in Chihuahua cheese, a study was conducted to search for the antimicrobial and antioxidant capabilities of LAB isolated from cheeses made in Chihuahua [17]. The aim is to identify the molecules responsible for the functional capabilities for subsequent use.

Regarding the use of lactic acid bacteria in biopreservation, there are different views on their use. Although lactic acid bacteria are considered generally recognized as safe (GRAS), bioactive compounds can also be produced through fermentation to purify them and then added to foods. For instance, commercial production of antimicrobial compounds like nisin showcases the possibilities. However, using protective cultures offers added advantages by imparting additional sensory characteristics to foods.

3.2 Probiotics in the production of functional foods

The consumption of fermented dairy products has demonstrated a beneficial effect on health. Elie Metchnikoff's observation, proposing that the consumption of fermented dairy, particularly sour milk, contributed to the longevity of inhabitants in Eastern European mountainous regions, is widely acknowledged. In the twentieth century, the term "probiotics" was coined to refer to live microorganisms that, when administered in adequate quantities, confer health benefits to the host [18].

Dairy products are most frequently used to isolate probiotic strains, although they can also be obtained from less typical products such as fruits, vegetables, or meat products [19]. The selection of probiotic strains involves gastrointestinal survival criteria, antimicrobial activity, antibiotic resistance, lysozyme resistance, adhesion to intestinal epithelial cells, acid tolerance, and bile salt tolerance [20]. In addition to surviving gastrointestinal conditions, probiotics must also endure the conditions encountered during food processing and shelf life to be successful candidates for functional food. Bacterial cells of probiotics must remain viable through their passage through the gastrointestinal tract at populations of 106–107 CFU/g [19]. Remarkably, probiotics exhibit transient colonization in the gastrointestinal tract, with fecal detection typically lasting only a few weeks post-consumption. The list of benefits attributed to probiotic strains is extensive, although some effects are still under study to determine the mechanism of action. These benefits include reducing lactose intolerance, improving immune function, managing inflammatory bowel diseases, preventing and reducing diarrhea associated with rotavirus and antibiotics, and producing vitamins [21].

Regarding functional food formulation, prebiotics—non-digestible compounds such as lactulose, inulin, and oligosaccharides—serve as preferred substrates for probiotic growth, enhancing their viability and activity during processing and transit through the gastrointestinal tract [22]. While the health benefits of prebiotics and probiotics have been extensively documented through *in vitro* and clinical studies, elucidating their mechanisms of action remains a subject of ongoing research (**Figure 2**). A recent conceptual addition, "postbiotics," denotes the biologically active metabolites or cell remnants of probiotics, which traverse the intestinal barrier and exert beneficial effects upon various host sites *via* the circulatory system [21].

Postbiotics have been shown to benefit conditions such as irritable bowel syndrome [22] and other gastrointestinal problems [21], suggesting their potential therapeutic relevance. It has been observed that using probiotics in the elderly can improve

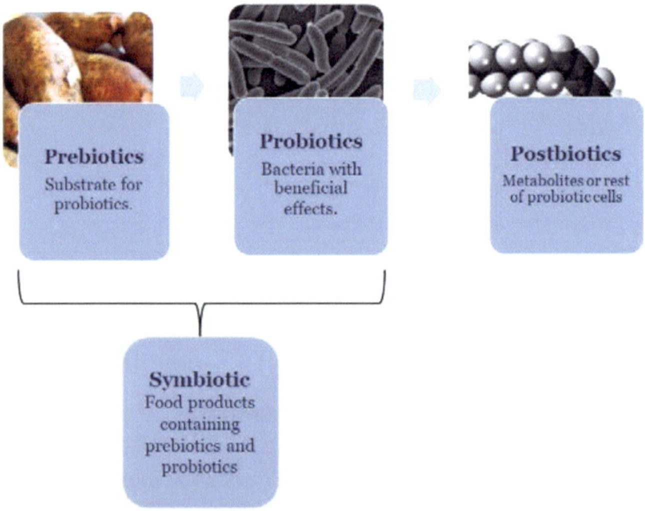

Figure 2.
Functional food products containing probiotics and their health effects.

many health conditions related to advanced age, such as constipation, malnutrition, and susceptibility to opportunistic pathogens [23].

4. Food loss and food waste

As mentioned before, food production has environmental impacts at different levels, from using almost half of habitable land to consuming almost three-quarters of the freshwater available [6]. Moreover, the transportation of food from production sites to end consumers imposes additional environmental burdens, which have intensified in recent years due to shifting global dietary patterns.

The issue's magnitude is further compounded by the revelation that approximately one-third of globally produced food remains unconsumed. This wastage may arise from physical, chemical, or microbiological degradation or production inefficiencies. It is imperative to distinguish between the juncture of food disposal and the concepts of food loss and waste (FLW) [18].

Food loss denotes the diminution of edible food quantities occurring from primary production stages through post-harvest handling, processing, and across the supply chain. On the other hand, food waste refers to the discarding of food from retail outlets to consumer households and waste from consumption practices (**Figure 3**). Notably, food loss includes provisions intended for human consumption and encompasses inedible portions of food items [18].

In a study conducted at a municipal landfill in Mexico, nearly 50% of the waste was organic matter, mainly consisting of fruit and vegetable peels and leftovers from prepared foods [24]. This highlights another severe environmental problem of food loss and waste: the generation of greenhouse gases by decomposing food waste. It is estimated that one-quarter of greenhouse gas pollutants worldwide result from the decomposition of unconsumed food [18].

Several strategies have been proposed to reduce food loss. Within the concept of a circular economy, the use of food unsuitable for human consumption or parts discarded during processing is proposed to generate other products. The valorization of food waste or its by-products presents a viable solution to alleviate the strain on supply chains at micro- and macro-levels within the global food system [3, 25]. Non-edible waste from food production and processing can be used as raw material

Figure 3.
Food loss and food waste.

in producing high-value-added components for food. Applications may include the preparation of flours, extraction of colorants, enzymes, or antioxidant compounds. These applications may be in the food and pharmaceutical or cosmetic industries [26]. One proposed approach for valorizing food waste is using microorganisms to obtain value-added products. Solid-state fermentation is an environmentally friendly methodology with many applications in obtaining high-value-added compounds [27, 28]. Efforts to reduce food loss and waste through circular economy and biorefinery strategies should consider the differences in their causes in developing and developed countries [25].

An interesting concept is food upcycling, which involves using products that might otherwise be destined for waste to generate products suitable for consumption. A distinction is made between two types of upcycled foods: those that have been incorporated into the production chain and are used in alternative ways to preserve their use and those that involve the use of food resources in novel ways [29].

It is commonly perceived that reducing food loss and waste would linearly mitigate the environmental impact associated with this issue. However, as posited by Cattaneo et al. [30], the geographical and socioeconomic context of FLW management significantly influences their efficacy, as the reduction of environmental footprints attributed to food loss centers upon production, processing, and transportation dynamics. On the other hand, the impact of reducing environmental impact by reducing consumer waste is similar worldwide. Therefore, consumer education strategies are crucial. Environmental education strategies can be implemented through formal education or awareness campaigns while reducing the impact of soil erosion and water pollution through government policies or international organizations.

In any case, addressing the complex issue of reducing food loss and waste requires the involvement of various disciplines, including microbiology. Improved food preservation schemes, the use of microorganisms in the valorization of agroindustrial waste, and fermentation as a method of biopreservation are contributions to addressing the issue.

5. The complexity of sustainable, safe food: What is next

The significance of sustainable food production cannot be overstated, necessitating an interdisciplinary approach to comprehend its complexity. Employing a multidisciplinary perspective, we must examine environmental ramifications, economic feasibility, and societal obligations. Undertaking measures toward sustainability and equity is imperative for shaping a more promising future for all.

An often-overlooked facet of food security pertains to the nutritional quality of available provisions for the human population. A considerable segment of the global population suffers from moderate to severe malnutrition, necessitating dietary supplementation for improved well-being. It is crucial to acknowledge that food security constitutes a fundamental human right. Nevertheless, the effort to provide nutritious food for all can impose additional burdens on the food production system. A necessity arises to redistribute food resources to mitigate the substantial percentage of unutilized food that increases the proportion of food loss and waste.

Despite numerous technological advancements in food processing and distribution, valuable insights remain to gather from traditional wisdom. Traditional culinary practices incorporate locally sourced ingredients that have demonstrated nutritional efficacy over decades. Antimicrobials derived from plants, such as those present in spices, not only help increase shelf life or reduce foodborne outbreaks but enhance the food's sensorial properties. The same can be mentioned for traditional fermented foods.

Alongside contemporary developments in food science and technology, educational initiatives at all levels are of paramount importance. Cultivating healthy dietary habits is decisive in mitigating metabolic diseases such as diabetes. Furthermore, adherence to sound food preparation practices can mitigate the food loss and waste problem. Ultimately, responsibly utilizing sustainably produced foods is a collective venture, and we all need to do our small part.

Conflict of interest

The authors have no conflict of interest to declare.

Author details

Guadalupe Virginia Nevárez-Moorillón*, Dayanira Morales-Corral and Daniela Sánchez-Aldana
School of Chemical Sciences, Autonomous University of Chihuahua, Circuito Universitario s/n, Campus Universitario II, Chihuahua, Mexico

*Address all correspondence to: vnevare@uach.mx

References

[1] UN General Assembly. Transforming our World: The 2030 Agenda for Sustainable Development. UN General Assembly; 2015, A/RES/70/1, Available from: https://www.refworld.org/docid/57b6e3e44.html [Accessed: January 28, 2024]

[2] FAO/WHO. Probiotics in Food. Health and Nutritional Properties and Guidelines for Evaluation - Report of a Joint FAO/WHO Expert Consultation on Evaluation of Health and Nutritional Properties of Probiotics in Food Including Powder Milk with Live Lactic Acid Bacteria. Rome, Italy: Food and Nutrition paper No. 85. FAO/WHO; 2006. 86 p

[3] Socas-Rodríguez B, Álvarez-Rivera G, Valdés A, Ibáñez E, Cifuentes A. Food by-products and food wastes: Are they safe enough for their valorization? Trends in Food Science & Technology. 2021;**114**:133-147. DOI: 10.1016/j.tifs.2021.05.002

[4] Roser M. Life Expectancy Increased in all Countries of the World. 2015. Available from: 'https://ourworldindata.org/life-expectancy-increased-in-all-countries-of-the-world' [Accessed: January 28, 2024]

[5] Ritchie H, Rodés-Guirao L, Mathieu E, Gerber M, Ortiz-Ospina E, Hasell J, Roser M. Population Growth. 2023. Available from: https://ourworldindata.org/population-growth [Accessed: January 28, 2024]

[6] Ritchie H, Rosado P, Roser M. Environmental Impacts of Food Production. 2022. Available from: 'https://ourworldindata.org/environmental-impacts-of-food' [Accessed: January 28, 2024]

[7] Network GF. Open Data Platform. York Univ. 2022. Available from: https://data.footprintnetwork.org/index.html#/ [Accessed: January 28, 2024]

[8] Martínez-Gracía C, González-Bermúdez CA, Cabellero-Valcárcel AM, Santaella-Pascual M, Frontela-Saseta C. Use of herbs and spices for food preservation: Advantages and limitations. Current Opinion in Food Science. 2015;**6**:38-43. DOI: 10.1016/j.cofs.2015.11.011

[9] Rastogi YR, Thakur R, Thakur P, Mittal A, Chakrabarti S, Siwal SS, et al. Food fermentation–significance to public health and sustainability challenges of modern diet and food systems. International Journal of Food Microbiology. 2022;**371**:109666. DOI: 10.1016/j.ijfoodmicro.2022.109666

[10] Ayivi RD, Gyawali R, Krastanov A, Aljaloud SO, Worku M, Tahergorabi R, et al. Lactic acid bacteria: Food safety and human health applications. Dairy. 2020;**1**(3):202-232. DOI: 10.3390/dairy1030015

[11] Abdul Hakim BN, Xuan NJ, Oslan SN. A comprehensive review of bioactive compounds from lactic acid bacteria: Potential functions as functional food in dietetics and the food industry. Food. 2023;**12**(15):2850. DOI: 10.3390/foods12152850

[12] Licitra G. World wide traditional cheeses: Banned for business? Dairy Science & Technology. 2010;**90**(4):357-374. DOI: 10.1051/dst/2010016

[13] Linares-Morales JR, Gutiérrez-Méndez N, Rivera-Chavira BE, Pérez-Vega SB, Nevárez- Moorillón GV. Biocontrol processes in fruits and fresh

produce, the use of lactic acid bacteria as a sustainable option. Frontiers in Sustainable Food Systems. 2018;**2**:50. DOI: 10.3389/fsufs.2018.00050

[14] Venegas-Ortega MG, Flores-Gallegos AC, Martínez-Hernández JL, Aguilar CN, Nevárez-Moorillón GV. Production of bioactive peptides from lactic acid bacteria: A sustainable approach for healthier foods. Comprehensive Reviews in Food Science and Food Safety. 2019;**18**(4):1039-1051. DOI: 10.1111/1541-4337.12455

[15] Sánchez-Gamboa C, Hicks-Pérez L, Gutiérrez-Méndez N, Heredia N, García S, Nevárez-Moorillón GV. Microbiological changes during ripening of Chihuahua cheese manufactured with raw milk and its seasonal variations. Food. 2018;**7**(9):153. DOI: 10.3390/foods7090153

[16] Tunick MH, Van Hekken DL, Molina-Corral FJ, Tomasula PM, Call J, Luchansky J, et al. Queso Chihuahua: Manufacturing procedures, composition, protein profiles, and microbiology. International Journal of Dairy Technology. 2008;**61**(1):62-69. DOI: 10.1111/j.1471-0307.2008.00359.x

[17] Venegas-Ortega MG, Flores-Gallegos AC, Aguilar CN, Rodríguez-Herrera R, Martínez-Hernández JL, Nevárez-Moorillón GV. Multi-functional potential of presumptive lactic acid bacteria isolated from Chihuahua cheese. Food. 2020;**9**(3):276. DOI: 10.3390/foods9030276

[18] FAO. Global Food Losses and Food Waste – Extent, Causes, and Prevention. Rome: FAO; 2011. Available from: http://www.fao.org/docrep/014/mb060e/mb060e00.pdf [Accessed: January 29, 2024]

[19] Pavli FG, Argyri AA, Papadopoulou OS, Nychas GJE, Chorianopoulos NG, Tassou CC. Probiotic potential of lactic acid bacteria from traditional fermented dairy and meat products: Assessment by *in vitro* tests and molecular characterization. Journal of Probiotics & Health. 2016;**4**:157. DOI: 10.4172/2329-8901.1000157

[20] Rajoka MS, Hayat HF, Sarwar S, Mehwish HM, Ahmad F, Hussain N, et al. Isolation and evaluation of probiotic potential of lactic acid bacteria isolated from poultry intestine. Microbiology. 2018;**87**:116-126. DOI: 10.1134/S0026261718010150

[21] Li HY, Zhou DD, Gan RY, Huang SY, Zhao CN, Shang A, et al. Effects and mechanisms of probiotics, prebiotics, synbiotics, and postbiotics on metabolic disease targeting gut microbiota: A narrative review. Nutrients. 2021;**13**(9):3211. DOI: 10.3390/nu13093211

[22] Martyniak A, Medyńska-Przęczek A, Wędrychowicz A, Skoczeń S, Tomasik PJ. Prebiotics, probiotics, synbiotics, paraprobiotics and postbiotic compounds in IBD. Biomolecules. 2021;**11**(12):1903. DOI: 10.3390/biom11121903

[23] Ale EC, Binetti AG. Role of probiotics, prebiotics, and synbiotics in the elderly: Insights into their applications. Frontiers in Microbiology. 2021;**12**:631254. DOI: 10.3389/fmicb.2021.631254

[24] Gomez G, Meneses M, Ballinas L, Castells F. Characterization of urban solid waste in Chihuahua, Mexico. Waste Management. 2008;**28**(12):2465-2471. DOI: 10.1016/j.wasman.2007.10.023

[25] Dora M, Biswas S, Choudhary S, Nayak R, Irani Z. A system-wide interdisciplinary conceptual framework for food loss and waste mitigation

strategies in the supply chain. Industrial Marketing Management. 2021;**93**:492-508. DOI: 10.1016/j.indmarman.2020.10.013

[26] Osorio LL, Flórez-López E, Grande-Tovar CD. The potential of selected agri-food loss and waste to contribute to a circular economy: Applications in the food, cosmetic and pharmaceutical industries. Molecules. 2021;**26**(2):515. DOI: 10.3390/molecules26020515

[27] Leite P, Belo I, Salgado JM. Co-management of agro-industrial wastes by solid-state fermentation for the production of bioactive compounds. Industrial Crops and Products. 2021;**172**:113990. DOI: 10.1016/j.indcrop.2021.113990

[28] Izábal-Carvajal AL, Sepúlveda L, Chávez-González ML, Torres-León C, Aguilar CN, Ascacio-Valdés JA. Extraction of bioactive compounds via solid-state fermentation using *Aspergillus niger* GH1 and Saccharomyces cerevisiae from pomegranate peel. Waste. 2023;**1**(3):806-814. DOI: 10.3390/waste1030047

[29] Aschemann-Witzel J, Asioli D, Banovic M, Perito MA, Peschel AO, Stancu V. Defining upcycled food: The dual role of upcycling in reducing food loss and waste. Trends in Food Science & Technology. 2023;**132**:132-137. DOI: 10.1016/j.tifs.2023.01.001

[30] Cattaneo A, Federighi G, Vaz S. The environmental impact of reducing food loss and waste: A critical assessment. Food Policy. 2021;**98**:101890. DOI: 10.1016/j.foodpol.2020.101890

Chapter

Pharmaceutical Particle Engineering: Technology That Contributes to Drug Therapeutic Efficacy

Sergio Alcalá-Alcala

Abstract

Particle engineering (PE) is an interdisciplinary science that focuses on designing, controlling, and manipulating particles in a range of micro-nanometric sizes. It uses manufacturing methods under *bottom-up* and *top-down* approaches to create particles with specific properties and characteristics. Its applications in medicine and pharmacy impact the therapeutic efficacy of drugs, with important implications that encompass controlled drug release, tissue-specific targeting, more precise diagnostic methods, encapsulation of insoluble drugs, improvements in bioavailability, personalization of treatments, and the creation of advanced materials under a technological application approach that is designed for formulation or route of administration, such as pulmonary inhalation, for instance. However, PE still faces important challenges, such as standardization of production methods, minimization of toxicity, adequate quality control, and proper regulation of products. Despite these challenges, the potential benefits are significant, as it can improve the therapeutic efficacy of medications and reduce side effects; in addition, it can open new possibilities in the treatment of diseases and in the design of novel pharmaceutical dosage forms.

Keywords: particle engineering, micro-nanotechnology, powders, drug delivery systems, drug inhalation, pharmaceutical technology

1. Introduction

Particle engineering (PE) is an emerging discipline that combines interdisciplinary sciences such as engineering, physics, chemistry, formulation, colloids and interfaces, solid-state physics, aerosols and powders, and nanotechnology to control and manipulate materials with the aim of generating and/or modifying particles to give them desired properties of shape, size, size distribution, density, surface chemistry, inner structure, composition, and/or powder properties [1]. This area offers a rational strategy with scientific bases to design well-structured particles with well-defined intentions of use and applications, which requires having solid knowledge about the mechanisms by which particles are formed and the physics around their properties. PE stands at the forefront of obtaining pharmaceutical particles ranging from the nanoscale to the microscale.

In the realm of healthcare, medicine, and pharmaceutical technology, this burgeoning field has revolutionized the therapeutic efficacy of medications by offering innovative solutions for drug administration, delivery, and controlled release. The research, innovation, and development of new medicines using particle engineering had a significant boom in the last two decades, coinciding with the advances in the understanding of the mechanisms of pulmonary particle deposition and formulation strategies for this route of administration [2], especially in the development of pharmaceutical powders for inhalation.

The therapeutic efficacy of drugs experiences a profound enhancement through advances in PE. This ability to govern the functionality of the particle properties enables the optimization of drug bioavailability, tissue distribution, and stability. Moreover, PE unlocks novel prospects in targeted drug delivery to specific tissues or cells, thereby mitigating systemic toxicity and augmenting clinical efficacy [3]. In this regard, the application of nanoparticles, microspheres, and controlled-release systems designed through PE has transformed the way in which medications are administered and utilized. These systems facilitate sustained drug release, shield against enzymatic degradation, surmount biological barriers such as the blood-brain barrier, or can achieve deep zones like alveolar sacks, thereby extending the reach of treatments for previously untreatable ailments, or improving them by increasing the therapeutic efficacy and reducing the side effects [4].

Furthermore, PE significantly contributes to personalized medicine by tailoring drug dosing and release according to the specific needs of each patient. This fosters an improvement in therapeutic efficacy and a reduction in adverse side effects, culminating in more precise and less invasive medical care. In this context, we can explore how PE has revolutionized the therapeutic efficacy of medications, paving the way for the treatment of diseases and enhancing the quality of life for patients.

2. Aspects of safety and efficacy in medications

Health protection is a right of all human beings, and its care requires health services to which a patient can go and access to different supplies that are used to diagnose, treat, prevent, rehabilitate, or cure a condition that affects health; medications are the most used therapeutic technology. Each medication must achieve three basic characteristics or principles: (a) safety, as a measure of risk-benefit, where the difference between an effective dose and a dose that causes serious adverse reactions (therapeutic window or index) has been evaluated; (b) efficacy, known as the ability to produce the desired therapeutic effect; and (c) quality, a concept related to the control of its manufacturing process, which guarantees satisfaction of the therapeutic need for which it is intended [5]; see **Figure 1**. Thus, small errors in dosage and the use of an inappropriate administration route can have harmful clinical effects, which are also highly related to the type of system (medicine or formulation) that has been administered and its effectiveness [6].

When we talk about the effectiveness of a drug, we take into account two essential biopharmaceutical premises: first, it is necessary for the active pharmaceutical ingredient (API) to reach its site of action or target to achieve the therapeutic effect, and second, in order for it to be absorbed and interact with the therapeutic target, it

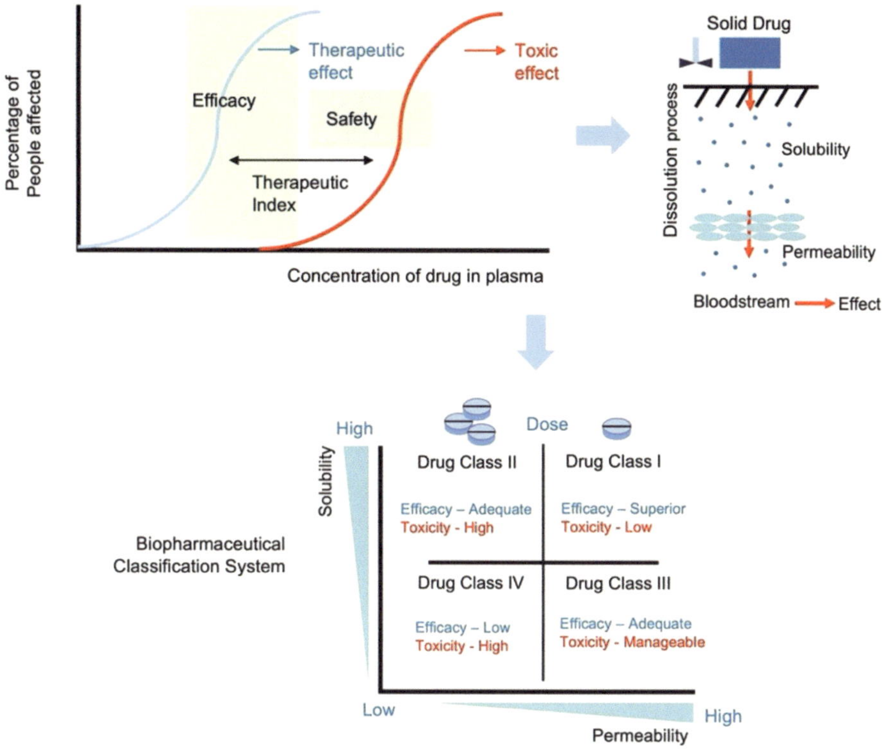

Figure 1.
Aspects of safety and efficacy in drugs.

must be dissolved in the fluids of the microenvironment and permeate. Thus, solubility is critical, both to define the dose to be used and the associated efficacy-toxicity; the Biopharmaceutical Classification System allows us to classify APIs in this sense (**Figure 1**). To meet the above implies having a vehicle or transporter (drug delivery system or pharmaceutical dosage form) that facilitates achieving both objectives [6]. However, a current problem is the fact that 90% of new APIs that are discovered or developed have poor aqueous solubility, which reduces bioavailability; that is why much of the research into new medications focuses on an approach to improving the solubility, dissolution rate, and the drug-controlled release [7].

In this context, the low aqueous solubility of APIs has led to the administration of drugs at high doses and with a greater frequency of administration, without taking into account the fact that the administration route is not always ideal for enhancing the therapeutic application; for example, many drugs that are administered orally (which is by far the most used route) are affected by pH changes in the gastrointestinal tract, enzymatic degradation, low permeability, hepatic first-pass metabolism, systemic effect that leads to various side effects, and an inefficient residence time on the site of action, which is necessary to achieve a therapeutic effect more efficiently [8]. These elements in some way violate the safety of medications, compromise effectiveness, waste resources, generate detachment from the treatment by the patient, and neglect vulnerable populations such as pediatric and geriatric patients.

What has research done in pharmaceutical technology? Considering that the conventional dosage form and administration can partially reach the objective of an efficient and safe therapeutic response and that, in the last 30 years, a better

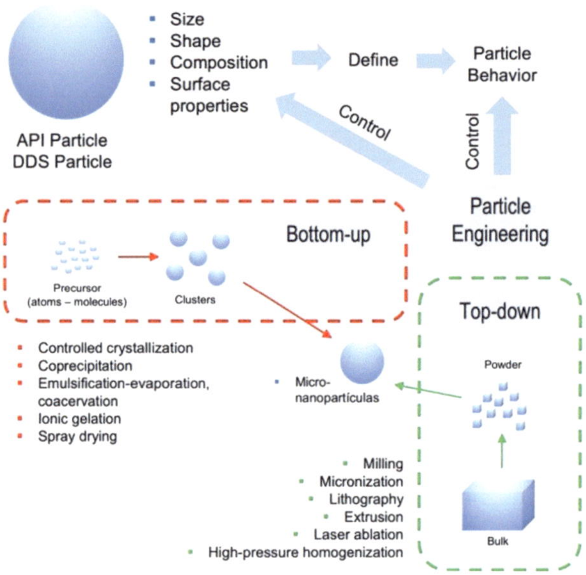

Figure 2.
Particle engineering; manufacturing approaches.

understanding of drug transport through membranes and tissues has been generated, new materials for pharmaceutical application have been developed and characterized, and emerging technologies such as micro-nanoencapsulation have been applied. Thus, research around drug development has focused its efforts on two important approaches: (a) The use of alternative routes of administration such as transmucosal (like pulmonary and nasal routes) and transdermal routes and (b) The development of pharmaceutical dosage forms with new and improved technological properties, either through improvements in existing technologies or in the development of novel drug delivery systems (DDS) [9]. Therefore, in addition to the physicochemical and biopharmaceutical properties, the shape, size, and composition of the API particles are determining factors in the physical, chemical, mechanical, and biological behavior of the drugs (**Figure 2**). Here, particle engineering has had a positive impact on both approaches.

3. Particle engineering (PE)

As mentioned above, PE is responsible for the design, manufacturing, and manipulation of particles, with the aim of controlling their physical, chemical, and biological properties to make them more useful and with better performance, so that, to understand PE, it is necessary to understand how particles form [10].

The formation of a particle occurs through a general process, which begins with nucleation, where precursor molecules group together to form small aggregates called nuclei, which are the basis on which particles develop. Subsequently, comes the growth stage, which is characterized by aggregating additional precursor molecules and by adding layers of material around the initial core. After that, it follows the agglomeration stage, where precursor particles tend to join. And finally, the stabilization stage, where the formation process could stop. It is important to highlight that the particle formation process can be highly influenced by a variety of factors, such as the nature of the materials, the kinetics of the reaction, the environmental conditions, and the presence of stabilizing agents [11].

3.1 Manufacturing methods

Currently, there are two main approaches in the application of particle engineering to obtain particles on the micro-nanometer scale: the *bottom-up* approach and the *top-down* approach (**Figure 2**). The *bottom-up* approach involves the construction of particles from the assembly of simpler components, at the atomic or molecular level, which allows fine control over the properties of the resulting particles; techniques such as controlled crystallization, coprecipitation, emulsification-solvent evaporation or solvent displacement, coacervation, ionic gelation, and spray drying are common. On the other hand, in the *top-down* approach, it starts with larger structures or materials that are gradually reduced to smaller sizes, by fragmentation; for example, it is similar to the idea of carving a piece of wood, where you start with a large block and then the material is removed to obtain the desired shape; techniques such as milling, micronization, lithography, extrusion, laser ablation, and the high-pressure homogenization are common [1, 12, 13].

Beyond the material used to obtain the particle, depending on the technique applied for its development and manufacturing, it is possible to observe differences in the particles regarding size distribution, surface properties (porosity), morphology, mechanical behavior, solid state properties, and even biointeraction properties [14].

3.2 PE applications

Since the last 20 years, the application of PE in drug development has been a rapidly growing field, as it is the key to the compression of various unit operations carried out in the pharmaceutical industry. In addition, PE has offered numerous benefits to pharmaceutical development, in terms of therapeutic efficacy, bioavailability, controlled drug delivery, and reduction of side effects [15].

When it comes to particles that are used in therapy as medications, there are particle engineering factors that influence on their effectiveness, as follows:

- Optimized manufacturing processes—shape and size particles determine the flow of materials through production equipment.

- A better administration—the size of the particles determines the ease and comfort when they are introduced into the body.

- Permanence of the therapeutic system at the site of application—the functionalization of the surface of the particle helps to identify tissues or adhere to the organism, for example, the use of mucoadhesion phenomena.

- Stability of the formulation in the body—the form and composition of particles reduce elimination or degradation in the site.

- Reach of the site of action—size determines if particles reach the tissues; for example, in the case of solid aerosols, particles need to get deeper into the lungs.

Some important application contexts for PE include controlled drug release, as it is possible to design systems that can deliver drugs sustainably over long periods, reducing the need for frequent dosing and minimizing adverse reactions. It also represents improvements in bioavailability, since modifying the shape and size of the particles can increase the solubility and permeability of drugs, facilitate the absorption of drugs through biological membranes, and improve their distribution in specific tissues. On the other hand, the encapsulation of drugs in particles using PE can protect them from enzymatic degradation, rapid metabolism, and environmental conditions that impact their storage. However, with PE, particles can be designed to target specific tissues, cells, or intracellular compartments, improving therapeutic efficacy and reducing systemic toxicity. Finally, PE allows for the personalization of treatments by adapting drug dosing, release, and administration for the specific needs of each patient [15–17].

There are different solid and liquid pharmaceutical dosage forms in the pharmaceutical and biotechnology market that are based on micro-nanoparticles that have been obtained through PE. Companies such as Pfizer, Novartis, GP Pharm, Astra Zeneca, Mannkind, Alnylam, Genentech, Vertx, Nova Laboratories, Janssen, among others, have placed medications on the market for oral administration (capsules and tablets), for inhalation, parenteral (injectable), and patches for transdermal administration, all of them with drugs that are used to treat diseases such as asthma, cancer, diabetes, acquired immunodeficiency syndrome (AIDS) caused by human immunodeficiency virus (HIV), cystic fibrosis, chronic obstructive pulmonary disease (COPD), hepatitis C, immunosuppressants, antibiotics to treat infections, among others, whose benefits have generated a growth of the pharmaceutical market and a better quality of life for patients [1, 14].

Thus, the application of particle engineering in drug development offers new opportunities to improve the effectiveness and safety of treatments, as well as to address complex challenges in the field of medicine and the pharmaceutical industry.

4. Case studies: On-site developments

In the Pharmaceutical Technology Research Laboratory at the Faculty of Pharmacy of the Autonomous University of the State of Morelos, in Cuernavaca, Mexico, different projects have been developed in which PE is applied to design, obtain, optimize, characterize, formulate, and evaluate *in vitro* and *in vivo* micro-nanoparticulate systems, as proposals for the development of new pharmaceutical dosage forms for modified drug release. The above, with the aim of offering improved alternatives to current treatments for diseases that afflict the population, such as hypercholesterolemia, infections, arterial hypertension, diabetes, asthma, and lung cancer.

The research strategies have been defined based on the following intentions: obtain pharmaceutical systems that are purposeful in improving treatments for diseases of current interest, generate new technologies (micro-nanoparticles or composites), protect the integrity of the drug (stability), search for control in the drug release rate, establishment of manufacturing processes achievable on an industrial scale, and finally, create drug delivery systems that can be administered by alternative routes, mainly by inhalation, such as the pulmonary and nasal routes. In the following sections, some successful study cases in the development of a drug delivery system (DDS) applying PE are summarized.

4.1 Microparticles to treat pneumonia

Pneumonia is an infectious disease that affects the lungs, specifically at the bronchiole zone. When the infection is caused by *Streptococcus pneumoniae*, symptoms such as cough, fever, difficulty breathing, and general malaise appear. Treatment includes the use of penicillin antibiotics such as amoxicillin, in combination with macrolides such as azithromycin. The latter was the drug of interest in the study, as it has low aqueous solubility, which leads to low bioavailability (barely 37%), and therefore to the use of high doses (500 mg every 12–24 hours). These drugs are commonly administered and formulated in tablets or capsules, by the oral route, which has generated resistance, low efficacy, side reactions, and low concentrations of antibiotic at the site of action (into the lungs) [18, 19].

In this way, applying PE and using a microencapsulation method by emulsification-solvent evaporation, mucoadhesive microparticles of a polymer called (poly) methyl vinyl ether maleic anhydride (PVM/MA) were designed and obtained. The particles obtained had a spherical, solid shape and presented an average size of 1.62 ± 0.47 μm, measured as mean aerodynamic mass diameter (MMAD) and a density of 0.36 ± 0.02 mg/mL, which were ideal for pulmonary administration. It is known that only particles between 1 and 3 μm reach the deep areas of the lung and formulations with bulk density < 0.4 mg/mL can be aerosolized (**Figure 3**). The particles deliver the antibiotic in a sustained drug release for 24 hours, using a dose 10 times lower compared to oral dose, when it is administered directly to the lung, suggesting a reduction in the amount and frequency of administration. The mucoadhesive particle dissolves in the lungs, while remaining attached to perform its function.

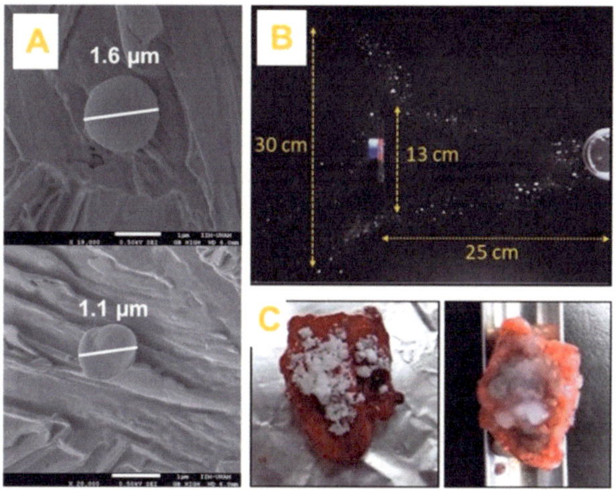

Figure 3.
Microparticles of PVM/MA loaded with azithromycin. (A) Micrographs by scanning electron microscope (SEM), (B) aerosolization capacity, and (C) mucoadhesion in pig lungs. Taken and modified from [20].

These particles were formulated with excipients for inhalation (four types of lactose) and were applied to bacterial cultures, observing that the drug concentration necessary to kill them is half of that required if applied conventionally [20].

4.2 Micro-nanocomposites to treat arterial hypertension

According to the WHO, high blood pressure affects more than 30% of the world's adult population and is the main risk factor for suffering from cardiovascular diseases. There are several classes of antihypertensive medications that are used to treat it, including diuretics, angiotensin II blockers, angiotensin-converting enzyme inhibitors, beta-blockers, and calcium channel blockers, which include drugs such as felodipine. This drug is prescribed as the first choice in antihypertensives; however, it has a water solubility of <1 μg/mL (low solubility), with an oral bioavailability of only 16%, which leads to the administration of 5–20 mg every 12–24 hours. It is found in 2.5, 5, and 10 mg tablets that are administered orally. Its side effects generate a high detachment from the treatment on the part of the patient [21, 22].

With this background in this regard, composites with aerodynamic properties were designed to treat hypertension by administering a dry powder through the lungs. A composite is a system formed by two or more materials or structures that are perfectly identifiable; in this development, biodegradable nanoparticles loaded with felodipine were embedded into mucoadhesive microparticles. In recent years, the pulmonary route has been proposed to treat systemic diseases in a safer, more effective, and economical way. The development of composites was designed using PE by a method that consists of the covering of nanoparticles in suspension.

Felodipine was encapsulated with 99.5% efficiency in nanoparticles of polylactic-glycolic acid (PLGA), a biodegradable polymer. The nanoparticles were obtained with an average size of 270 ± 20 nm, ideal for reaching the alveoli and thus favoring their arrival into the bloodstream; they were spherical and solid in shape. These nanoparticles were coated with a mucoadhesive polymeric material (PVM/MA), obtaining a spherical micro-nanocomposite, with an MMAD of 1.07 ± 0.026 μm, that is, with adequate properties to be formulated as a dry powder and administered by inhalation. The system released the drug for 30 days. To evaluate its effectiveness, it was administered by nebulization to hypertensive rats, showing a reduction in blood pressure in the first 3 hours, maintaining it in control for 5 days, with a single administration (**Figure 4**). The above suggests a great reduction in the frequency of administration and the dose to be administered without going through the systemic effect of oral administration, which will reduce side effects and improve the quality of life of patients suffering from hypertension [23, 24].

4.3 Lipid nanocarriers to treat lung cancer

Lung cancer is the second cause of cancer death worldwide and is one of the six most common types of cancer, with non-small cell lung cancer (NSCLC) being the most common (80% incidence). Its treatment includes surgery, when possible, radiotherapy, and chemotherapy above all when the disease is advanced stage (metastasis); it is known that 70% of patients are diagnosed in an advanced stage. Among the drugs used in chemotherapy, it is found that vinorelbine, a high molecular weight molecule, is sensitive to light and temperature, which is administered in a usual dose of 25–60 mg/m^2/week. It is formulated as an intravenous solution that is irritating and has generated trauma in the blood vessels, in addition to the systemic effect that is

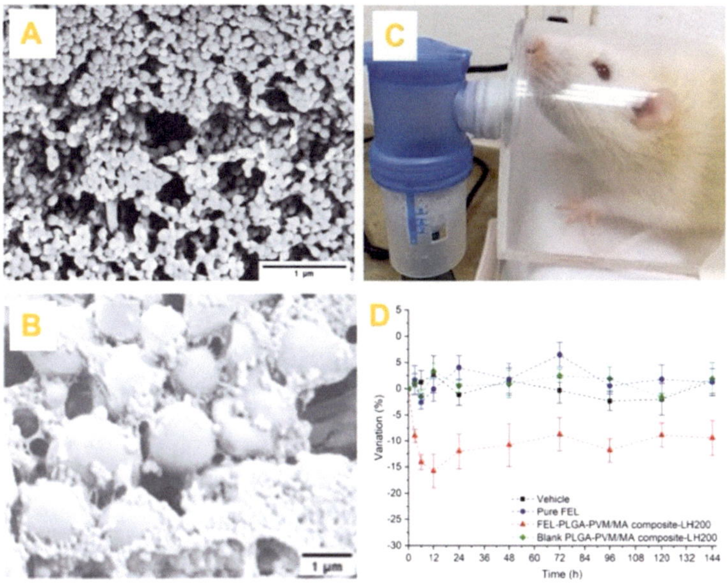

Figure 4.
Micro-nanocomposites of PLGA-PVM/MA loaded with felodipine. (A) Micrograph of PLGA nanoparticles by scanning electron microscope (SEM), (B) micrograph of PLGA-PVM/MA composites (SEM), (C) administration of composites by pulmonary route, and (D) variation in percentage (%) of systolic blood pressure in rats after pulmonary administration; red line shows the pressure reduction effect due to composites. Taken and modified from [23, 24].

common in chemotherapy, also affecting healthy cells. Therefore, a more targeted and localized administration in the cancerous nodules would be more convenient for the patient [25, 26].

The drug delivery system developed using PE was a nanostructured lipid carrier (NLC), which was designed with technological properties to be aerosolized and administered directly to the lung by inhalation. NLCs are biodegradable systems that provide high stability to drugs and offer sustained release. Using an emulsification-solvent evaporation method, spherical and homogeneous NLCs were obtained, with a size of 0.93 ± 0.011 μm and an encapsulation efficiency of 27%, which is equivalent to a load of 60.95 mg of encapsulated vinorelbine. This size was already ideal for pulmonary administration; however, to favor its permanence in the site, the surface was functionalized with 0.16% chitosan (a biodegradable mucoadhesive polymer), giving an MMAD of 1.6 ± 0.30 μm (**Figure 5**), suitable for reaching the areas deep and middle parts of the lung, which is where the nodules are located. The NLCs were dried with the spray-drying technique and formulated with excipients for inhalation (lactose), achieving a sustained release of a dose of 10 mg of drug for 48 hours (2 days). Currently, the system is in a stage of evaluating its effectiveness in a lung cancer cell model [27].

4.4 Lipid nanocarriers to treat chronic asthma

Asthma is a chronic disease characterized by inflammation and narrowing of the airways, making it difficult to breathe. Treatment usually involves a combination of control measures and rescue medications to relieve acute symptoms. Available

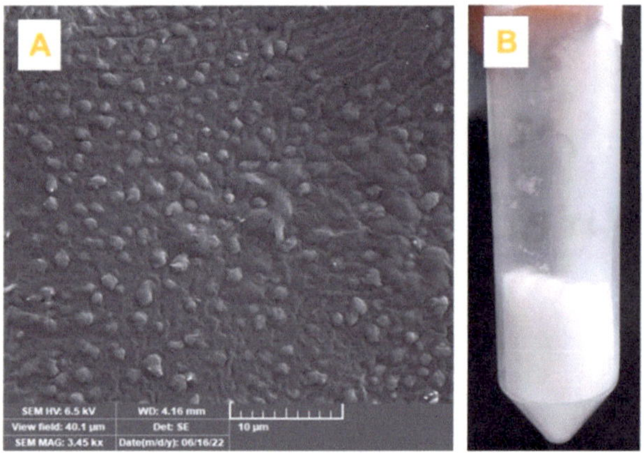

Figure 5.
NLC loaded with vinorelbine. (A) Micrograph of NLC by scanning electron microscope (SEM), (B) NLC dry powder for inhalation. Taken and modified from [27].

medications include anti-inflammatory and bronchodilator drugs that are formulated in solution and inhaled using a pressurized cartridge, for example, albuterol and salbutamol. However, these medications in most cases are inefficient, as the formulation and the patient's coordination with the action of the device limit the dose that reaches the bronchi, leading to frequent doses, swallowing the dose and generate side effects, without considering that they are only administered in acute situations. This has led to the proposal of new molecules, such as flavones of natural origin, which have been shown to have an efficient bronchodilator effect, but they have the problem of being highly insoluble. Likewise, the formulation of a dry powder system, in contrast to a pressurized cartridge, eliminates the negative effect of patient coordination with their device and provides greater stability to the drug, while prolonged release offers alternatives to address the effects of chronic asthma [28].

To provide an alternative that overcomes the above concerns, a dry powder based on NLC was developed, loaded with 6-hydroxyflavone, an insoluble flavone with a short half-life of just 6 hours. Through an emulsification-solvent evaporation process, NLCs were obtained with an average size of 0.62 ± 0.02 μm, spherical in shape and with a smooth surface, with a drug encapsulation efficiency of 83%.

These particles were also functionalized on their surface with chitosan to give them retention properties in the bronchi, generating an MMAD of 2.14 ± 0.6 μm when formulated with excipients for inhalation. These already functionalized particles generate a prolonged release of the drug for 72 hours (3 days), which suggests a reduction in the frequency of administration (**Figure 6**). To support this finding, the efficacy of the particles was evaluated by applying them to the mouse trachea, evidencing a constant bronchodilation effect during the study time, compared to a maximum effect of 1 hour for the unformulated molecule. These results allowed us to offer an alternative to the treatment of asthma, especially in patients who have chronic conditions [29].

4.5 Nanoplexes to treat hypercholesterolemia

Hypercholesterolemia is a condition characterized by elevated levels of blood cholesterol, specifically low-density lipoprotein (LDL), commonly known as

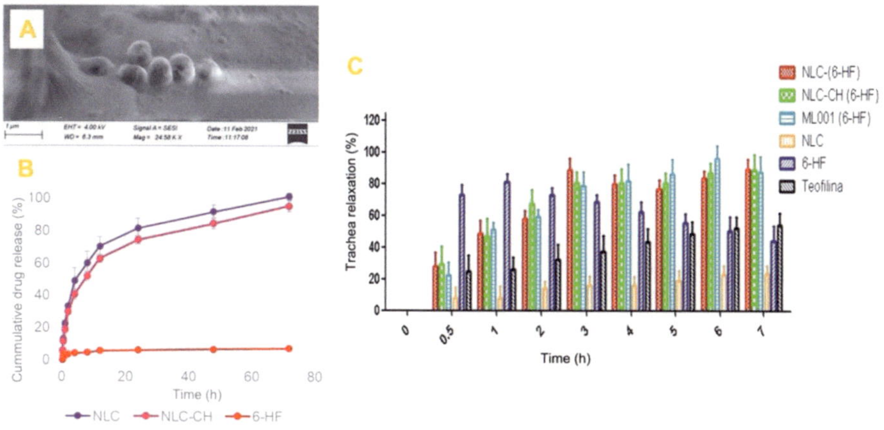

Figure 6.
NIC loaded with 6-hydroxyflavone. (A) Micrograph of NIC by scanning electron microscope (SEM), (B) drug release profile, (C) relaxation in percentage (%) of lise muscle of rat trachea; red, green and blue bars shows the bronchorelaxant effect using the NLC-formulations. Taken and modified from [29].

"bad cholesterol," and is a major risk factor for the development of cardiovascular diseases. It has been reported that 5–20% of the population has cholesterol levels above 240–250 mg/dL. Treatment for hypercholesterolemia focuses on reducing LDL cholesterol levels in the blood; among the drugs available are statins, the first-line medications. Statins are molecules that inhibit the enzyme HMG-CoA reductase, which is responsible for cholesterol synthesis in the liver. Within statins atorvastatin is found, a drug that has low aqueous solubility <0.01 mg/mL, with an oral bioavailability of only 12% and 14%. The normal starting dose of atorvastatin is 10 mg, available as tablets for oral use, and may reach a maximum of 80 mg per

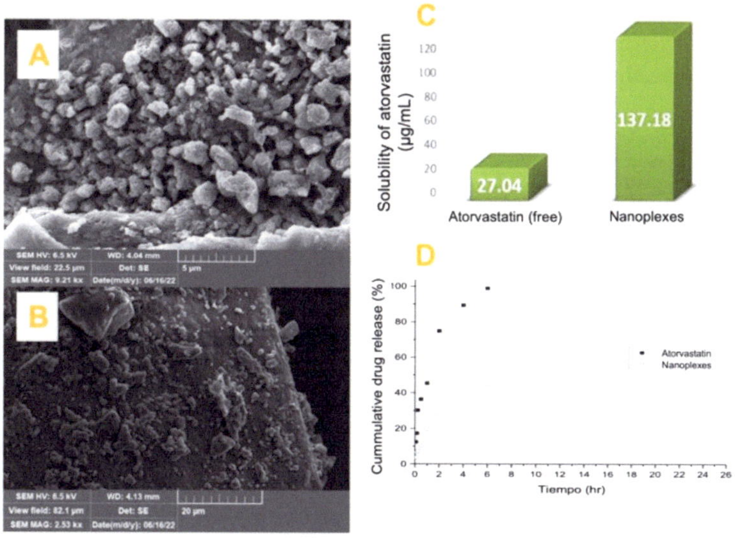

Figure 7.
Nanoplexes of atorvastatin. (A) Micrograph of nanoplexes by scanning electron microscope (SEM), (B) micrograph of nanoplexos formulated with lactose (SEM), (C) atorvastatin solubility improvement, (D) drug release profile. Taken and modified from [31].

day. Atorvastatin can cause side effects such as diarrhea, heartburn, gas, and joint pain, causing the patient to neglect treatment [30].

To solve the problem of low solubility, polymeric nanoplexes were developed using the PE method called controlled nanoprecipitation, which are micro-nanoparticles that are generated by interaction between the charges of the molecule of interest and a polyelectrolyte (charged polymer). In this study, the polymer used was chitosan, which is a biodegradable and mucoadhesive material that has a positive charge when found in a slightly acidic medium. Particles with a slightly ovoid morphology, with a rough surface, with an average size of 0.98 ± 0.02 μm and a complexation efficiency of 94.5% were generated. These nanoplexes release atorvastatin sustainably for more than 24 hours, at an initial dose of 10 mg. For its part, the aqueous solubility of the drug went from 27.04 to 137.18 μg/mL, that is, a radius of improvement of up to five times more (**Figure 7**). These polymeric micro-nanoparticles were formulated with lactose for inhalation, presenting an MMAD of 1.15–1.72 μm, which, together with the density (0.25 g/mL), give these particles technological capabilities to be aerosolized and administered through the lungs, offering an alternative to the administration of atorvastatin. This project is currently in the testing phase in rats with high cholesterol levels, being administered by nebulization [31].

5. Conclusions

In conclusion, the application of particle engineering in drug development and medicine represents a significant advance that promises to transform the way drugs are administered, delivered, and used. Throughout this writing, we have explored the various benefits that PE brings to the pharmaceutical industry and the field of medicine. From controlled drug delivery to improving bioavailability and personalizing treatments, PE offers innovative solutions to address complex challenges in disease treatments. The ability to engineer particles with specific properties, target them to specific sites in the body, and control their release has revolutionized the therapeutic efficacy of drugs and expanded the scope of treatments for previously intractable diseases. Thus, the future in the application of this science is related to the perspectives in the development and obtaining of molecules with therapeutic activity, as a large percentage of them present solubility and bioavailability problems, as well as to the perspectives of optimizing formulations and manufacturing processes that lead to the generation of more effective and accessible medicines.

However, it is important to highlight the need to address the challenges and concerns related to the safety, quality, toxicity, and regulation of products developed through PE. It is necessary for continuous evaluation of the risks and benefits to ensure that these technologies are used ethically and responsibly for the benefit of public health. That is why particle engineering represents a powerful tool that is transforming the landscape of modern drug development, providing hope for more effective, safe, and personalized treatments that will improve the quality of life of millions of people around the world.

Conflict of interest

The author declares no conflict of interest.

Thanks

I would like to thank my collaborators for making it possible to acquire experience and knowledge in this area, and especially to my students because without them, we would not have results and contributions in this science.

Author details

Sergio Alcalá-Alcala
Autonomous University of State of Morelos, Cuernavaca, Mexico

*Address all correspondence to: sergio.alcala@uaem.mx

References

[1] Cun D, Zhang C, Bera H, Yang M. Particle engineering principles and technologies for pharmaceutical biologics. Advanced Drug Delivery Reviews. 2021;**174**:140-167. DOI: 10.1016/j.addr.2021.04.006

[2] Vehring R. Pharmaceutical particle engineering via spray drying. Pharmaceutical Research. 2008;**25**(5):999-1022. DOI: 10.1007/s11095-007-9475-1

[3] Mitchell MJ, Billingsley MM, Haley RM, Wechsler ME, Peppas NA, Langer R. Engineering precision nanoparticles for drug delivery. Nature Reviews. Drug Discovery. 2021;**20**(2):101-124. DOI: 10.1038/s41573-020-0090-8

[4] Di Stefano A. Nanotechnology in targeted drug delivery. International Journal of Molecular Sciences. 2023;**24**(9):8194. DOI: 10.3390/ijms24098194

[5] Ross JS, Dzara K, Downing NS. Efficacy and safety concerns are important reasons why the FDA requires multiple reviews before approval of new drugs. Health Affairs (Millwood). 2015;**34**(4):681-688. DOI: 10.1377/hlthaff.2014.1160

[6] Garjón Parra J. Evaluation and selection of medicines. Farmacéuticos de Atención Primaria. 2011;**9**(3):89-94. Available from: https://www.elsevier.es/es-revista-farmaceuticos-atencion-primaria-317-pdf-X2172376111012370

[7] Sun D, Gao W, Hu H, Zhou S. Why 90% of clinical drug development fails and how to improve it? Acta Pharmaceutica Sinica B. 2022;**12**(7):3049-3062. DOI: 10.1016/j.apsb.2022.02.002

[8] Bhalani DV, Nutan B, Kumar A, Singh Chandel AK. Bioavailability enhancement techniques for poorly aqueous soluble drugs and therapeutics. Biomedicine. 2023;**10**(9):2055. DOI: 10.3390/biomedicines10092055

[9] Mignani S, El Kazzouli S, Bousmina M, Majoral JP. Expand classical drug administration ways by emerging routes using dendrimer drug delivery systems: A concise overview. Advanced Drug Delivery Reviews. 2013;**65**(10):1316-1330. DOI: 10.1016/j.addr.2013.01.001

[10] Schenck L, Koynov A, Cote A. Particle engineering at the drug substance, drug product interface: A comprehensive platform approach to enabling continuous drug substance to drug product processing with differentiated material properties. Drug Development and Industrial Pharmacy. 2019;**45**(4):521-531. DOI: 10.1080/03639045.2018.1562467

[11] Ismat SS, Hassnain J, Yassitepe E, Ali B. Evaporation: Processes, bulk microstructures, and mechanical properties. In: Peter MM, editor. Handbook of Deposition Technologies for Films and Coatings. 3rd ed. William Andrew Publishing; 2010. pp. 135-252. DOI: 10.1016/B978-0-8155-2031-3.00004-1

[12] Abid N, Khan AM, Shujait S, Chaudhary K, Ikram M, Imran M, et al. Synthesis of nanomaterials using various top-down and bottom-up approaches, influencing factors, advantages, and disadvantages: A review. Advances in Colloid and Interface Science. 2022;**300**:102597. DOI: 10.1016/j.cis.2021.102597

[13] Biswas A, Bayer IS, Biris AS, Wang T, Dervishi E, Faupel F. Advances

in top-down and bottom-up surface nanofabrication: Techniques, applications & future prospects. Advances in Colloid and Interface Science. 2012;**170**(1-2):2-27. DOI: 10.1016/j.cis.2011.11.001

[14] Arpagaus C. Pharmaceutical particle engineering via nano spray drying—Process parameters and application examples on the laboratory-scale. International Journal of Medical Nano Research. 2018;**5**:026. DOI: 10.23937/2378-3664.1410026

[15] Thomas F. Putting a spotlight on particle engineering. Pharmaceutical Technology. 2022;**46**:24-25. Available from: https://www.pharmtech.com/view/putting-a-spotlight-on-particle-engineering

[16] Sun CC, Davé RN. Crystal and particle engineering—An indispensable tool for developing and manufacturing quality pharmaceutical products. Pharmaceutical Research. 2022;**39**(12):3041-3045. DOI: 10.1007/s11095-022-03449-x

[17] Tahara K. Pharmaceutical formulation and manufacturing using particle/powder technology for personalized medicines. Advanced Powder Technology. 2020;**31**(1):387-392. DOI: 10.1016/j.apt.2019.10.031

[18] Yang W, Peters JI, Williams RO. Inhaled nanoparticles-a current review. International Journal of Pharmaceutics. 2008;**356**(1-2):239-247. DOI: 10.1016/j.ijpharm.2008.02.011

[19] Bakheit AH, Al-Hadiya BM, Abd-Elgalil AA. Azithromycin. In: Profiles of Drug Substances, Excipients and Related Methodology. Vol. 39. 2014. pp. 1-40. DOI: 10.1016/B978-0-12-800173-8.00001-5

[20] Mastache-Juárez A. Desarrollo y evaluación de un micro-nanosistema biodegradable y bioadhesivo para administración pulmonar en el tratamiento de la neumonía [thesis]. Faculty of Pharmacy: Autonomous University of the State of Morelos; 2019

[21] Okur ME, Karantas ID, Okur NU, Siafaka PI. Hypertension in 2017: Update in treatment and pharmaceutical innovations. Current Pharmaceutical Design. 2017;**23**(44):6795-6814. DOI: 10.2174/1381612823666170927123454

[22] Sahu BP, Das MK. Preparation and in vitro/in vivo evaluation of felodipine nanosuspension. European Journal of Drug Metabolism and Pharmacokinetics. 2014;**39**(3):183-193. DOI: 10.1007/s13318-013-0158-5

[23] Isabel CZ, Luis AD, Samuel ES, Elizabeth PS, Dea HR, Sergio AA. Novel mucoadhesive PLGA-PVM/MA micro-nanocomposites loaded with felodipine intended for pulmonary administration by nebulization. International Journal of Pharmaceutics. 2022;**628**:122295. DOI: 10.1016/j.ijpharm.2022.122295

[24] Cruz-Zazueta I. Obtención de micro-nanocompósitos mucoadhesivos para la administración pulmonar de delodipino: Caracterización, Formulación y Bioevaluación [thesis]. Faculty of Pharmacy: Autonomous University of the State of Morelos; 2022

[25] Lee WH, Loo CY, Traini D, Young PM. Inhalation of nanoparticle-based drug for lung cancer treatment: Advantages and challenges. Asian Journal of Pharmaceutical Sciences. 2015;**10**(6):481-489. DOI: 10.1016/j.ajps.2015.08.009

[26] Bahadori F, Topçu G, Eroğlu MS, Onyüksel H. A new lipid-based nano

formulation of vinorelbine. AAPS PharmSciTech. 2014;**15**(5):1138-1148. DOI: 10.1208/s12249-014-0146-3

[27] Salinas-Cortes R. Desarrollo farmacéutico de un polvo seco multiparticulado inhalable a base de NLC-vinorelbina como potencial tratamiento para cáncer pulmonar [thesis]. Faculty of Pharmacy: Autonomous University of the State of Morelos; 2022

[28] Flores-Flores A, Estrada-Soto S, Millán-Pacheco C, Bazán-Perkins B, Villalobos-Molina R, Moreno-Fierros L, et al. Functional mechanism of tracheal relaxation, antiasthmatic, and toxicological studies of 6-hydroxyflavone. Drug Development Research. 2019;**80**(2):218-229. DOI: 10.1002/ddr.21484

[29] Ahedo-Arenas MF. Desarrollo y evaluación in vitro y ex vivo de un polvo inhalable de NLC y 6-hidroxiflavona con potencial uso antiasmático [thesis]. Faculty of Pharmacy: Autonomous University of the State of Morelos; 2021

[30] Mathur M, Devi VK. Investigation of different types of nano drug delivery systems of atorvastatin for the treatment of hyperlipidemia. Drug Development and Industrial Pharmacy. 2018;**44**(12):2048-2060. DOI: 10.1080/03639045.2018.1508225

[31] Torres-Mojica JR. Desarrollo de un polvo seco inhalable de nanoplexos de atorvastatina-chitosán con potencial uso en la hipercolesterolemia [thesis]. Faculty of Pharmacy: Autonomous University of the State of Morelos; 2022

Chapter

Computational Tools to Convert Rainfall to Direct Runoff in Small Watersheds

Sergio Ignacio Martínez-Martínez

Abstract

A refinement of a simplified method to estimate direct runoff hydrographs for ungagged small watersheds is presented. One of the uses of these hydrographs is the design of spillways for reservoirs. Frequently, it is sufficient to apply simplified methods to estimate hydrographs; one of those is the triangular empirical hydrograph method. This method does not consider watershed shape, stream network properties other than the main stem, and spatial variation of rainfall depth and soil-coverture. But nowadays, thanks to geoinformatics software it is possible to discretize the watershed to approximately consider these additional characteristics of the watershed and the storm to which it is submitted. The data needed is a digital elevation model, several points that represent the outlets of watershed and sub-watersheds, and rainfall depth and soil-coverture information of the watershed and neighboring region. Also, it considered the attenuation of flood waves when they pass by reservoirs and stream channels. To build the watershed hydrograph, it was needed to generate a methodology and then to implement it in software. Its development is briefly described. The software was implemented as a plug-in for MapWindow Geographic Information System (GIS) (version 1.4.8). A case study is presented showing the hydrographs produced for several watershed configurations.

Keywords: hydrograph, direct runoff, hydrograph estimation, small watershed, GIS

1. Introduction

A watershed is a part of the earth's surface that collects storm's rainfall; a portion of rainfall infiltrates, and the other runs off over the terrain, eventually arrives to the drainage system, and flows through it, until reaching the outlet as a flood wave. Flooding can happen depending on the volume and form of the wave and affect populated, agricultural, or industrial areas that need to be protected. That protection should be consistent with the possible damages caused by the event, and it is not the same to protect a densely populated town than a farm field; the former must have better protection to avoid human life and property losses, and the latter only needs to protect cultivated crops. The protection level is usually determined using design storms of different return periods or frequencies. The time in years that a given hydrology event is equaled or exceeded is called the return period. The greater the storm, the greater the return period; the smaller the storm, the smaller the return period. To avoid flooding downstream of the outlet of the watershed sometimes a dam or another waterwork can be built. The dam is provided with a spillway that discharges the excess flow from the lake created upstream of the dam. A lot of times, it could be advantageous to subdivide or discretize the watershed to include more of its details. The subdivision of the watershed could be performed from points of interest located over some stream, such as the location of a dam or a point where the soil-coverture changes. To design the spillway of the dam, it is necessary to estimate the hydrograph produced by a design storm. In the case of watersheds that are sensible to intense storms, the hydrograph can be estimated with simplified methods, such as the triangular empirical hydrograph method (MHET for its acronym in Spanish) [1, 2]. This method needs, for the watershed, a few parameters: area, time of concentration, rainfall depth (generally associated with a return period), and soil-coverture data, but lacks a way of considering the form of the watershed, the topology of the stream network, and the effect of attenuation of the flood waves by inner reservoirs and streams connecting upstream to downstream sub-watersheds. A sub-watershed can be named headwater or interior. A headwater watershed has no tributary sub-watersheds; an interior sub-watershed has at least one. The consideration of all these characteristics of a subdivided watershed tends to produce smoother hydrographs more like those measured in hydrographic stations than simple triangular hydrographs. The availability of geoinformatics software and geographical data in the form of Digital Elevation Models (DEM) and vectorial datasets representing points, lines, and polygons made possible the development of a series of computational tools useful to consider the new parameters transforming MHET to a new method, the triangular empirical hydrograph with attenuation method (MHETA for its acronym in Spanish).

2. Theoretical background

Throughout the last several years, various approaches for obtaining unit and direct runoff hydrographs have been proposed, including those presented in Refs. [3–9]. However, none of them has the characteristics of the method described in this work.

For the development of the methodology to be proposed, it is necessary to consider several parts: type of data, specialized software, a programming language, and hydrological models to be applied.

2.1 Geographic information systems applied to the problem

Surface water hydrology in many engineering disciplines has a geographic component that can be helpful to represent in a geographic information system (GIS). Inside geoinformatics software, geographic objects can be represented in several formats. A grid or raster is used to represent values that change continuously over a geographic area; a particular type of grid is a digital elevation model (DEM) that represents the terrain elevation of an area. Another is the vectorial format, which consists of points, lines, or polygons. Commonly, vectorial data are stored in shapefiles. A shapefile of points can represent, for example, dams or hydrometric stations; a shapefile of lines can represent a stream or an entire drainage network; and one of the polygons can represent watersheds, sub-watersheds, or water bodies. In the case of this work, a DEM is needed, the watershed exit in a point shapefile, and another shapefile with the exit point and one or more points inside the watershed to define sub-watersheds. The soil-coverture can also be represented in a polygon shapefile and the variation of rainfall with a polygon shapefile with Thiessen (or Voronoi) polygons of climatologic stations populated by the design rainfall depth of the waterwork needed at the exit point of the watershed.

Inside a GIS, not only geographic objects are represented, but they can also be used to extract information to achieve a particular goal, such as delineate a watershed or find the greater water body of a region among many things. Some of these operations/processes may be already implemented in the used geoinformatics software, but others need to be implemented *via* plug-ins or extensions. In our problem, several processes needed to be implemented.

2.2 Hydrologic models

Simple, but still useful hydrologic models were selected to construct the algorithms to finally build hydrographs. They are the following:

- An approximate way to estimate the time of concentration, t_c, which is the time that water travels from the farthest point in the watershed until its exit point. An empirical formula, the Kirpich formula [10], is useful to compute t_c, in km:

$$t_c = 0.0663\left(\frac{L}{\sqrt{s_c}}\right)^{0.77} \tag{1}$$

where L is the main steam length, km; $s_c = H/L$, the main steam slope, dimensionless; and H, difference in height between the beginning and the end of the main steam, km. For this work, it can be assumed that the Kirpich formula can be used to compute the time of travel of flood waves in stream reaches. Where L is the length of the reach, in km, and so on.

- The estimation of the excess rainfall depth, P_e, in mm, with the Natural Resources Conservation Service (NRCS) formula [11]:

$$P_e = \begin{cases} 0; & \text{if:} P - \dfrac{5080}{N} + 50.8 \leq 0 \\ \dfrac{\left(P - \frac{5080}{N} + 50.8\right)^2}{P + \dfrac{20320}{N} - 203.2}; & \text{if} : P - \dfrac{5080}{N} + 50.8 > 0 \end{cases} \tag{2}$$

Where P is the rainfall depth in mm, N is the curve number function of the soil-coverture and soil moisture of the watershed.

- To build hydrographs for small watersheds the MHET is used (**Figure 1**):

$$Q_{max} = \frac{2V_e}{t_c + t_r} \tag{3}$$

where Q_{max} is the maximum flow rate, in m^3/s; $t_r = x\, t_c$, t_r is the recession time of the hydrograph, and x is a function of the watershed area, A, according to **Table 1** [2]; $V_e = P_e A$ is the direct runoff volume, in m^3.

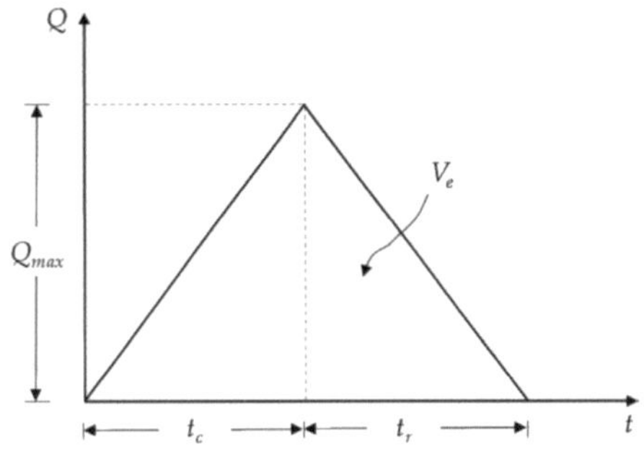

Figure 1.
Triangular empirical hydrograph.

A (km^2)	*X*
< 500	1.0
500–5000	1.5
< 5000	2.5

Table 1.
MHET x-parameter.

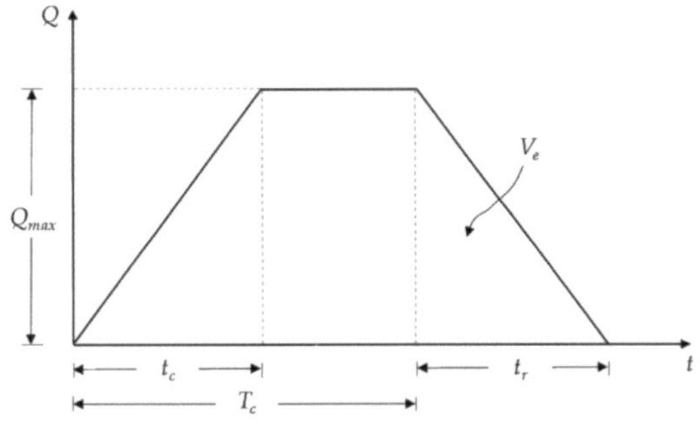

Figure 2.
Triangular empirical hydrograph converted in a trapezoidal hydrograph for a sub-watershed.

When the watershed is divided into sub-watersheds, the hydrograph of each sub-watershed will have a trapezoidal form in which the maximum flow rate is reached in the t_c corresponding to the sub-watershed, then, the flow rate is kept constant and, when the time is equal to the time of concentration of the watershed, the flow rate begins to fall linearly during the time of recession corresponding to the sub-watershed, until it reaches zero. This is illustrated in **Figure 2**, where all variables correspond to the sub-watershed, while T_c is the concentration time of the watershed. The maximum discharge under these conditions is:

$$Q_{max} = \frac{2V_e}{(2T_c - t_c + t_r)} \tag{4}$$

- Flow routing in reservoirs. Flood waves suffer an attenuation when they arrive at a reservoir, that is a reduction of its maximum discharge and an amplification of its duration. The evolution of the outflow discharge of the reservoir, $O(t)$, is obtained solving the continuity equation applied to the reservoir considering that the water surface is horizontal. This process is known as the level pool routing [12]. The equation can be written:

$$\frac{dE}{dt} = \frac{I(t) - O(t)}{A(t)} \tag{5}$$

Additionally, there are needed the inflow hydrograph, and water surface elevation—outflow discharge and water surface elevation—water surface area relationships (given in this work as tables). In the Eq. (5) t is time since the beginning of the simulation, h; E is the water surface elevation in the reservoir, m; $I(t)$ is the inflow discharge, m^3/s; $O(t)$ is the outflow discharge, m^3/s; and $A(t)$ is the area of the water surface, m^2. Several numerical schemes have been proposed to solve this equation. Here, it is used the third-order Runge-Kutta method [12]. To integrate this equation, time is discretized and proceeds step by step from the initial time $t = 0$ to the time at which the simulation ends. A sufficiently small time increment Δt must be used for the integration to be successful.

- Flow routing in channels. A method useful to estimate the evolution of the outflow in a channel reach is the Muskingum Method [13]. It is a simplified model that can be used with the data that can be extracted from a DEM. This method is based on the continuity equation like the flood routing in reservoirs (change of volume in the channel reach with respect to time, $(\Delta V/\Delta t)$, is equal to the difference between mean inflow and mean outflow, $(\overline{I} - \overline{O})$, during the increment of time):

$$\frac{\Delta V}{\Delta t} = \overline{I} - \overline{O} \tag{6}$$

and, in the storage equation in the channel reach:

$$V = KO + Kx(I - O) \tag{7}$$

where K is storage constant, h; and x is a dimensionless weight factor of inflows and outflows. Eq. (7) has a physical interpretation [13], the term KO is prism storage, and the term $Kx(I - O)$ is wedge storage.

Depending on the length of the stream channel, the solution of Eq. (7) can consider three cases: If the length is too long, divide the channel and perform successively the flood routing in as many reaches as needed. If the length is between an interval of appropriate lengths, the flood routing simulation is performed once, and if the channel reach is too short, simply translate the flood wave from the beginning of the reach to its end.

3. Project methodology

To generate the methodology to build direct runoff hydrographs for discretized small watersheds from one DEM, geographic coordinates of the watershed and sub-watershed outlets, and soil-coverture, rainfall, and reservoir data; it is needed to consider the mentioned hydrologic models, the design of several algorithms, the coding of procedures based on such algorithms, and the implementation of the code in computational tools to be run in a specific system. The implementation of procedures was performed using the integrated development environment (IDE) Visual Studio 2015 of Microsoft. The programming language VB.Net was used to create the DACHydro Tools, a plug-in that runs on the MapWindow GIS (version 4.8.8) program [14] for Windows. The software development involves an iterative refining process of algorithm-code-test-debug and going back to one or more steps until the required result is attained, as happened in this study. Several cases of the study were used to test the geoinformatics tools until the results obtained were correct.

3.1 Processes needed to apply MHET

The first set of procedures implemented in the tools are those needed to apply MHET. That is, all those needed to obtain watershed area, and main stem and its physical properties.

The tools were created sequentially, beginning with the (FL, WP, and FD) tools. This tool was developed using the algorithm proposed in Ref. [15]; with it, the Filled Hole DEM (FLGrid), Boundary Watershed Grid (WSGrid), and Flow Direction Grid (FDGrid) are produced. All other algorithms were generated for this work except where indicated. In general, new tool(s) to be developed need as input data some of the files created in the previously developed tool(s). The input files needed in some of the tools are mentioned. The second was the (FA) tool where it is created the Flow Accumulation Grid (FAGrid). The input file needed for this tool is the FDGrid. The third was the (Dir-DW) tool. This tool needs as input the Watershed Exit Point (or Outlet) shapefile (ExitShp), FAGrid, FLGrid, FDGrid, and FAGrid; in it, the Delineated Watershed DEM (DWGrid) is obtained. The next tool (SNG) was written to create the Stream Network Grid (SNGrid). Once the previous Grid was created, the (SNS) tool was written to create the Stream Network shapefile (SNShp). Then, it was possible to create the (Identify MS) tool, which produces the Main Stem shapefile (MSShp). Although defining a sub-watershed for each stream segment is not necessary to apply MHET, the tool to create sub-watersheds would be helpful later to code a tool to create custom sub-watersheds. Therefore, the (SWhG) tool was developed to create the Sub-watershed Grid (SWhGrid). Once the previous tool was ready, the (SWhS) tool useful to create the Sub-watershed shapefile (SWhShp) was developed (the core of this tool is a procedure already built in MapWindow GIS).

3.2 Additional processes to apply MHETA

Several additional tools were needed to build the Discretized-Watershed Hydrograph. Some of the previous procedures were used to develop these tools. The first one was the creation of the Custom Sub-watershed Grid (CSWhG) tool. This tool needs as input data the Custom Sub-watershed Exit Point (or Outlet) Shapefile (CSWhEPShp). The next was the (CSWhS) tool in which it is produced the Custom Sub-watershed shapefile (CSWhShp). This tool was developed analogously to the (SWhS) tool. At this point, it was necessary to develop the (CSWhSNG) tool to generate the Custom Sub-Watershed Stream Network Grid (CSNGrid). Afterwards, the (CSWhSNS) tool is used to build the Custom Sub-Watershed Stream Network shapefile (CSNShp). In this shapefile, there are attributes to characterize each stream, such as length, height, order, and the code of the custom sub-watershed to which it belongs, among others. The next tool to be created was the computation of time of travel and time of concentration of the Custom Sub-watersheds, that is, the

(CSWhTtTcComp) tool. In this tool, the Custom Sub-watershed Main stems are identified; then are pinpointed the streams from upstream Custom Sub-watershed Exit Points to adjacent downstream Custom Sub-watershed Exit Points; after that, the Custom Sub-Watershed times of concentration as well as the time of travel of streams from upstream Custom Sub-watershed Exit Points to adjacent downstream Custom Sub-watershed Exit Points are computed. For the general case of the presence of reservoirs in the Sub-watershed Exit Point, the procedure to introduce water surface elevation—water surface area and water surface elevation—spillway discharge tables were developed and named (Introduce Reservoir Data) Tool. Finally, the most important tool of the set was developed as the Custom Sub-watershed Hydrograph Building (CSWhHydrBuild) tool. This tool builds the Discretized-Watershed Hydrograph from the properties of the Custom Sub-watersheds, the topology of the Custom Sub-watershed Stream Network, and the presence of reservoirs inside the watershed.

4. Results

The results of this project are theoretical and practical. The former are the algorithms, and the latter are the implemented procedures based on the algorithms.

4.1 Theoretical results

The more important theoretical result obtained is the algorithm to build the hydrograph for the discretized watershed. This process is carried out after all the GIS operations are performed. The building discretized-watershed hydrograph algorithm is presented in **Table 2** in pseudocode form. All interior computational processes are indented. In **Table 2** appear for the first time the term junction. A junction is the exit of a sub-watershed where it accumulates the discharge of all its upstream sub-watersheds including the discharge of its own sub-watershed. The definitions of some variables in that table are sub-watershed: custom sub-watershed; nJuncs: number of junctions; i: index of junctions and sub-watersheds; nProcessedJuncs: number of junctions processed. Once all the junctions are processed, the discretized-watershed hydrograph is completed.

4.2 Practical results

The practical results are the implemented procedures in the DACHydro Tools. Its menus and tools are briefly described in **Table 3**. For this work, the more important are those of Menu **<Build CSWhHydrographs… >** and the most important tool there is the ***<Execute CSWhHydrBuild Process >*** tool. The workflow of the DACHydro Tools is shown in **Figure 3**.

4.3 A case of study

One of the several watersheds that were studied during the development of the Building Hydrographs Toolset was the watershed of Ordeña Vieja dam. This dam is in the State of Aguascalientes, México; its coordinates are latitude 21°56′50″ N, longitude 102°43′20″ W. Elevation 1780 mamsl. This watershed has an area of 75.33 km^2; inside the watershed there is another dam, El Temazcal dam, as is shown in **Figure 4**, along

```
Begin
  nJuncs = number of sub-watersheds
  For all sub-watersheds
    obtain the hydrograph with MHET considering its properties, such as area, runoff number, rainfall
    depth, time of concentration and time of recession.
  i = 1
  nProcessedJuncs = 0
  Do
  If junction(i) is not processed, then:
    If the sub-watershed(i) is not interior, then:
      If exists a reservoir at junction(i), then:
        Perform the level pool flood routing in the reservoir.
      Mark the junction(i) as processed.
      nProcessedJuncs = + 1
    Else
      If the upstream(s) junction(s) of junction(i) is/are processed, then:
        For all upstream processed junctions
          Perform the Muskingum flood routing in the stream linking the upstream junction and junction(i).
        Summation of hydrograph from sub-watershed(i) and hydrographs routed from upstream junctions.
        If exists a reservoir at junction(i), then:
          Perform the level pool flood routing in the reservoir.
        Mark the junction(i) as processed.
        nProcessedJuncs = + 1
  If nProcessedJuncs > = nJuncs, then:
    Exit Do
  i = + 1
  If i > nJuncs, Then:
    i = 1
 Loop
End
```

It is presented in pseudocode.

Table 2.
Algorithm for building the discretized-watershed hydrograph.

with the DACHydro Tools Menu installed in MapWindow GIS. The DEM for this case of study was downloaded from the website "Mexican Elevation Continuum" (CEM for its acronym in Spanish) of the National Institute for Statistics and Geography (INEGI for its acronym in Spanish) [16].

To review the effect of the application of MHETA on the outflow hydrograph of the watershed, several configurations of the basin can be considered, for example, the basin without reservoirs, with the El Temazcal dam, and with the El Temazcal and the Ordeña Vieja dams. At the time of the storm, both reservoirs are full to the crest of the spillway. The data considered are shown in **Tables 4–6**.

After running the necessary tools in the three configurations, the results presented graphically in **Figure 5** were obtained. The first hydrograph is the one obtained with MHET. The following three hydrographs were obtained with MHETA. In the hydrograph for the configuration without reservoirs, it is noted how the consideration of two sub-watersheds makes the hydrograph flatten and longer. In addition, the hydrograph tends to be smooth, and it is no longer polygonal. Hydrographs of the watershed with reservoirs show that they have a larger base time and an even smaller peak discharge. The hydrograph with only the El Temazcal dam is the hydrograph at

Menu/tool	Description
Introduce data...	
• *Default Result Name Files*	Creation of Default Data and Result Name files
• *File with File Names*	Loading of File with Data and Result Name files
• *Execute FL, WP, and FD Processes*	Creation of Fill depressions, boundary Watersheds, and Flow Direction Grids
• *Execute FA process*	Creation of Flow Accumulation Grid
Delineate single watershed ...	
• *Execute Dir-DW Process*	Watershed delineation with the direct method
• *Execute Iter-DW Process*	Watershed delineation with the iterative method
Build streams...	
• *Execute SNG Process*	Creation of Stream Network Grid
• *Execute SNS Process*	Creation of Stream Network Shapefile
• *Identify MS*	Identification of Main Stem
Build Sub-watersheds...	
• *Execute SWhG Process*	Creation of Sub-watershed Grid
• *Execute SWhS Process*	Creation of Sub-watershed Shapefile
Build custom subwatersheds...	
• *Execute CSWhG Process*	Creation of Custom Sub-watershed Grid
• *Execute CSWhS Process*	Creation of Custom Sub-watershed Shapefile
Build CSWhHydrographs...	
• *Execute CSWhSNG Process*	Creation of Custom Sub-watershed Stream Network Grid
• *Execute CSWhSNS Process*	Creation of Custom Sub-watershed Stream Network Shapefile
• *Execute CSWhTtTcComp Process*	Calculation of concentration times of Custom Sub-watersheds and travel time of streams that link adjacent upstream and downstream Custom Sub-watershed outlets
• *Introduce Reservoir Data*	Introduction of the tables of elevations-areas and elevations-discharges of the reservoirs of the watershed
• *Execute CSWhHydrBuild Process*	Calculation of hydrographs for Custom Sub-watersheds and the entire Watershed

Table 3.
Description of the DACHydro tools.

the entrance to the Ordeña Vieja dam, while the hydrograph for both reservoirs is the outflow hydrograph of the Ordeña Vieja dam.

Several numerical results are shown in **Table** 7. From the hydrograph of MHET to that of the two reservoirs, peak times increase, peak discharges decrease, and base times increase. A review of these results consists of checking the volume under the hydrograph or direct runoff volume. All volumes must be very similar, which is true

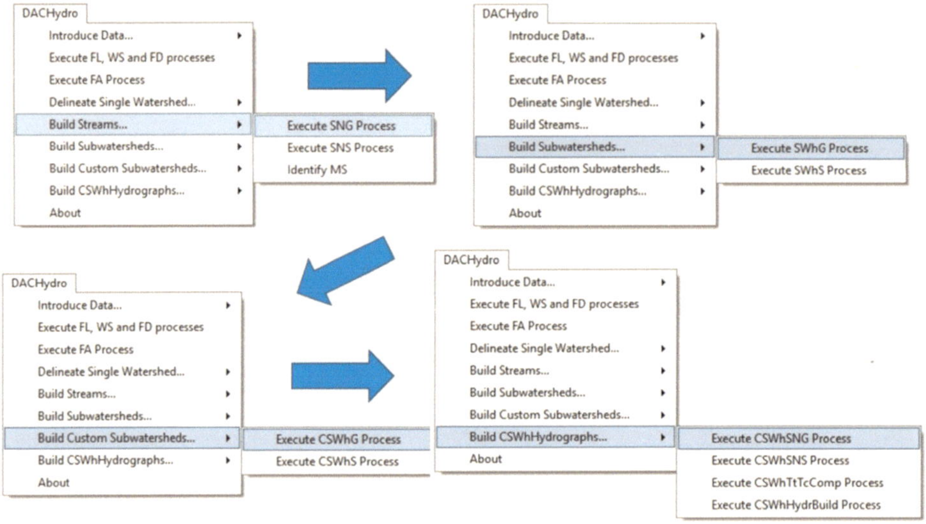

Figure 3.
Workflow of the DACHydro tools. Several toolsets including the build custom sub-watershed hydrographs toolset are shown.

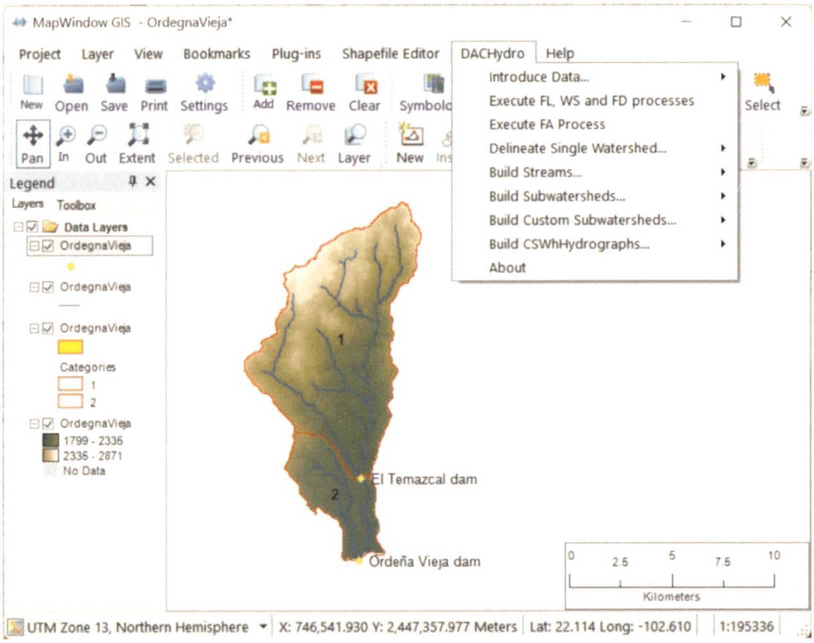

Figure 4.
Watershed of Ordeña Vieja dam along with the DACHydro tools menu inside MapWindow GIS.

Sub- watershed	A (km^2)	P (mm)	N
1. El Temazcal	60.80	65.5	67.3
1. Ordeña Vieja	14.53	56.8	69.8

Table 4.
Sub-watershed's: area (A), rainfall depth (P), and runoff number (N).

E (mamsl)	A_w (m^2)	E (mamsl)	Q (m^3/s)
1917	0	1920	0
1918	136,178	1920.1	9.795
1919	164,593	1920.2	27.705
1920	173,350	1920.3	50.897
1921	177,103	1920.4	65.08
		1920.5	109.513
		1920.6	143.959

Table 5.
El Temazcal dam. $E - A_w$ and $E - Q$ tables, where E is water surface elevation, A_w is water surface area, and Q is spillway discharge.

E (mamsl)	A_w (m^2)	E (mamsl)	A_w (m^2)	E (mamsl)	Q (m^3/s)
1778	0	1789	255,064	1796.5	0
1779	41,380	1790	278,802	1796.6	6.324
1780	66,027	1791	304,800	1796.7	17.889
1781	83,670	1792	326,080	1796.8	32.863
1782	100,969	1793	347,384	1796.9	50.596
1783	116,769	1794	367,755	1797	70.711
1784	135,371	1795	390,460	1797.1	92.952
1785	154,783	1796	424,518	1797.2	117.132
1786	179,602	1797	454,988	1797.3	143.108
1787	206,239	1798	480,986	1797.4	170.763
1788	229,165	1799	501,185	1797.5	200.000

Table 6.
Ordeña Vieja dam. $E - A_w$ and $E - Q$ tables; where: E is water surface elevation, A_w is water surface area, and Q is spillway discharge.

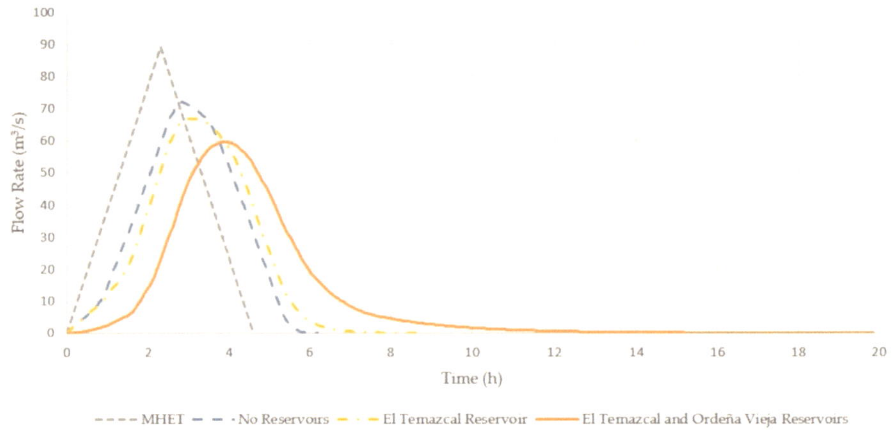

Figure 5.
Hydrograph comparison. HET, watershed without reservoirs, with El Temazcal reservoir, and with El Temazcal and Ordeña Vieja reservoirs.

Configuration	t_p (h)	Q_p (m^3/s)	t_b (h)	V_e (m^3)
MHET	2.30	89.29	4.60	739,313
No reservoirs	2.85	71.91	6.20	738,562
With El Temazcal Reservoir	3.10	66.91	8.75	738,542
With both reservoirs	3.85	59.58	19.85	738,486

T_p*: peak time,* Q_p*: pick flow rate,* t_b*: base time, and* V_e*: direct runoff volume.*

Table 7.
Some properties of hydrographs for different watershed configurations.

in this case. All direct runoff volumes remain within −0.11% of the volume of the hydrograph obtained with MHET. This difference is because the receding part of the hydrographs has many too small flow rates that the software does not consider for computing volume.

5. Conclusions

The work carried out represents a refinement of the Triangular Empirical Hydrograph Method (MHET). Given that in the new method, based on MHET, the watershed is discretized, and the flood waves suffer attenuation in the streams and reservoirs located within the watershed it was named Triangular Empirical Hydrograph with Attenuation method (MHETA). From a practical point of view, the benefit provided by the implementation of MHETA in the DACHydro tools is the automatic computation of the direct runoff hydrograph at the watershed outlet, considering flood routing if reservoirs are located inside the watershed and in the stream channels that link adjacent upstream-downstream sub-watersheds. This software was originally intended to improve the teaching of surface hydrology to civil engineering undergrade students. The idea was to provide them with a resource that could be used to finish faster and systematically projects involving the design of water works to control flood waves at the outlet of small watersheds using geographical data. This software can also be useful to practicing engineers solving the same type of problem.

However, sometimes DEMs, hydrologic models, and/or additional data used may not adequately reflect the physical characteristics of the watershed and its surface flow. This would result in less approximate hydrographs. Although MHETA is a method to design spillways for ungagged watersheds, in future research work, it can be proposed the implementation of tools to check hydrographs for storm hyetographs with variable rainfall depths in gauged watersheds. It could be done using some type of optimization to calibrate the parameters to compute the watershed hydrograph and compare it with the measured hydrograph. Those parameters could be, that is, times of concentration, times of travel, and runoff numbers, among others. Thus, once the values of the parameters have been calibrated and with any new storm occurring over the watershed, a hydrograph can be obtained that better reflects the watershed response to rainfall. The later application would be very useful in the case of real-time flood forecasting.

Acknowledgements

The author is grateful to the Autonomous University of Aguascalientes for the resources to carry out the research projects PIIC12-1, PIIC14-2, PIIC15-3, and PIIC-17-1, in which a good part of the DACHydro tools was developed. The author would also like to thank the students Gonzalo E. Espinoza-Dávalos, Diego A. García-Gómez, and Juan A. Hernández-Andrade who, in different periods of time between 2009 and 2017, helped in the development of the tools; after 2017 the author has continued developing of them.

Conflict of interest

This work does not have any conflict of interest.

Author details

Sergio Ignacio Martínez-Martínez
Autonomous University of Aguascalientes, Aguascalientes, Mexico

*Address all correspondence to: simartin@correo.uaa.mx

References

[1] Klemes V. Applications of Hydrology to Water Resources Management (Planning and Design Level). Operational Report No. 4, WMO – No. 356. Geneva: World Meteorological Organization; 1973. 122 p

[2] Campos-Aranda DF. Manual para la estimación de avenidas máximas en cuencas y presas pequeñas. México: Secretaría de Agricultura y Recursos Hidráulicos; 1982. 225 p

[3] Muzik I. Derivation of unit and flood hydrographs using a GIS. Environmental Monitoring and Assessment. 1992;**23**: 45-56. DOI: 10.1007/BF00406951

[4] Hellweger F. TABHYD - TR55 Tabular Hydrograph Method in ArcView. 1996. Available from: http://www.ce.utexas.edu/prof/maidment/gishydro/ferdi/research/tabhyd/tabhyd.html [Accessed: April 8, 2014]

[5] Kilgore JL. Development and evaluation of a GIS-based spatially distributed unit hydrograph model [thesis]. Blacksburg: Virginia Polytechnic Institute and State University; 1997

[6] Györi M-M, Haidu I. Unit hydrograph generation for ungauged subwatersheds. Case study: The Monoroştia river, Arad County, Romania. Geographia Technica. 2011;**2**:23-29

[7] Dawod GM, Koshak NA. Developing GIS-based unit hydrographs for flood management in Makkah metropolitan area, Saudi Arabia. Journal of Geographic Information System. 2011;**3**: 153-159. DOI: 10.4236/jgis.2011.32012

[8] Ogunlela AO, Adewale PO, Adamowski JF. Developing design storm hydrographs for small tropical catchments with limited data. Ethiopian Journal of Environmental Studies and Management. 2012;**5**:356-365. DOI: 10.4314/ejesm.v5i4.4

[9] Foda RF, Awadallah AG, Gad MA. A fast semi distributed rainfall runoff model for engineering applications in arid and semi-arid regions. Water Resources Management. 2017;**31**:4941-4955. DOI: 10.1007/s11269-017-1787-2

[10] Kirpich ZP. Time of concentration of small agricultural watersheds. Civil Engineering. 1940;**10**:362

[11] Natural Resources Conservation Service NRCS. Part 630 Hydrology. Chapter 10: Estimation of direct runoff from storm rainfall. In: National Engineering Handbook. United States Department of Agriculture; 2004

[12] Chow VT, Maidment DR, Mays LW. Applied Hydrology. New York: McGraw-Hill; 1988 580 p

[13] Chow VT, editor. Handbook of Applied Hydrology. New York: McGraw-Hill; 1964. 1495 p

[14] MapWindow. Download of MapWindow GIS v4.8.8 - Stable Release - 32Bit. 2013. Available from: https://mapwindow4.codeplex.com/releases/view/110244 [Accessed: August 15, 2013]

[15] Wang L, Liu H. An efficient method for identifying and filling surface depressions in digital elevation models for hydrologic analysis and modelling. International Journal of Geographical Information Science. 2006;**20**:193-213. DOI: 10.1080/13658810500433453

[16] INEGI. Mexican Elevation Continuum 3.0 (CEM 3.0). 2013. Available from: https://www.inegi.org.mx/app/geo2/elevacionesmex/index.jsp [Accessed: August 21, 2017]

Chapter

Lovecraft's Horror Legacy and Its Implications in the Twenty-First Century

Antonio Alcalá González

Abstract

Through his view of humanity as something insignificant when contrasted with a cosmic background, Lovecraft anticipated the kind of concerns over uncertainty proper to the twenty-first century. In the current context of the Anthropocene, we are not threatened by the physical hyperobjects shaped in the bodies of his monsters; however, we struggle harder every year to accept that no matter how much we pretend to be in control of what happens on Earth, we cannot escape the destructive combined effects of mere natural events which we insist on calling catastrophes and the consequences of our own intervention on the planet surface, atmosphere and waters. Moreover, the drastic changes brought to our lives due to the presence of a virus, COVID-19, have provoked a reflection on the fragility of our supposed position as owners of the planet. In relation to this and for the purpose of this article, I propose to consider that microscopic life still represents challenges in the face of attempts to control it through medical and chemical action; besides, there is the fact that this kind of life is permeated with a halo of mystery given the impossibility to interact directly with it. Said microscopic scenario is influential to *The Color Out of Space* by Richard Stanley. This film is a reinterpretation of Lovecraft's story "The Color Out of Space", published in 1927, in which vaporous entities of an indescribable color arrive to Earth in a meteorite and unleash a polluting process that ends up diluting and extracting the color of all life around its point of arrival. The events faced by Stanley's characters remind us that our dominance over our surroundings is but an arbitrary illusion. My intention here is to scrutinize Stanley's film appropriations of Lovecraftian literary events and creatures to express a view of humanity as a species with a very limited role on the planet and evaluate the relevance of Lovecraft's legacy in the current context of uncertainty around human existence.

Keywords: Anthropocene, adaptation, horror of contagion, ecohorror, lovecraft, Richard Stanley

1. Introduction

Before COVID-19, we tended to consider ourselves safe and secure thanks to our apparent control of the planet. It is true that we have not eradicated many diseases or prevented the aging of our bodies, nor have we escaped the effects of hurricanes, earthquakes or other types of disasters, but as the current century approached the end of its second decade, we felt prepared to control and even to predict Nature. However, the pandemic put us on a path to reconsider our apparent position as masters of the planet. In relation to this and for the purpose of this paper, I propose to consider the following scenario: we know that microscopic life represents challenges in attempts to control it through medical and chemical action, in addition to the fact that it is impossible to interact with it directly, which permeates it with an aura of mystery. In relation to the above, the film, *The Color Out of Space* [1] by Richard Stanley, is a reinterpretation of Lovecraft's story "The Colour Out of Space", published in 1927, in which what seems to be vaporous entities of an indescribable color arrive on Earth in a meteorite and unleash a polluting process that ends up diluting and extracting the color of all life around its point of arrival. The intention of this text is to reflect on Stanley's appropriation of Lovecraftian creatures to raise concerns of EcoHorror proportions and to scrutinize how their presence places us as a species with a very limited role on the planet.

2. Horror of contagion

When literary elements are incorporated into film, according to Linda Hutcheon, we move from imagination to the realm of direct perception [2]. Instead of the call to our imaginative powers from the words on a page and the conceptual reading experience, cinematic resources allow to produce a sense of depth on a two-dimensional surface that drags the viewer into a perceptual experience where we accompany the characters and their stories as they move and develop within the screen through multiple spaces and temporalities. Given these and other differences between film and literature, my attention will focus on the monster as a crucial element in the two narratives included in this analysis. I propose to consider the connection between Stanley's film to Lovecraft's fiction as an example of what Linda Hutcheon would call "extended intertextual engagement with adapted work" [3].

Stanley adapts "The Colour Out of Space" for the screen and the twenty-first century while changing the spelling from the word "color" to the American dialect (without the "u"), marking his appropriation of the plot. The monstrous, unmaterial entities that appear in the source text evidence the writer's view of humanity as insignificant and ephemeral compared to the universe. The attitude of such presences toward humans constantly confirms that they find us irrelevant. In fact, when they are described as evil, it is the result of imposing on them a human perspective that they lack, for, from their point of view, we are mere insignificant presences on their path. Human understanding of what they are is so vague that neither the source text nor the film confirms at all whether they are several entities or a unified one. Their presence reminds us that existence and corporeality transcend our ability to comprehend space and time. Hutcheon proposes that "stories travel to different cultures and diverse mediums, adapting to the context that receives them" [2]. What adapts to our moment are the monsters created by Lovecraft, which I consider from an ecocritical perspective because their presence reminds us of our trivial position on Earth, and questions our irruption into the dark corners of Nature for extractive purposes. As for *The Color* in film, the story is adapted to the context of the twenty-first century to demonstrate that, almost a hundred years later, humanity is just as vulnerable to the unknown and our bodies lie defenseless against multiple microscopic threats that enter and leave them permanently.

DOI: http://dx.doi.org/10.5772/intechopen.115472

In horror narratives, characters who are confronted with monsters experience two interrelated moments. First, they suffer a devastating confrontation with the monstrous entity. Second, comes a more heartbreaking event: they have no choice but to accept that the irrefutable existence of the monster proves that our environment contains more than we could ever know. In *The Color Out of Space*, Ward (Elliot Knight) lives these two moments in a scenario of horror in the face of the risk of a contagion that imperceptibly bursts into living organisms. He is left with the feeling that the pollution left by color on the desolate wasteland will remain transmissible through the water of the dam that covers the place. This echoes what happens in Lovecraft's story where color does not completely escape into space, as a part of it returns to the well of the Gardners' farm. The result in both narratives is an annihilating fear generated by the mere suspicion that what is left by the extraterrestrial presence will enter human bodies after ingesting water from the dam. Ward and Lovecraft's narrator undergo a prevalent fear that the events of organic deformation experienced by the Gardners and all life on their farm will be replicated in a greater scale. The story is anti-pastoral in the sense that Nature around the Gardners becomes strange; it turns into something beyond what it originally was after being infected by an extraterrestrial presence that absorbs humanity and all other forms of organic life. The entire family is affected and destroyed because of their contact with the color. The bodies of Theresa (the mother) and Jack (one of the sons) merge into a horrific mass that threatens to kill Lavinia (the daughter, played by Madeleine Arthur), who is later apparently possessed by the alien energies from the well (where the color hides) and transported to the outer dimension where the color comes from. As for Benny (the younger brother) and Nathan (the father, played by Nicholas Cage), the former is sucked into color, while the latter dies accidentally, but then returns, just like the rest of the family members, in seemingly color-reanimated forms before the house is ultimately destroyed and a barren land of only gray remains in what used to be the farm and its natural surroundings. The first 15 minutes, before the meteorite's arrival, show us the Gardners as a family in control of their farm and the nature around it. However, that is preceded by Ward's epilogue about the warnings given to him about the area; he describes the place as a wild region where: "There are valleys with deep forests that no axe has ever cut." [00.01.43] In this sense, the farm must be considered to occupy a liminal zone since, around it, everything seems to be wild nature. On the savage as an opposite, Berenice Murphy reminds us that the natural had long served as the sinister "other" of civilization's supposed security [4]; in this context, controlling and cultivating such territory, and of course, subjugating or exterminating the "savage" people who lived there, was not only desirable but necessary to ensure the security of civilization. Lavinia seems to be aware of that from her first appearance, right after the credits, when she shows her respect for the unknown forces that humanity cannot control at all, in addition to her desire to leave the place.

As Simon Estok states, human history reflects a constant attempt to control the natural environment; we take rocks and turn them into tools or weapons to modify or kill parts of the natural environment; we build shelters to protect ourselves from weather and predators [5]. However, the first two decades of the twenty-first century show an even stronger impact of the Anthropocene on the planet than that witnessed in the previous two centuries. The footprint left by our activity on land, in the atmosphere and in water is becoming more evident as time progresses. The fears of Nature's possible responses to our excessive manipulation of the planet are expressed in literary manifestations of Ecophobia, which, as defined by Estok, is an irrational and unfounded hatred of the natural world, so present and subtle in our daily lives.

Said hatred justifies our plundering of animal and non-animal resources [5]. In *The Color Out of Space*, Ward gradually comes to understand that the degradation of life around the Gardner farm trumps any attempt to tame the unknown forces in the universe. Everything in the area is absorbed by the color. The family's bodies are in turn manipulated or absorbed by it; Nathan's own mind also goes crazy as a result of this process. A similar fate befalls the family's only neighbor, Ezra, whose body is found by Ward and the local sheriff drained and about to be consumed by color. All these cases illustrate the fragility of human bodies and remind us that we cannot protect them from unknown threats that challenge our supposed control of Nature. The impossibility of escaping Stanley's version of Lovecraft's alien color represents the fact that humanity has forgotten to think of itself and all other beings on the planet as what Timothy Morton calls a *mesh* or *net* that extends to all living entities [6]. Within this mesh, all creatures coexist in a complex web of intertwined shared relationships that emphasize their differences and bind them together. Morton's view is echoed in Donna Haraway's argument about *tentacular thinking* [7] which refers to the interconnectedness of life that governs all organic interactions on our planet. The artificial denial of our belonging to this network led to the end of the Holocene and the beginning of the Anthropocene.

A key facet of horror is being confronted by creatures that annihilate our faculties and destroy all knowledge, perceptions, and convictions. Eugene Thacker's definition of horror reinforces this conceptualization of the term; for him, horror is the ever-potential threat of the senses being overwhelmed by perceiving something that exceeds the classificatory mechanism of understanding and the synthetic function of reason [8]. When a horror encounter is related to environmental concerns, some critics have applied the term *Ecohorror*. Christy Tidwell, Stephen A. Rust, and Carter Soles have previously provided explanations of this genre that are useful here. For Tidwell, the word applies to narratives in which elements of the natural world attack humanity [9, 10]. Rust and Soles situate EcoHorror in texts where environmental disruption is tormenting humanity's relationship with the non-human world [10]. Both definitions propose texts belonging to EcoHorror as those in which the damage caused by humans to the ecology is responsible for an attack by natural forces. On the basis of these ideas, my intention here is to continue discussions around a significant branch of EcoHorror: the Horror of Contagion, which frames anxieties that result from the irrefutable truth that our bodies are inhabited by a multiplicity of microscopic entities that can carry disease and death imperceptibly to our senses; This reaffirms the impossibility of humanity to rise above other forms of life since we are internally cohabited by non-human presences. We label the attack of microbial forms as a violent irruption of catastrophic proportions for us. However, we are the ones who have tried, unsuccessfully, to ignore that the microbial life within us is part of Nature and when we remember it, it is to manipulate it by introducing chemicals into our bodies. In the end, the amount and mutability of microbial life within us is of such magnitude that it is equally uncontrollable as well as extremely unpredictable.

In the case of the color impossible to describe in Lovecraft's tale, it is also impossible to escape in both, literary and film versions. It becomes fundamental in this Contagion of Horror narrative since it consists of a life form that escapes our tangible perception of the Earth (due to its lack of materiality), and it has the capability to unleash contagion of uncontrollable proportions as it invades human bodies. In his version of the story, Stanley uses a pink-violet color, as well as the multiple tentacles that emerge from the world from which color seems to have come to dramatize the irruption of something alien to the known and that surpasses any attempt to stop its colonization of the human body. The intertextual engagement of Lovecraft's creatures with the film is not only visual but also auditory: Ward identifies the presence of nocturnal and strange sounds around the area affected by the color, and even his cell phone and Lavinia's receive inexplicable static when they are in the area affected by the presence of color. Human understanding and technology are challenged and intervened by the overwhelming reach of color.

In Lovecraft's original tale, it is the unbridled curiosity of its narrator that leads him to discover the existence of color that challenges the idea of human supremacy on the planet. Likewise, scientific curiosity leads Ward to inquire about the events that take place around the Gardners after the meteorite falls. Like its counterpart in the original text, he knows that the water from the dam should not be ingested by humans and, at least in his personal case, both narrators decide that they will not consume it for anything in the world. However, they choose not to make their story public because they know it will be considered as an example of madness by those who have not lived what they did. The degradation of life at Gardner Farm witnessed by Ward, a man with a scientific background, can be summed up in the recorded words of Ezra that Ward hears as the final collapse of the place of contagion approaches: "That lives in the well. He grew up down there poisoning everything; changing everything into something resembling the world he came from and what he knows. We all know it's coming, but we can't escape. It takes over everything that lives" [1.36.40]. The "It" he refers to is the color, Stanley's proposal for what Nature might be like outside of Eart. Its invasion of organic life cannot be stopped, let alone understood by civic authority, the sheriff, or Ward's scientific mind. Close to its presence, humans have two choices: flee as the narrator does or decompose into color itself and its lack of materiality.

3. Conclusion

To sum up, *The Color Out of Space* is a Horror of Contagion film that draws on monstrosities adapted from Lovecraftian creatures that go beyond the merely organic as we know it to shatter our assumptions of dominance in ocean waters. The presence of these creatures in the film extends the view of humanity as insignificant left by Lovecraft in his fiction. When the color takes over the Garners mentally and physically and ends up disintegrating them, we remember that countless forms of microscopic life surround us without us being able to guarantee our safety from the risks generated by their entry into our bodies. Having analyzed Stanley's inability to control microscopic life, this article seeks to encourage others to explore similar manifestations of EcoHorror experiences that can be examined as reflections on the consequences and dangers of our anthropocenic attempts to assign ourselves an exaggeratedly prominent role on the planet we inhabit, especially in times like the present when a simple virus has shown that we are not in control not even from our own flesh-and-blood ecosystems.

Conflict of interest

No conflict of interest was created by producing the present text.

Author details

Antonio Alcalá González
Tecnológico de Monterrey, Mexico

*Address all correspondence to: antonio.alcala@tec.mx

References

[1] Color Out of Space. [Film]. Directed by Richard Stanley. USA: RLJE Films; 2019

[2] Hutcheon L. A Theory of Adaptation. New York: Routledge; 2006. 296 p

[3] Lovecraft HP. The New Annotated H. P. Lovecraft. New York: W.W. Norton & Company; 2014. 928 p

[4] Murphy BM. The Rural Gothic in American Popular Culture: Backwoods Horror and Terror in the Wilderness. New York: Palgrave; 2013, 2013. 265 p

[5] Estok SC. Theorizing in a space of ambivalent openness: Ecocriticism and Ecophobia. ISLE. 2009;**16**(2)

[6] Morton T. The Ecological Thought. Cambridge. MA: HUP; 2010. 163 p

[7] Haraway DJ. Staying with the Trouble: Making Kin in the Cthulucene. Durham, NC: Duke University Press; 2016. 313 p

[8] Thacker E. Tentacles longer than Night: Horror of Philosophy. Vol. 3. London: Zero Books; 2014. 219 p

[9] Tidwell C. Ecohorror. In: Braidotti R, Hlavajova M, editors. Posthuman Glossary. London: Bloomsbury; 2018. pp. 115-117

[10] Rust SA, Soles C. Ecohorror special cluster: Living in fear, living in dread, pretty soon We'll all Be dead. Interdisciplinary Studies in Literature and Environment. 2014;**21**(3)

Chapter

Managed Aquifer Recharge: An Alternative for Long-Term Water Sustainability

Adriana Serrano-Santiago and Adriana Palacios

Abstract

The industrial, economic, and social development of a community can affect its natural resources, especially water security. Managed aquifer recharge (MAR) is one method that can help achieve long-term water sustainability by increasing the availability of groundwater reservoirs. This review provides an overview of the MAR methodology and its benefits to a groundwater-dependent community. Different types and techniques for implementing MAR are listed, as well as some of the challenges that need to be addressed for MAR development. Some countries already have regulations in place for the monitoring and installation of MAR. However, there is a need for a risk-based regulatory framework that considers changes in the quality and quantity of the groundwater and water to be recharged, as well as the inclusion of continuous monitoring that supports confidence in the sustainability of the MAR. Different MAR systems adapt to different conditions. Therefore, a detailed assessment of local conditions and the characteristics of available technologies, as well as a study of legal, environmental, and budgetary constraints, is required to select the most appropriate one for a particular site and purpose. MAR contributes to the achievement of the 6th Sustainable Development Goal, particularly concerning the protection and restoration of aquifers.

Keywords: artificial recharge, groundwater sustainability, managed aquifer recharge (MAR), review, water

1. Introduction

Global urbanization has accelerated with the economic boom, particularly in developing countries. As the world's population grows, there is an increasing need to reconcile competition for environmental resources between commercial demands so that communities have what they need to meet their needs. The use of these resources can be seen as a matter of fundamental rights but also of responsibilities. Although the industrial, economic, and social development of a community is carried out in harmony with the environment, the lifestyle of the population has a profound impact on the planet and affects its natural resources [1].

IntechOpen

Environmental issues occupy a prominent place on the international agenda, not only for scientists, academics, and researchers of various professions, but also for politicians and the population in general, due to the multiple, derived, interrelated, or isolated causes that negatively affect the ecosystem. In 2010, the United Nations (UN) General Assembly recognized the right to water and sanitation, reaffirming that the liquid is fundamental for the development of human rights, indicating that it must be sufficient for domestic and personal use, safe, acceptable, and physically accessible. Likewise, in 2015, the 2030 Agenda for Sustainable Development was approved, with 17 Sustainable Development Goals (SDGs), ranging from the eradication of poverty to the fight against climate change. SDG 6 proposes to ensure the availability of water, its sustainable management and sanitation, including equitable access to the resource and improving quality, reducing pollution, eliminating the discharge of untreated wastewater, and increasing reuse [2].

According to the Food and Agriculture Organization of the United Nations (FAO), the amount of precipitation falling on the Earth is almost 110,000 km^3/year. 56% of this is evapotranspired by natural landscapes, and about 5% is used for agriculture. The remaining 39% (43,000 km^3/year) becomes surface runoff, feeding renewable freshwater resources (rivers, lakes, and aquifers) that can be extracted through the installation of appropriate infrastructure. Most water withdrawals will be returned to the environment sometime after they have been used. However, its quality is lower than the original. The FAO estimates three types of abstraction worldwide: (a) agricultural (irrigation, livestock, and aquaculture), accounting for 72%; (b) municipal (domestic), 16%; and (c) industrial, 12% of the total [3].

Surface water harvesting, groundwater filtration and wastewater reuse are essential to ensure irrigation water security for the development of green agriculture and the protection of agricultural lands. They also aim to reduce the impact of soil pollution [4].

The aquifers are recharged by the precipitation that percolates through the soil and reaches the rock material. When the amount of water pumped out of the underground system is greater than the amount that is infiltrated, the water table drops and can change direction. An example of this is that some of the water that used to flow into a stream no longer does, and the stream now feeds the system, reducing the flow of the stream in the future. Water resources are influenced by factors such as population, land use, exploitation, wastewater reuse, economic/social value, accessibility, and quality [5].

Knowledge of the aquifer system is essential for the sustainable development of groundwater resources and their management. In this case, the term "sustainable" relates to groundwater replenishment; risk security; political, economic, and social challenges; improving water quality; groundwater biogeochemical processes and, in short, water security [6]. Water security is based on multidisciplinary and environmentally sound approaches and studies that include quantity and quality, relationships between groundwater and surface water, and between dependent ecosystems. These studies are based on a participatory approach that involves scientists, water professionals, administrators, and planners, to include, in addition to technical concepts, the ethical, religious, and cultural traditions of society.

In response to the need for high-quality water for agricultural production and human consumption, there has been a global increase in groundwater resource pumping, an activity that has led to a worrying decline in water availability. River ecosystems are threatened by reduced rainfall, global warming and, in the long term, population growth. According to studies [7], hydrographic basins around the world have shrunk by between 17 and 21% in the last decade. Increasing available water resources by artificially recharging aquifers, a technique called Managed Aquifer Recharge (MAR), is an ideal option.

MAR has been practiced for decades and in some places for centuries, so it is not a new technique. Artificial groundwater recharge has been practiced using ancestral methods such as sowing and harvesting (also known as "*careos*" or "*amunas*"), initiated several centuries ago in Europe and South America [8]. Since the 1960s, implementation has accelerated at a rate of 5% per year. However, it has not kept pace with increasing resource extraction [9]. MAR slows surface flows and/or facilitates the infiltration of water into the soil at specific locations through infrastructure such as ponds, ditches, injection wells, and others. Recharge water can be stored in a wide range of aquifer types (confined and unconfined), from unconsolidated alluvial deposits to karst and fractured rocks [9, 10]. The current focus in many areas is to increase and enhance MAR to manage water supplies during droughts and to provide greater reliability of water supplies as demand increases and hydrologic changes occur due to the climate and human alteration of the land surface.

MAR is the "managed" intentional recharge of water into aquifers for subsequent recovery or environmental benefit [11]. The flexibility and different configurations that MAR has adopted internationally is a strengths that will see its increasing adoption in sustainable water management. This technology is being recognized for its underground water storage and distribution capabilities, which are increasingly essential for managing risks to human health and the environment. This engineering approach plays an increasingly important role in adapting to climate change by increasing water supply and environmental issues and reusing treated wastewater. However, in some cases, the interaction of "unmanaged" recharge with anthropogenic activities has led to widespread declines in groundwater quality [12, 13].

Rapid urbanization or natural landscape development can alter urban water hydrology, affecting groundwater quality and quantity. While increased runoff over impervious surfaces can decrease direct groundwater recharge, increased artificial, indirect, and localized recharge from leaking pipes can increase recharge. Furthermore, climate change is expected to intensify hydrological events [14], further reducing recharge. Although MAR is applied in water-stressed areas, in some places, it is associated with improving water quality and diluting contaminated groundwater [8].

Most MARs are carried out in water-stressed areas, with the primary objective of groundwater storage. In contrast, MARs to improve water quality occur in areas with relatively high water availability through precipitation, indicating that high precipitation events can produce contaminated stormwater runoff from a watershed, either agricultural or urban [8]. **Figure 1** shows some of the case studies, which are in the most arid regions of the world, where the problems of water quantity and water quality are linked [15, 16].

The purpose of this review is to provide an overview of the types of MAR and some of the challenges faced in implementing MAR projects. It also lists some of the countries that have regulations related to the establishment, monitoring, and implementation of MAR systems. MAR projects are adapted to the hydric and hydrogeologic conditions of each area of interest. Therefore, it is necessary to design MAR systems according to the specific characteristics of the area. A study is currently underway to develop a dynamic simulation model to evaluate the response of a free aquifer to managed artificial recharge. The aquifer is in the central region of Mexico. It is in an area of increasing productive activity, and therefore intensive groundwater exploitation is recorded.

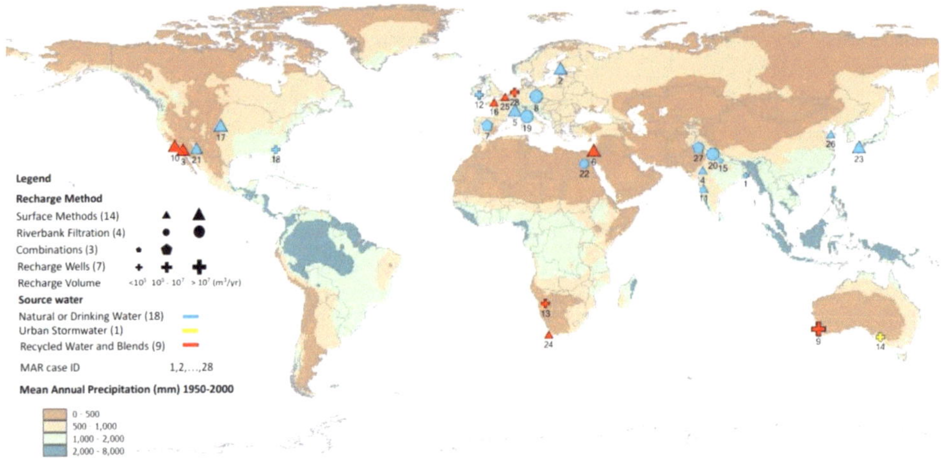

Figure 1.
Some case studies of MAR projects. Adapted from [16].

2. MAR information resources

MAR is defined as the technique of intentional aquifer recharge to stabilize water storage and improve water quality. This technique is increasingly recognized as an important mechanism to address water quantity and/or quality problems [13]. Alternatively, the International Association of Hydrogeologists Commission on Managing Aquifer Recharge (IAH-MAR) defines MAR as the intentional addition of water to aquifers for subsequent recovery or environmental benefit [11]. Managed aquifer recharge is one of the options to provide solutions to water scarcity, water security, declining water quality, declining groundwater levels, and endangered ecosystems that depend on groundwater, as noted by Dillon et al. [17].

In the past, the term “artificial recharge” was used when the goal was to increase the quantity of recharge, but much less attention was paid to water quality management. MAR means that both quantity and quality are effectively managed [9]. MAR can make a significant contribution to the solution of the myriad of local water problems, because it includes projects where water recharged to unconfined aquifers is of the same quality as natural recharge, or where water quality is managed before recharge or during recovery to ensure public and environmental health protection.

Recharge is the process of adding water to an aquifer. Recharge occurs naturally from precipitation and stream flow, incidentally from various human uses such as irrigation or leaks in water pipes, and finally intentionally from facilities or projects developed to recharge an aquifer. It is this latter type of recharge that is considered managed aquifer recharge [15].

MAR has five main objectives: (1) water system management, (2) water storage maximization, (3) physical aquifer management, (4) improving water quality, and (5) achieving ecological benefits. Other benefits include improving soil conditions and preventing land subsidence. Important co-benefits, such as protection against seawater intrusion into coastal aquifers, improvement of environmental flows, water storage for drought relief, and water purification through natural processes, add to the popularity and success of MAR programs. In addition, managed aquifer recharge has the objectives of restoring contaminated systems, minimizing the process of land subsidence, increasing the availability of resources, and reducing the risk of flooding [8, 16, 18]. Therefore, well-conceived and executed MAR projects offer water managers the opportunity to realize water resilience benefits [9].

Intentional aquifer recharge is popular due to its low implementation costs, low evaporative losses, and ability to infiltrate large volumes from various sources such as rivers, treated wastewater, and urban and agricultural runoff, making it a suitable strategy for water resource management [8, 9]. The use of groundwater has increased significantly in the last 50 years due to its widespread presence, high reliability in times of drought, good quality in most cases, advances in hydrological knowledge, and development of modern drilling and pumping technologies. However, these challenges require in-depth studies and research, the implementation of new scientifically based methodologies, and the adoption of principles for integrated management and an environmentally sound protection system for water resources.

The water cycle, the scale and jurisdictional characteristics of the geographic setting, and the location of the water supply and demand are critical factors in determining the solutions to water challenges, including climate change, poverty, and geopolitics. The publication of MAR case studies internationally has contributed to the understanding of how this technique can be part of the set of strategies that can address these problems [15].

Improving water security is perhaps the greatest goal and benefit of MAR, which is understood as an Integrated Water Resources Management (IWRM) technique. MAR is an adaptation mechanism to climate change to ensure the long-term availability of groundwater. It also facilitates the transfer of water through an aquifer to users' wells, reduces pumping energy costs due to rising groundwater levels, and can help mitigate flooding in karst areas by infiltrating water into the aquifer, increasing storm concentration time, and reducing erosion [17].

2.1 Types and processes for MAR implementation

An understanding of hydrogeology is essential to the determination of the options available and the technical feasibility of projects. MAR methods are designed to intentionally infiltrate water from various sources into an aquifer and extract the treated water for later use [8, 9]. Implementing or selecting MAR is challenging because it depends on many interrelated factors, including the availability of water sources for recharge, soil types for infiltration of water sources, the capacity of the aquifer to store and extract enough water, as well as the quality of the water source, the capacity of the infrastructure to remove contaminants, and the cost of implementing and maintaining the projects [9].

A variety of methods, including aquifer storage and recovery, aquifer soil treatment, percolation tanks, bank filtration, infiltration basins, and subsurface dams, can be used to achieve the objective of a MAR project. However, these methods can be grouped into four broad categories: stream channel modification, bank filtration, water distribution, and recharge wells [17, 19]. A variety of recharge methods and water sources using various aquifers for water storage and treatment are shown in **Figure 2**.

Most MAR projects are selected based on water quantity challenges, rarely considering MAR meeting water quality requirements or other potential secondary benefits that would implement the economically viable technique at some sites or capital investment and maintenance costs. Understanding the difference in MAR design and how these designs help infiltrate water or remove contaminants would help in

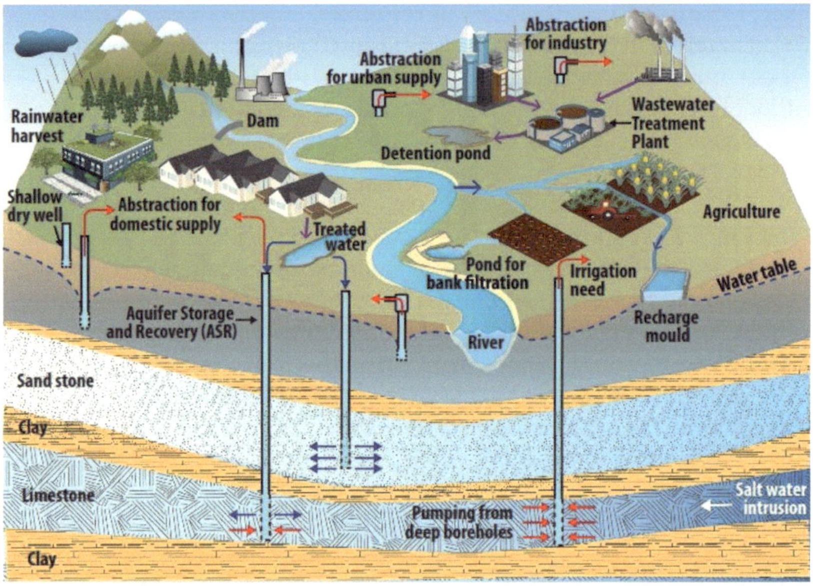

Figure 2.
Schematic of managed aquifer recharge techniques based on the type of aquifer, topography, land use, and intended uses of the recovered water. Adapted from [8].

MAR selection and implementation based on site-specific conditions and needs. However, it must be remembered that the choice of the type of MAR depends, among other limitations, on on-site conditions such as hydraulic conductivity, texture, and mineral properties of the soil and water source, because chemistry can affect both the infiltration capacity as well as the contaminant removal capacity of the MAR [8]. In conclusion, the selection of MAR at a site depends on four main criteria: (a) the capacity of the soil to infiltrate water sources, (b) the scarcity or need of water, (c) the availability of water sources according to the climate, and (d) water quality needs or challenges.

DINA-MAR, TRAGSA, and Fernández Escalante et al. [20] proposed a MAR classification scheme that includes the categories proposed by Gale [19], plus an additional category for accidental (unmanaged) recharge sources. **Table 1** summarizes this classification system, including the main categories and specific types of MAR that can provide benefits in terms of water quantity, quality, or both. Different MAR systems are appropriate for different conditions. Therefore, the selection of the most appropriate system for a particular site and purpose requires a detailed evaluation of local conditions and characteristics of available technologies, as well as a study of regulatory, environmental, and budgetary constraints.

Higher volume infiltration projects involve capturing and recharging flow diversions from perennial streams, as this is the highest volume water source available. The second largest source of water identified is treated wastewater. This is followed by stormwater runoff [21]. However, multiple water sources also add complexity to water quality risk management and MAR regulations, so risk-based management is essential to ensure public health and environmental protection. While fully utilizing the potential of MAR to provide natural treatment and facilitate recycling and reuse [13].

N°	MAR Category	System	MAR Device
1 2 3 4 5	Water spreading	Dispersion	Infiltration ponds wetlands Channels and infiltration ditches Soil ridges and aquifer treatment techniques Infiltration fields (flood and controlled) Bofedal wetlands
6 7 8 9 10 11	River channels	Channels	Reservoir dams Permeable dams Levees Riverbed scarification Subsurface underground dams Drilled dams
12 13 14 15 16 17 18	Targeted recharge	Wells	Qanats (underground gallery) Open infiltration wells Deep wells and boreholes Sinkholes, collapses Aquifer storage and recovery (ASR) Aquifer storage, transfer, and recovery (ASTR)
19 20 21	Filtration	Filtration	River bank filtration (RBF) Interdune filtration Underground irrigation
22 23	Rain based systems	Rain	Rainwater harvesting Sustainable urban drainage systems (SUDS)
24 25	Accidental recharge		Accidental recharge from pipes and sewer system Accidental recharge by irrigation return

Source: Adapted from [20].

Table 1.
Summary of MAR categories and systems.

2.2 Challenges for MAR implementation

Although a site may satisfy primary and secondary criteria, the implementation of MAR can be challenging due to other design and policy constraints [8]. Some challenges to be faced for the implementation of MAR projects, according to different research projects, are listed below.

2.2.1 Clogging

The primary cause of the failure of MAR systems has been that clogging control methods have been inadequate at some sites, primarily in infiltration basins and injection wells [17]. A decrease in infiltration of the water source due to clogging could render the recharge project unusable by causing waterlogging. Clogging can occur due to deposition of suspended sediment in the water source and compaction of the soil layer that allows infiltration. Clogging can also occur due to biofilm formation, especially if the source water has high levels of dissolved organic carbon (DOC) or nutrients [12]. Pretreatment strategies based on water source and site conditions, as well as maintenance regimens to eliminate blockages and extend the useful life of facilities, often require studies to quantify their cost-benefit ratio and improve design and maintenance practices [8, 21].

2.2.2 Groundwater pollution

In some cases, MAR sites may also serve as a source of contaminants or influence the removal of some contaminants. Currently, the natural attenuation or assimilation capacity of aquifers is relied upon for the degradation of organic contaminants. A site that allows rapid infiltration of the source water may increase the risk of groundwater contamination by decreasing sorption, colloid retention, and biodegradation, and pose significant risks to human health. Therefore, a cost-benefit analysis is necessary in cases where additional pre-treatment or post-treatment is required [8, 22]. In urban areas, the stored water is at risk of contamination from the urban land uses above it. Confined aquifer systems are inherently protected from contamination but require wells to be accessed [23]. Tools are needed to enable real-time monitoring, data assimilation, and reactive transport modeling to predict the fate of chemicals and pathogens and assess risks to human health and aquifer integrity [13].

2.2.3 Damage to crops due to waterlogging

Excessive groundwater recharge can raise the water table near the root zone of plants and cause waterlogging. This can reduce aeration and increase plant susceptibility to disease, although the extent of the damage depends on the geology, soil, topography, aquifer characteristics, and drainage of the recharge area. In addition, due to the interconnected nature of the aquifer, waterlogging can occur at a location that is far away from the recharge area [8]. The criteria of interest for implementing a type of MAR should include environmental characteristics such as precipitation, vegetation index, drainage density, and subsurface conditions such as groundwater depth and geology. Mouhoumed et al. [24] emphasized the importance of precipitation, slope, and vegetation index as the most influencing decision parameters, while drainage density had the lowest influence.

2.2.4 Lack of adequate infrastructure and public acceptance

MAR implementation may require additional infrastructure to transport surface water to the recharge site. Depending on the type of MAR desired, a detailed site- or location-based economic/social/technical analysis is required to justify the implementation of MAR. Specifically, using treated wastewater in a MAR project can pose a significant challenge, not only for infrastructure costs but also for public acceptance of the use of that type of effluent [8].

2.2.5 Groundwater governance and socioeconomic challenges

In addition to the technical issues that MAR projects typically address, the lack of provisions for the recovery of stored water [18] makes the regulatory framework a challenge that must be addressed to scale up MAR facilities. Existing water rights, rules, regulations, and policies are important constraints that could limit additional water diversions for the implementation of MAR projects. Rules and regulations can vary significantly from region to region. Therefore, it may be difficult to develop a set of recommendations that can be applied to all regions. Therefore, solutions must be developed based on local water constraints and needs [8].

Science-policy collaboration between water management sectors and institutions, such as researchers, water administrators, NGOs, and consultants, has been put in place for the development of MAR projects. However, structural barriers have been found due to a mismatch between long-term and short-term goals, which results in the availability of funding, which in turn discourages collaboration between science and policy [25].

2.3 Guideline and regulation formulation

Compared to the number of countries where MAR systems are already operating, there are few countries and organizations that have regulations related to MAR directly or indirectly. Countries such as Belgium, Brazil, Chile, China, United States, India, Italy, Mexico, Namibia, Netherlands, New Zealand, Portugal, South Africa, Spain, and Thailand have formulated their own regulations and guidelines regarding the study, implementation, and MAR project monitoring. However, there is no explicit guidance on the minimum requirements to publish a robust regulatory framework and essential aspects that it must address; therefore, it is necessary to strengthen institutional capacity for water allocation, permitting, and protecting water quality [13, 20].

Although most guidelines vary in their level of development and detail, they share a common objective. **Table 2** shows the classification of MAR regulations in some countries, including the scope (international, national, regional, and local), the level of compliance (hard or soft), the type of document (guidelines, regulations, and operator rules), and the year of publication.

Fernández et al. propose nine factors to be considered when drafting any future comprehensive MAR policy or regulation. These factors are grouped under the conceptual model of controlled and intentional recharge. The nine blocks that make up the MAR conceptual model are (a) water sources; (b) environmental conditions, including climate, aquifer type, geology, surface water basin, groundwater body, and purification capacity; (c) technology; (d) sensors to monitor the system and collect data; (e) guidelines for monitoring water quantity and quality; (f) final use of the recovered water; (g) analytical aspects, such as maximum allowable concentrations of potential contaminants; (h) risk assessment, which details some of the risks to be evaluated; and finally, (i) others, including relevant issues that are difficult to include in the previous factors, such as emerging contaminants, economic aspects, and public participation [20].

2.4 MAR projects worldwide

The wide variety of MAR project experiences around the world means that many countries have undertaken initiatives to advance aspects of aquifer recharge. They also recognize the need for MAR regulations to protect aquifers. Some countries with emerging and growing economies are seeing real opportunities to make valuable water supplies more resilient and are moving more quickly than others with more stable economies. These experiences include the 500,000 barraginhas (infiltration ponds) in Brazil, the injection of highly treated recycled water into crystalline rock in Namibia to provide drinking water, and the restoration of wetlands due to rising water levels after MAR in Spain. On the other hand, regulations in Australia have ensured sustainable operations and supported the adoption of MAR. In contrast, existing water allocation policies in Mexico do not provide an incentive to invest in MAR [17].

Country	Type	Scope	Severity	Year
Belgium (Torreele)	Operator rule	Local	Hard	2012
Brazil	Regulation	National	Soft	2019
Chile	Regulation	National	Soft	2013
China	Guideline draft	National	Soft	2014
United States	Guideline	Regional	Hard	1994
Arizona				
California	Guideline	Regional	Hard	2012
Florida	Guideline	Regional	Soft	1999
India	Guideline draft	National	Soft	2014
Italy	Regulation	National	Hard	2016
Israel (Shafdan)	Operator rule	National	Hard	1966
Mexico	Regulation	National	Hard	2003–2009
Namibia (Windhoek)	Guidelines. Regulation proposal	National	—	2004
Netherlands	Regulation	National	Hard	1993
New Zealand	Technical guidance	National	Soft	2017
Portugal	Regulation	National	Hard	2000
South Africa	Regulation draft	National	Hard	2004
Spain	Regulation	National	Hard	2007
Thailand	Guideline	National	Hard	2022
USA (ASCE)	Regulation	National	Soft	1974–2019
WFD (EU)	Regulation	International	Soft	2000
WHO	Guideline	International	Soft	2001

Source: Adapted from [20].

Table 2.
MAR guidelines classified according to their focus, content, and severity.

A global inventory of MAR projects can be seen on the IGRAC MAR portal, which includes 1136 pilot and large-scale schemes. In this portal, the different implementation categories were grouped into five methods: (a) induced bank filtration; (b) in-channel modification; (c) spreading methods; (d) rainwater & runoff harvesting; and (e) wells, shafts, and boreholes. **Figure 3** shows the global inventory of MAR schemes registered on the online portal, which continuously updates its database [26].

The analysis of 1127 aquifer recharge projects showed that they were mainly carried out in sites with loamy-clay-sandy soils and access to river water for infiltration. Spatial analysis showed that many depleting regions have opportunities for such projects [8].

2.5 MAR projects in Mexico

Mexico is on the brink of a water crisis with depleted aquifers and a lack of universal access [25]. The amount of water in Mexico varies due to its geographical position and seasonal cycles that alternate drought and rain. The typical seasonality

DOI: http://dx.doi.org/10.5772/intechopen.115472

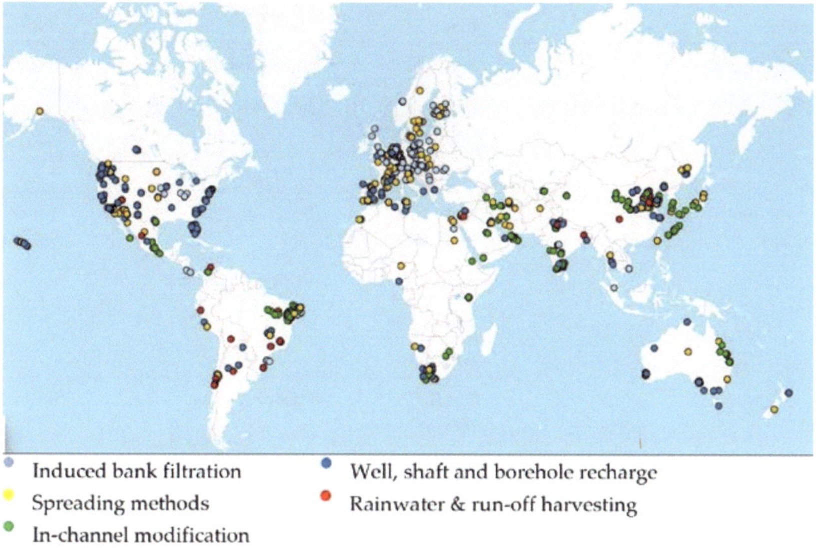

Figure 3.
Global inventory of MAR schemes presented as an on line portal. Adapted from [26].

in tropical-subtropical latitudes are factors that determine the country's climate and the availability of water resources. Two-thirds of the territory is considered arid or semi-arid, with annual rainfall at less than 500 mm, while one-third, in the southeast, is humid, with annual rainfall exceeding 2000 mm torrential rain [26]. Existing conflicts over water can be intensified in generally dry countries where production processes are concentrated in arid regions. In the arid zone of northern and central Mexico, the largest population volume is concentrated, and productive processes are developed, which require a large amount of water for agroindustry. In the northern states, the overexploitation of aquifers is fundamentally related to export agriculture and dairy production; it has been observed that the agricultural districts of Sonora and Baja California have caused marine intrusion [27, 28].

MAR implementation is progressing apace, driven by necessity and with management aspects supported by research that improves risk assessment on resource sustainability and water quality. Unlike many other countries, Mexico is one of the few that has regulations for the development and implementation of projects to recharge underground bodies of water. Two official standards establish the guidelines and water quality parameters that these types of projects must comply with: NOM-014-CONAGUA-2003: Requirements for the artificial recharge of aquifers with treated wastewater [29], and NOM-015-CONAGUA-2007: Artificial infiltration of water into aquifers – Characteristics and specifications of the works and water [30]. Within the requirements, it is specified that the treated wastewater used must meet the permissible limits of quality for human use and consumption set out in NOM127-SSA1-2021, published in May 2022.

Aquifer recharge projects in Mexico have been designed and implemented by researchers, consultants, and local governments since the 1950s. Small-scale plants currently exist in rural communities and medium-sized cities. The approximate total volume of annual water recharge is estimated at 156 million m^3 [31]. **Table 3** presents the list of artificial recharge projects carried out in Mexico, especially in arid and semi-arid regions. These projects were designed to restore aquifers and prevent

Objective	Place	Device	Source
Repair the aquifer	Sonora	Injection wells	Rainfall
	Baja California Sur	Reservoir dams	
	Querétaro	Underground dams	
Repair the aquifer and increase availability	Baja California	Injection wells	Treated wastewater
	Sonora	Infiltration ponds/ wetlands	
	Chihuahua		
	Sonora		
	State of México	Injection wells	Rainfall
	Oaxaca	Underground dams and injection wells	
	Chihuahua		
Repair the aquifer, reduce land subsidence, and increase availability	México	Reservoir dams	Rainfall
	México	Injection wells	Effluent
Repair the aquifer and reduce land	State of México	Injection wells	Rainfall
Flood management and repair the	San Luis Potosí	Injection wells	Rainfall
Repair the aquifer	Coahuila	Injection wells	Dam

Source: Adapted from [18].

Table 3.
MAR guidelines classified according to their focus, content, and severity.

flooding [18]. In the early 2000s, plants using treated wastewater were proposed in the states of Chihuahua, Sonora, and Mexico.

3. Conclusions

Just as the population increases, the demand for water to meet needs also increases. Water resources are under increasing pressure not only from population growth and rapid urbanization but also from the impact of climate change. These factors must be studied to guarantee sustainable management of groundwater since they alter rainfall patterns, humidity, the balance of glacier masses, river flow, and changes in the sources of aquifers.

On the planet, groundwater accounts for 98% of unfrozen freshwater, making it a significant component of the hydrogeological cycle. Groundwater must be integrated into the economic, social, and environmental dimensions of water resources. Many people depend on this resource as a supply of drinking water and to ensure food security and sustainable living. Groundwater is a critical component of water supply for agriculture, urban areas, industry, and ecosystems, but its management is challenging because it is difficult to quantify and evaluate groundwater quality. The groundwater crisis in several countries is especially aggravated by the average annual precipitation has decreased in the long term, which means limited water resources, an increase in population and its respective demand, intensification of droughts and floods, acceleration of changes in land use and land cover, and finally poor management and lack of adequate water governance.

Groundwater is a finite and invisible resource. Surface water availability is being negatively affected by climate change. This article is designed to provide an overview of the Managed Aquifer Recharge (MAR) technique worldwide. Although not exhaustive, information was provided about the types and processes for implementing the technique, as well as a few challenges that the water community has faced. The countries that have certain regulations for the implementation and operation of MAR projects were also listed, and finally, the projects developed in Mexico.

Managed aquifer recharge is tailored to the hydrology and hydrogeology of each area, and is generally governed by aquifer type, topography, land use, environmental groundwater quality, and intended uses of recovered water.

The growth of applied research in the development, analysis, and implementation of MAR projects has allowed us to consolidate the fundamentals for the achievement of reliable MAR systems that will produce resilient water supplies. Structural implementation alternatives have been proposed, including water sources for recharge, reducing water treatment costs, and improving the quality and quantity of water for recharge, among others. Despite this progress, there are still several steps that would improve the efficiency of MAR investments and support the adoption of MAR as a water supply option.

Innovations in MAR methodologies and monitoring are constantly under development and implementation. However, in most countries, governance frameworks need to be strengthened to ensure that MAR is sustainable, protects groundwater quality, and provides benefits to all community members so that the MAR system is reliable.

Acknowledgements

This study is part of the Ph.D. thesis of the first author, developed during the Doctoral Program in Water Sciences, Universidad de las Américas Puebla (UDLAP), with the support of the National Council of Science and Technology (CONAHCYT).

Conflict of interest

The authors declare that they have no competing interests.

Author details

Adriana Serrano-Santiago[1*] and Adriana Palacios[2]

1 Departamento de Ingeniería Civil, Universidad de las Américas Puebla, Puebla, Mexico

2 Departamento de Ingeniería Química, Alimentos y Ambiental, Universidad de las Américas Puebla, Puebla, Mexico

*Address all correspondence to: adriana.serranoso@udlap.mx

References

[1] UN-Water. Summary Progress Update 2021: SDG 6—Water and Sanitation for all. Switzerland; 2021

[2] United Nations. Goal 6: Ensure Access to Water and Sanitation for all 1999. Available from: https://www.un.org/sustainabledevelopment/water-and-sanitation/ [Accessed: December 05, 2023]

[3] FAO. AQUASTAT—FAO's Global Information System on Water and Agriculture. 2021. Available from: https://www.fao.org/aquastat/en/overview/methodology/water-use/ [Accessed: December 01, 2023]

[4] UNESCO. Water Security and the Sustainable Development Goals Training Manual for Decision Makers. Uruguay; 2020. Available from: https://unesdoc.unesco.org/ark:/48223/pf0000374917.locale=es

[5] Srinivasa K, Nagesh D. Impact of Climate Change on Water Resources with Modeling Techniques and Case Studies. Springer Climate; 2018. Available from: http://www.springer.com/series/11741

[6] Fundación Aquae. Las aguas subterráneas y peligroso descenso. 2021. Available from: https://www.fundacionaquae.org/wiki/el-preocupante-descenso-de-las-aguas-subterraneas/ [Accessed: December 01, 2023]

[7] Fernández, Escalante E, Stefan C, Brown CJ, Hutchinson A. Managed aquifer recharge: A key to sustainability. In: Water (Switzerland). Vol. 15. Multidisciplinary Digital Publishing Institute (MDPI); 2023

[8] Sarfaraz A, Annesh B, Sujith R, Mekonnen G, Sanjay KM. Managed Aquifer Recharge Implementation Criteria to Achieve Water Sustainability, Science of the Total Environment. Vol. 768. Elsevier B.V.; 2021

[9] Dillon P, Stuyfzand P, Grischek T, Lluria M, Pyne RDG, Jain RC, et al. Sixty years of global progress in managed aquifer recharge. Hydrogeology Journal. 2019;**27**(1):1-30

[10] Parker TK, Alley WM, Job CA. Managed Aquifer Recharge: A Proven Technology for Water Supply Resilience, Groundwater. Vol. 60. John Wiley and Sons Inc; 2022. pp. 577-578

[11] IAH. Commission on Managing Aquifer Recharge. Available from: Welcome—Managed Aquifer Recharge (iah.org) [Accessed: December 01, 2023]

[12] Song Y, Du X, Ye X. Analysis of potential risks associated with urban stormwater quality for managed aquifer recharge. International Journal of Environmental Research and Public Health. 2019;**16**(17)

[13] Zheng Y, Vanderzalm J, Hartog N, Escalante EF, Stefan C. The 21st century water quality challenges for managed aquifer recharge: Towards a risk-based regulatory approach. Hydrogeology Journal. 2023;**31**(1):31-34

[14] IPCC. Summary for policymakers. In: Lee H, Romero J, editors. Climate Change 2023: Synthesis Report. A Report of the Intergovernmental Panel on Climate Change. Contribution of Working Groups I, II and III to the Sixth Assessment Report of the Intergovernmental Panel on Climate Change Core Writing Team. Geneva, Switzerland; 2023. Available from: https://www.ipcc.ch/ar6-syr/

[15] Megdal S. MAR as a mechanism to advance water policy goals: A perspective.

The Water Report. 2022. Available from: www.TheWaterReport.com

[16] Zheng Y, Ross A, Villholth K, Dillon P. MANAGING AQUIFER RECHARGE a Showcase for Resilience and Sustainability. Paris: UNESCO; 2021. Available from: http://www.unesco.org/open-access/terms-use-ccbysa-en

[17] Dillon P, Escalante EF, Megdal SB, Massmann G. Managed Aquifer Recharge for Water Resilience. Vol. 12. Water (Switzerland): MDPI AG; 2020

[18] Cruz-Ayala MB, Megdal SB. An overview of managed aquifer recharge in Mexico and its legal framework. Water (Switzerland). 2020;**12**(2)

[19] Gale I. Strategies for Managed Aquifer Recharge (MAR) in Semi-Arid Areas. UNESCO: Paris, France; 2005. p. 33

[20] Fernández Escalante E, Henao Casas JD, San Sebastián Sauto J, Calero Gil R. Monitored and intentional recharge (MIR): A model for managed aquifer recharge (MAR) guideline and regulation formulation. Water (Basel). 2022;**14**(21):3405. Available from: https://www.mdpi.com/2073-4441/14/21/3405

[21] Palma Nava A, Parker TK, Carmona Paredes RB. Challenges and experiences of managed aquifer recharge in the Mexico City metropolitan area. Groundwater. 2022;**60**(5):675-684

[22] Ajjur SB, Baalousha HM. A Review on Implementing Managed Aquifer Recharge in the Middle East and North Africa Region: Methods, Progress and Challenges, Water International. Vol. 46. Routledge; 2021. pp. 578-604

[23] Page D, Bekele E, Vanderzalm J, Sidhu J. Managed aquifer recharge (MAR) in sustainable urban water management. Water (Switzerland). 2018;**10**(3)

[24] Mouhoumed RM, Ekmekcioğlu Ö, Başakın EE, Özger M. Integrated fuzzy AHP-TOPSIS model for assessing managed aquifer recharge potential in a hot dry region: A case study of Djibouti at a country scale. Water (Switzerland). 2023;**15**(14)

[25] Cruz-Ayala MB, Soto JR, Wilder MO. On lessons from water recharge projects in Mexico: Science-policy collaboration and stakeholder participation. Water (Switzerland). 2023;**15**(1)

[26] UNESCO. MAR Portal—IGRAC. 2023. Available from: https://ggis.un-igrac.org/view/marportal/ [Accessed: December 05, 2023]

[27] CONAGUA. Estadísticas del Agua en México. México: Secretaría de Medio Ambiente y Recursos Naturales; 2019

[28] Spring Ú. Water security and national water law in Mexico. Earth Perspectives. 2014;**1**(1):7

[29] Secretaría de Medio Ambiente y Recursos Naturales. NOM-014-CONAGUA-2003. Requisitos para la recarga artificial de acuíferos con agua residual tratada. Diario Oficial de la Federación; 2009a

[30] Secretaría de Medio Ambiente y Recursos Naturales. NOM-015-CONAGUA-2007. Infiltración artificial de agua a los acuíferos. Diario Oficial de la Federación; 2009b

[31] Bonilla Valverde JP, Stefan C, Palma Nava A, Bernardo da Silva E, Pivaral Vivar HL. Inventory of managed aquifer recharge schemes in Latin America and the Caribbean. Sustainable Water Resource Management. 2018;**4**(2):163-178

Chapter

A Comparison of GoogleNet Model Performance in RNN Cells for Breast Cancer Diagnosis from Thermographic Images

Andrés Munguía-Siu and J. Horacio Espinoza-Rodríguez

Abstract

Breast cancer is one of the most common diseases of death in women worldwide, and early diagnosis could reduce the death rate. Different medical imaging modalities have been used to detect breast cancer, but there is a novel noninvasive, non-radiative, and low-cost technique that has gained prominence recently, namely Infrared Thermography. The use of deep learning models has become relevant in the medical field and to improve the classifier performance. The aim of this work is to evaluate the GoogleNet architecture coupled with recurrent neural networks using breast thermal images obtained from the dataset Visual Lab DMR. A methodology was established to filter the data and obtain a balanced binary categorization, automatically segment the data with the U-Net network, apply Data Augmentation, and finally train GoogleNet coupled with Recurrent Neural Networks (RNN), Long-Short Term Memory (LSTM) and Gated Recurrent Unit (GRU). The results achieved an accuracy of 72.70% for the simple GoogleNet model and for the coupled models 77.30, 94.08 and 87.50% for RNN, LSTM and GRU, respectively. The results suggest that GoogleNet and LSTM increase the performance metrics to diagnose breast cancer in thermal images.

Keywords: breast cancer, thermography, DTI, GoogleNet, RNN

1. Introduction

In the last decade, there has been an interest in detecting diseases through non-invasive imaging techniques. One example has been the timely diagnosis of breast cancer at an early age to reduce the probability rate of death. There have been several imaging modalities to perform this task such as mammography, ultrasound, and magnetic resonance imaging (MRI); however, these screening techniques can become uncomfortable, with radiation risk, or high cost [1]. A novel imaging technique for disease detection has been infrared thermography (IT), which involves the detection of heat patterns on the surface of the body caused by blood perfusion and metabolism [2]. One of the advantages of using this modality is that it can diagnose breast cancer 8–10 years earlier than mammography, the standard technique for detecting breast cancer, in addition to being noninvasive, low-cost, non-radiative and with a higher detection rate in dense breasts [3].

For the study of these infrared images, the passive approach, also called Static Infrared Imaging (STI), was first proposed. In STI, only one image is captured without considering variations over time, trying to capture the heat patterns in the area of interest [4]. The problem with this approach is the high false negative rate, reaching an accuracy of 41% [5]. An example of STI image acquisition is shown in **Figure 1**.

One way to address the low sensitivity for detection of breast cancer disease in thermographic imaging was the development of the dynamic thermal imaging (DTI) protocol acquisition. In this modality, a thermal stimulus must be applied to the area of interest to reach a target temperature. Once this point is reached, the heat or cold source is removed and the capture of thermal images begins over the area to obtain thermal information until the basal temperature is reached [7]. This modality has demonstrated an increase in sensitivity of up to 82%, higher than that mentioned in the static approach [8]. **Figure 2** shows the two phases of the DIT approach, where first, a thermal stimulus (in this case cold) is applied until the target temperature is reached, and second, in the recovery stage, where the recording of the thermal behavior is captured until the basal temperature is reached.

Artificial Intelligence (AI) has been used for diagnosis in the medical field, either for segmentation, detection or classification of breast cancer. Deep learning is one of the most popular techniques for performing this task, and it is a sub-field of machine learning [9]. In deep learning, it has been shown that using ensemble models of Convolutional Neural Networks (CNNs) and Recurrent Neural Networks (RNNs) can improve performance metrics in imaging modalities for cancer detection [10, 11]. CNNs are mainly used for feature extraction in images, whereas RNNs allow feature

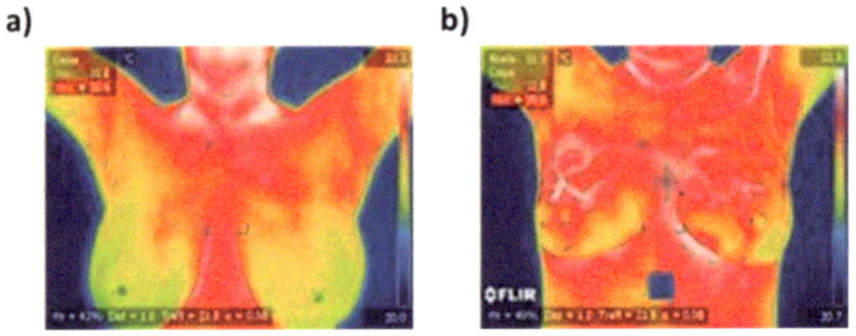

Figure 1.
Infrared images for breast cancer detection (a) healthy volunteer and (b) sick volunteer [6].

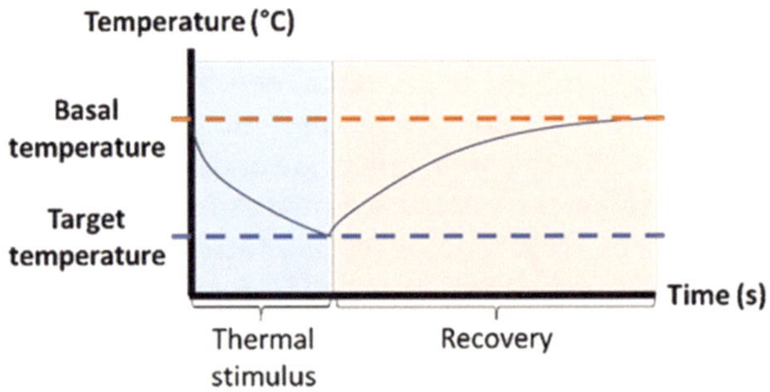

Figure 2.
Recovery rate of the DTI approach using a fan as a thermal stimulus to cool the area of interest.

extraction in the sequence obtained by the CNN. Therefore, the aim of this work is to evaluate the performance of the CNN architecture GoogleNet assembled with the RNN cells (RNN, Long-Short Term Memory (LSTM) and Gated Recurrent Unit (GRU)) in infrared thermography for breast cancer diagnosis using the DTI approach.

2. Methodology

2.1 Dataset

The Visual Lab DMR database is the only public database containing breast thermal images for both static and dynamic approaches [6]. For data acquisition, they used the FLIR SC-620 thermal camera, which contains a resolution of 640 × 320 pixels. Their protocol for image acquisition with STI, consists of having the volunteer resting for 10 minutes to reach its basal temperature; once the time has elapsed, one image is captured from the front, two at 45° and another two at 90° rotation. For data acquisition with the DTI protocol, a fan was used on the volunteer to cool the thorax until a target temperature of 30.5°C was reached. Once the target temperature was reached, the cold source was removed to capture a thermal image every 15 seconds for 5 minutes. All volunteers were labeled into two classes: 244 of them belonged to the healthy class and 44 of them belonged to the sick class.

To use a balanced database, the images were filtered in two steps. First, the samples were filtered so that there were 20 images in sequence per volunteer, and any blurred images were discarded. Second, we counted the resulting samples from each class and selected the same number of samples as the class with the fewest volunteers for the two groups. This resulted in a total of 38 volunteers for each class. Due to the small amount of data, it was decided to apply data augmentation on the samples to obtain a better overview in the training of the classifier. In total, four different transformations were used: horizontal flipping, rotation at 15° and 30°, and zoom.

2.2 Segmentation

For the automatic segmentation, a CNN architecture called U-Net was trained to discard the noisy areas of the thermal images, such as the armpits, stomach and neck, and only highlight the breast area. The advantage of using this network is that it allows obtaining a good performance in automatic segmentation with the training of few images, which makes it ideal for use in the biomedical area [12]. For the training of the network, 40 thermal images of the breasts were manually segmented with the ITK-SNAP software. **Figure 3** shows the representative image of a thermal image after being segmented by U-Net.

Additionally, the resulting images were resized to obtain network input tensors of size 128 × 128 × 20, where the first two values are the size of the image width and height, and the third value is the 20 images in sequence per volunteer.

2.3 Classification

In 2014, the ImageNet Large Scale Visual Recognition Challenge (ILSVRC) for image and video analysis was held as a competition to propose different approaches to classify or detect objects. In that year, Google implemented its CNN architecture called GoogleNet, which achieved results of 93.33% accuracy for classification [13].

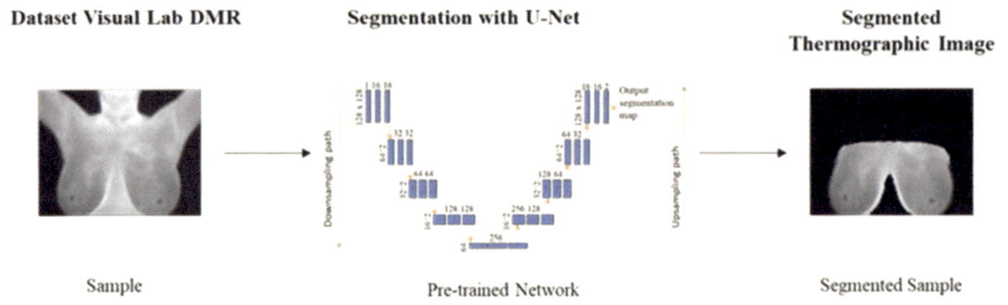

Figure 3.
Breast thermal image segmentation process using the U-Net pre-trained network.

This network introduced inception modules, which perform operations known as convolutions of size 1 × 1, 3 × 3, 5 × 5 and max pooling operations for reduction in dimensions of size 3 × 3. In addition, this architecture has considerable advantages over other CNN models because it is trained with fewer parameters and layers [14]. **Table 1** shows the operations performed by each layer of the GoogleNet network with their respective inception modules.

GoogleNet is used to assess the performance of the simple deep learning model and for the ensemble model, it is used for feature extraction in the segmented images. For its use in the ensemble model with RNN, the Fully Connected (FC) and softmax

Layer	Operation	Patch size/stride	Size
'data'	input		224x224x3
'conv1-7x7_s2'	convolution	7x7/2	112x112x64
'pool1-3x3_s2'	max pool	3x3/2	56x56x64
'conv2-3x3'	convolution	3x3/1	56x56x192
'pool2-3x3_s2'	max pool	3x3/2	28x28x192
'inception_3a'	inception		28x28x256
'inception_3b'	inception		28x28x480
'pool3-3x3_s2'	max pool	3x3/2	14x14x480
'inception_4a'	inception		14x14x512
'inception_4b'	inception		14x14x512
'inception_4c'	inception		14x14x512
'inception_4d'	inception		14x14x528
'inception_4e'	inception		14x14x832
'pool4-3x3_s2'	max pool	3x3/2	7x7x832
'inception_5a'	inception		7x7x832
'inception_5b'	inception		7x7x1024
'pool5-7x7_s1'	average pool	7x7/1	1x1x1024
'pool5-drop_7x7'	drop out		1x1x1024
'fc'	dense		1x1x1000
'inception_4e'	inception		14x14x832

Table 1.
GoogleNet model detailed in each layer.

layers were removed since they perform a logistic regression where the probability of belonging to some class is obtained. This last layer is where the RNN cells are coupled for the analysis of the features in sequence.

RNNs are a type of deep learning network whose main function is to analyze data in sequence, they are built from intermediate layers which represent hidden states where the activation of one depends on the previous state [15]. This network works well with short data sequences, and the problem occurs when the input data has a large amount of information. One way to solve the vanishing gradient problem is through two variations of the RNNs which are LSTM and GRU. These RNNs are the ones that were evaluated for the coupled GoogleNet + RNN model, where the convolutional network extracts the spatial features from the thermal image and the RNNs analyze those features obtained from the input sequence.

3. Results

A total of 304 samples of 128 x 128 x 20 were obtained after applying Data Augmentation to the filtered data, where 152 belong to the healthy group and the remaining to the sick group. **Figure 4** shows a sample of how the input data for the CNN is organized.

The way to evaluate the performance of the models with this data was through the accuracy, sensitivity and specificity metrics (see Eqs. (1)–(3)), which were calculated with the assistance of the confusion matrix (see **Figure 5**). Furthermore, the validation used was the Leave One Out Cross Validation (LOOCV) due to the small number of samples for model training.

$$ACC = \frac{TP + TN}{TP + TN + FP + FN} \tag{1}$$

$$Sens = \frac{TP}{TP + FN} \tag{2}$$

$$Sens = \frac{TP + TN}{TP + TN + FP + FN} \tag{3}$$

where TP is the prediction of a sample for the sick class when the real class is sick, TN for a healthy predicted class when the real class is healthy, FP is the prediction of a sick class when the real is healthy and FN for the prediction of a healthy class when the real class is sick. The results obtained by the different models are reported in confusion matrices (see **Figure 6**), where GoogleNet and FC were used to evaluate the simple deep learning model, and RNN, LSTM and GRU were used to evaluate the coupled models.

With the confusion matrices, metrics can be calculated using Eqs. (1)–(3), which are summarized in **Table 2**. From this table, it can be seen that by coupling the GoogleNet architecture with RNN, LSTM and GRU cells, the performance can be improved to classify volunteers as having or not having breast cancer, compared to only using a simple model with an FC layer. In addition, it is important to mention that of the coupled models, the one that performed best was GoogleNet + LSTM, which achieved 94.08%, 90.13% and 98.03% in accuracy, sensitivity and specificity, respectively.

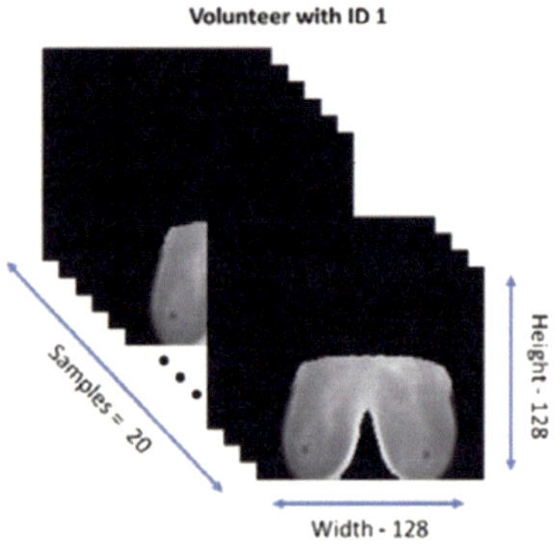

Figure 4.
Sample of the volunteer with ID 1 using images from the DTI approach and with automatic segmentation for deep learning model input.

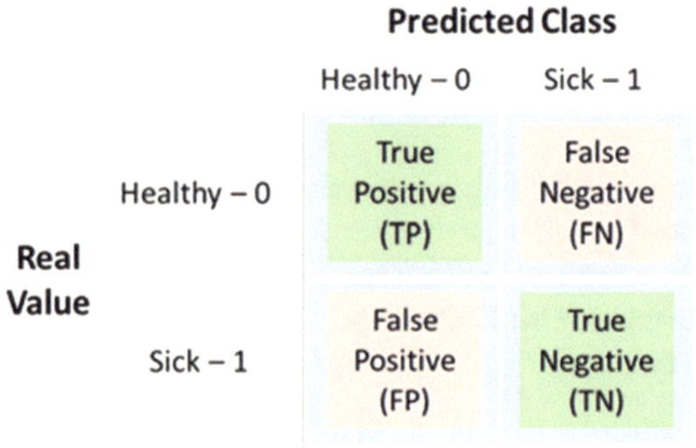

Figure 5.
Confusion matrix and the meaning of the data classified according to the real value and the predicted class.

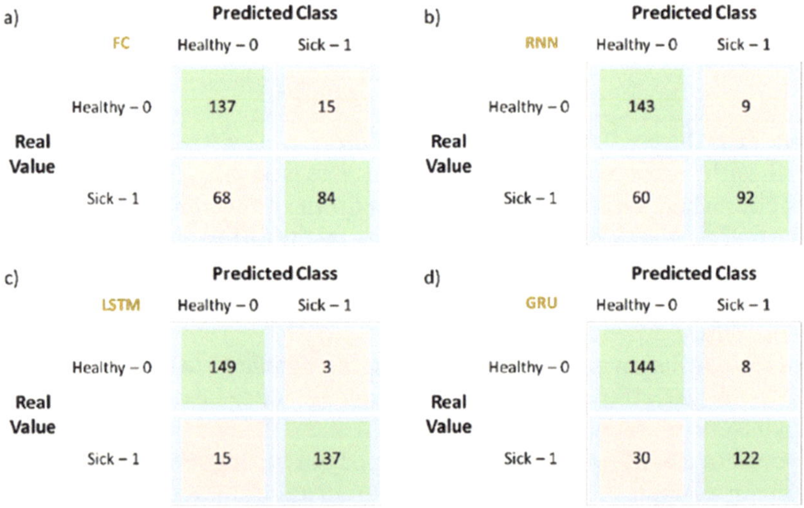

Figure 6.
Confusion matrices to classify volunteers for breast cancer diagnosis where the networks used were as follows: (a) GoogleNet + FC, (b) GoogleNet + RNN, (c) GoogleNet + LSTM, and (d) GoogleNet + GRU.

Individual #	Chromosome	Fitness	% from the total
1	11,010,110	254	24.5
2	10,100,111	47	4.5
3	00110110	457	44.1
4	01110010	194	18.7
5	11,110,010	85	8.2
TOTAL		1037	100.0

Table 2.
A hypothetical example of a population with five individuals having chromosomes of 8 bits in length. The third column shows the fitness values of each individual.

4. Conclusions

In this work, GoogleNet was used to classify breast cancer disease through thermal images obtained with the DTI acquisition protocol. This was done through the evaluation of a typical model and coupled models of GoogleNet + RNN (RNN, LSTM and GRU). It was demonstrated that using the coupled GoogleNet + RNN models can increase the performance metrics compared to the simple model, which has proven good results in the medical field for the early diagnosis of diseases in imaging.

Acknowledgements

The authors acknowledge the Universidad de las Americas Puebla (UDLAP) for supporting this publication. Andrés Mungua-Siu thanks CONAHCyT for a PhD scholarship.

Conflict of interest

The author declares there is no conflict of interest.

Author details

Andrés Munguía-Siu and J. Horacio Espinoza-Rodríguez*
Department of Computing, Electronics and Mechatronics, Universidad de las Américas Puebla, San Andrés Cholula, Puebla, México

*Address all correspondence to: juan.espinoza@udlap.mx

References

[1] Gonzales-Hernandez JL, Recinella AN, Kandlikar SG, Dabydeen D, Medeiros L, Phatak P. Technology, application and potential of dynamic breast thermography for the detection of breast cancer. International Journal of Heat and Mass Transfer. 2019;**131**:558-555

[2] Lozano A III, Hassanipour F. Infrared imaging for breast cancer detection: An objective review of foundational studies and its proper role in breast cancer screening. Infrared Physics and Technology. 2019;**97**:244-257

[3] Singh D, Singh AK. Role of image thermography in early breast cancer detection-past, present and future. Computer Methods and Programs in Biomedicine. 2020;**183**:105074

[4] Resmini R, Faria da SL, Medeiros Petrucio RT, Araujo Adriel S, Muchaluat-Saade Débora C, Aura C. A hybrid methodology for breast screening and cancer diagnosis using thermography. Computers in Biology and Medicine. 2021;**135**:104553

[5] Kandlikar Satish G, Isaac P-R, Raghupathi Pruthvik A, Jose-Luis G-H, Donnette D, Lori M, et al. Infrared imaging technology for breast cancer detection–current status, protocols and new directions. International Journal of Heat and Mass Transfer. 2017;**108**:2303-2320

[6] Silva LF, Saade DCM, Sequeiros GO, Silva AC, Paiva AC, Bravo RS, et al. A new database for breast research with infrared image. Journal of Medical Imaging and Health Informatics. 2014;**4**(1):92-100

[7] Elias da STA, Faria da SL, Christina M-SD, Aura C. A computational method to assist the diagnosis of breast disease using dynamic thermography. Sensors. 2020;**20**(14):3866

[8] Yasuhiko O, Isao U. Applying dynamic thermography in the diagnosis of breast cancer. IEEE Engineering in Medicine and Biology Magazine. 2000;**19**(3):42-51

[9] Ella M, Akhloufi MA. Applying deep learning for breast cancer detection in radiology. Current Oncology. 2022;**29**(11):8767-8793

[10] Xiaomei W, Ahmad I, Danish J, Armana ZS, Alotaibi FM, Ghoneim Mohamed E, et al. Intelligent hybrid deep learning model for breast cancer detection. Electronics. 2022;**11**(17):2767

[11] Kushangi A, Kumar SB, Bodhey Narendra K, Bilas PR. Mammography and ultrasound based dual modality classification of breast cancer using a hybrid deep learning approach. Biomedical Signal Processing and Control. 2023;**86**:104919

[12] Mohamed Esraa A, Rashed Essam A, Tarek G, Omar K. Deep learning model for fully automated breast cancer detection system from thermograms. PLoS One. 2022;**17**(1):e0262349

[13] Tiantian F. A novel computer-aided lung cancer detection method based on transfer learning from GoogLeNet and median intensity projections. IEEE International Conference on Computer and Communication Engineering Technology (CCET). 2018;**2018**:286-290

[14] Deepak S, Ameer PM. Retrieval of brain MRI with tumor using contrastive loss based similarity on GoogLeNet encodings. Computers in Biology and Medicine. 2020;**125**:103993

[15] Qiao P, Yuanyuan Z, Chen Dehua X, Guangwei. Character-based convolutional grid neural network for breast cancer classification. International Conference on Green Informatics (ICGI). 2017;**2017**:41-48

Chapter

Effect of Sample Distribution in Pulsed Light Equipment for the Inactivation of *Escherichia coli* ATCC 25922

Christelle Pihen, Aurelio López Malo and Nelly Ramírez Corona

Abstract

Pulsed light stands out as an emerging technology, proving to be an effective and rapid treatment method for preserving the shelf life of food. While previous assessments have explored the influence of vertical sample positioning on microbial inactivation efficiency, the impact of the horizontal plane, coupled with the photothermal effect of the equipment, in relation to changes in color, absorbance, and temperature remains unexplored. Thus, this study aims to analyze the alterations in color, absorbance, and temperature in the horizontally distributed irradiated dose within a solution of water and tartrazine, concurrently assessing the inactivation of *Escherichia coli* (*E. coli*) ATTC 25922 at two distances, 5.7 cm and 10.8 cm. The findings revealed a more uniform distribution of equipment fluence at a greater distance between the sample and the lamp compared to shorter distances (5.7 cm). Nevertheless, at 5.7 cm, a notable difference exists among horizontally distributed samples concerning color, absorbance, temperature, and *E. coli* inactivation. Furthermore, the difference in the horizontal plane's uniformity is more pronounced with longer treatment times, emphasizing the need for additional analysis regarding sample replicability in the horizontal plane.

Keywords: pulsed light technology, *Escherichia coli* inactivation, sample distribution, photothermal effect, fluence uniformity

1. Introduction

The preservation of food products and the extension of their shelf life in the food industry have historically relied on heat-based techniques such as pasteurization and sterilization. However, these thermal processing methods come with inherent challenges, particularly in terms of high energy requirements. Additionally, the elevated processing temperatures associated with these methods can cause damage to nutrients and alterations in sensory characteristics [1, 2].

The projected global population increase by 2050 is estimated to be around 2 billion individuals [3]. With consumers continually demanding enhanced food safety, superior food quality, and the preservation of nutritional properties [4, 5], food processing scientists are actively seeking faster and more environmentally friendly pasteurization methods capable of ensuring food safety while keeping nutritional properties intact.

Pulsed light emerges as a promising technology for food decontamination, providing an innovative approach to enhance food safety and prolong shelf life through rapid treatment [6–8]. This advanced technique utilizes high-intensity light within the wavelength range of 200-1100 nm and employs short exposure times for microbial inactivation [2, 9]. In photochemical inactivation (PL) treatments, electrical energy is accumulated in capacitors and then released in the form of high-intensity pulses. The inhibitory mechanism of PL is linked to photochemical activity that alters DNA through the absorption of ultraviolet light. Furthermore, a photothermal effect, involving localized heating of microbial cells, can result in cell rupture due to exposure to infrared light [10, 11]. Nevertheless, the efficacy of pulsed light equipment in food decontamination is intricately tied to the uniform distribution of light, as a critical element that directly influences food safety, product quality, and process efficiency [1, 7, 12, 13].

To determine the response of a microorganism to pulsed light (PL), it is essential to maintain a constant fluence rate for each microorganism within the tested food matrix. Consequently, ensuring a uniform distribution of radiation within the PL equipment is crucial to ensure the replicability of the experiment. Several authors have evaluated that closer distances between the irradiation source and the evaluated samples guarantee a higher dose emitted to the samples. However, improving the uniformity of radiation distribution within a Petri dish involves employing several methods, such as establishing a consistent radiation pattern on the surface of the dish and using a shallow liquid depth in the samples [14, 15]. However, achieving and quantifying perfect uniformity in irradiance distribution remains a challenge, even with a reduction of the distance between the radiation source and the sample.

Although analytical methods have been utilized to verify the fluence distribution emitted by light equipment, these methods encompass various approaches. This includes the use of a radiometer, which employs sensors to make instantaneous or cumulative measurements of ultraviolet energy. Actinometry involves exposing a radiation-sensitive chemical compound to light, inducing a photochemical change that allows for the analytical determination of the dose emitted by the light equipment [16, 17]. Biodosimetry is a technique that entails exposing a liquid inoculated with microorganisms to radiation and subsequently assessing the number of surviving microorganisms in relation to the average dose equivalent for microbial reduction [16, 18, 19]. The identification of fluence patterns in the distribution of radiation within the PL apparatus remains crucial.

Therefore, the aim of this research was to analyze the irradiated dose distribution in water solutions with tartrazine inoculated with *Escherichia coli* (*E. coli*) ATTC 25922 at a distance of 5.7 and 10.8 cm, evaluating the absorbance, color change and temperature difference and correlating the results with the inactivation of *E. coli* ATTC 25922.

2. Methodology

2.1 Pulsed light equipment

PL was applied to solutions by using SteriPulse ® Z-1000 (XENON Corp., Wilmington, MA). The system was a linear xenon flash lamp (Ø 2.54 × 40.6 cm, model LH-840, B-type, mercury-free) that could deliver high-intensity noncollimated white light (100-1100 nm) that generated 5.6 J/cm^2 per pulse on the strobe surface at an input voltage of 3 kV, with three pulse per second and 360 microseconds duration per pulse, with a fluence emitted by the equipment of 0.846 at 5.7 and 0.45 J/cm^2/pulse at 5.7 and 10.8 cm, respectively, at different treatment time. **Figure 1** shows the conformation and distribution of the samples in the equipment tray, where each number corresponds to a Petri dish (100×15 mm) with 10 mL of water with tartrazine (0.025 g/L) with a sample thickness of 0.5 cm.

2.2 Color determination

The color of the samples was determined with the CM-2300d colorimeter (Konica Minolta, The Netherlands). It calculated the total change of color (ΔE), using the L* values (luminosity), a* (red to green), b* (yellow to blue) obtained by the colorimeter.

$$\Delta E = \sqrt{\left(L_1^* - L_0\right)^2 + \left(a_1^* - a_0\right)^2 + \left(b_1^* - b_0\right)^2} \quad (1)$$

2.3 Absorbance

The absorbance of each sample at both distances was analyzed from a UV-Vis 1900 spectrophotometer (Shimadzu Spectrophotometer, Japan) and the results were reported by the number of each sample described in **Figure 1**.

2.4 Microorganism growth conditions

Escherichia coli ATTC 25922 was obtained from the bacteria collection from the Food Microbiology Laboratory of the Universidad de las Americas Puebla. Bacteria were grown in 50 mL of sterile trypticase soy broth (BD Bioxon, Estado de México, México) at 37°C for 18 h for each test and then inoculated in 10 mL tartrazine solution to reach a population of around 107 CFU/mL.

2.5 Statistical analysis

The experimental data were evaluated using Minitab 18 (Minitab Inc., State College, PA, ABD) for the experimental design and the analysis of variance (ANOVA).

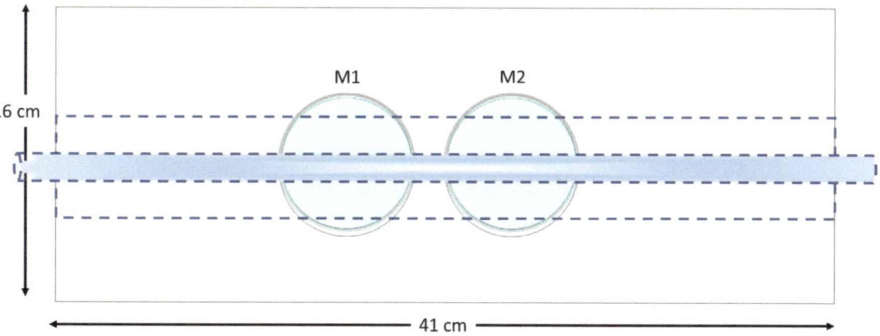

Figure 1.
Distribution of Petri dishes with tartrazine in PL equipment.

Tukey's pairwise comparison test was used to compare the means of data at a 95% confidence interval ($p \leq 0.05$).

3. Results

3.1 Color change

Color plays a crucial role among the essential quality attributes of food products, wielding substantial influence on consumer acceptance. Although the determination of color may not inherently provide insights into the nutritional or functional aspects of a product, it serves as an indicator of the overall dose distribution within an array of samples exposed to pulsed light, especially when considered alongside temperature gradients.

Table 1 report color change after 10s of treatment with irradiance of 0.846 J/cm^2/pulse at 5.7 cm and 0.45 at 10.8 cm. The findings indicate that, despite the proximity of the M1 and M2 samples, the most significant color change is observed in the M2 at the tested distance of 5.7 cm, registering a color change of 12.8 ± 0.23 compared to M1 at 5.7 cm, which shows a change of 11.21 ± 0.42. However, with an increase in the distance between the irradiation source and the samples at 10.8 cm, both tartrazine samples exhibit similar color change values. This suggests that the distribution of pulsed light (PL) becomes more uniform at greater distances, indicating a more consistent effect on the samples.

Despite findings by authors such as Garner et al. [20], who identified no significant differences in color change for apple and cranberry juices treated by pulsed light (PL), several researchers have reported notable color changes in other juice varieties. These include apple, melon, orange, and strawberry juices [14, 21], as well as milk [14], pineapple, amla juice, and coconut water [22] and verjuice [10]. The change in have been attributed to photodegradation of color pigments, light-induced Maillard reaction, and enzymatic browning in the sample after treatment. This phenomenon tends to occur more rapidly in foods with dark and opaque shades, which absorb more light energy and therefore reduce L^* value of Hunter scale [22], compared to lighter products, as noted by Fine and Gervais [23].

Regarding the alteration in color attributed to the varying distance between the irradiation source and the samples, studies by authors like Bianlka and Demirci [24], Chakraborty et al. [25], Gómez et al. [26], and Salazar-Zuñiga et al. [27] indicate that samples positioned closer to the lamp receive higher fluence compared to those situated farther away. Consequently, a more substantial color change is observed in the samples closer to the lamp.

Sample	ΔE		Absorbance (300 nm)		Temperature (°C)	
	5.7 cm	10.8 cm	5.7 cm	10.8 cm	5.7 cm	10.8 cm
Control	—	—	0.425 ± 0.003 a	0.425 ± 0.003 a	22.8 ± 1.3 a	23.4 ± 0.9 a
M1	11.21 ± 0.42 a	11.60 ± 0.34 a	0.1675 ± 0.017 b	0.026 ± 0.002 b	47.4 ± 0.8 b	38.8 ± 1.1 b
M2	12.8 ± 0.23 b	11.59 ± 0.14 a	0.1445 ± 0.025 c	0.024 ± 0.003 b	54.7 ± 1.3 c	39.9 ± 1.3 b

Table 1.
Color change, absorbance, and temperature in the PL-treated tartrazine solution for 10 s at 5.7 and 10.8 cm distance.

Nevertheless, even though the primary mechanism of microbial inactivation is the photochemical effect, primarily associated with the wavelength between 200 and 300 nm, there exists a second mechanism of inactivation by pulsed light (PL) known as the photothermal effect. This effect involves infrared radiation (800-1100 nm) and a portion of the visible range (400-700 nm), leading to an increase in the surface temperature of the samples. This temperature elevation could potentially induce non-enzymatic browning in samples exposed closer to the surface of the lamp, resulting in a change in color. Therefore, it is crucial to monitor the temperature both before and after PL treatment.

3.2 Absorbance

Absorbance pertains to the amount of light that a solution absorbs at a specific wavelength, and this absorbance is directly associated with the concentration of substances within the liquid [28]. Consequently, the examination of absorbance in pulsed light treatments for food serves as a crucial screening tool to comprehend and regulate the interactions between food and pulsed light, as well as to assess its impact on physical, chemical, and microbiological properties. Analyzing the absorbance spectrum at a designated wavelength allows for the evaluation of structural changes in compounds within the food [29, 30].

In **Table 1**, exhibits a pattern of results similar to those observed for color change (Section 3.1). Samples positioned at a closer distance to the irradiation source display a more pronounced change, even when placed at the same vertical distance. Where sample M1 exhibits a value of 0.1675 ± 0.017, while sample M2 demonstrates an absorbance of 0.1445 ± 0.025 and the distance between each other is 0.5 cm, indicating that M2 absorbs more irradiance compared to sample M1. In contrast, with samples placed at greater distances for both distributions, the absorbances of the samples is replicable (0.026 ± 0.002) between M1 and M2. Noticing that at a greater distance, the fluence from the xenon lamp becomes increasingly uniform concerning the samples.

The effect of distance on absorbance has not been reported until now; however, Kaya et al. [10] noted variations in the absorbance of verjuice and apple and grape juice pasteurized by pulsed light (PL). They emphasized that the change in absorbance results from the decomposition of colored compounds and the formation of dark-colored pigments, specifically melanin and melanoidins, due to a browning reaction [31, 32]. The absorption properties of melanin pigment and its impact on UV-C light irradiation (200-300 nm) have been documented in the literature [33]. The observed changes in absorbance reported by these authors align with alterations in the color of the samples.

3.3 Temperature analysis

The temperature analysis of the samples (**Table 1**) indicates the presence of a photothermal effect induced by pulsed light (PL), as evidenced by a significant temperature increase between the control samples and those treated with PL for 10 seconds. While the effect is more pronounced at a shorter distance between the samples (5.7 cm compared to 10.8 cm), it is noteworthy that, even at the same horizontal distance of 5.7 cm, sample M2 exhibits higher temperature values (54.7°C ± 1.3) than M1 (47.4°C ± 0.8). This observation aligns with the greater changes in absorbance and color observed in the analyzed samples. Similarly, the temperature rise at 10.8 cm coincides with changes in color and absorbance, with no significant difference observed between samples M1 (38.8°C ± 1.1) and M2 (39.9°C ± 1.3).

Previous investigations have emphasized the impact of distance on the temperature increase in samples treated with pulsed light (PL) equipment. The same pattern is evident as observed in this study, where shorter distances between the samples and the irradiation source result in higher temperature increases [21, 22, 34, 35]. Although it has been noted that this temperature rise is instantaneous, and samples return to near their initial temperature after processing; however, this thermal change does not counteract the chemical alterations in color and absorbance compounds, as well as inactivation. Nevertheless, the variations in dose distribution emitted by PL equipment, considering sample distribution or temperature gradient within the equipment, remain unexplained thus far.

3.4 Inactivation of *E. coli* ATTC 25922 in tartrazine solution

To validate the impact of distance and sample distribution within the pulsed light (PL) equipment, the inactivation of *E. coli* ATTC 25922 was assessed at various treatment durations—3, 5, 7, and 10 seconds—at both 5.7 and 10.8 cm distances from the irradiation source to the samples (**Figure 2**). The analysis revealed that there is an influence on inactivation when the samples are positioned closer to the xenon lamp. However, the fluence delivered to the samples at a greater distance is more uniform, independent of the distribution of the sample. When evaluating the M1 and M2 samples, a significant difference is observed between the two distributions at 5.7 cm compared to 10.8 cm. Similarly, it was observed that at 5.7 cm, the disparity in inactivation between sample M1 and M2 expands with longer treatment times, in contrast to the 10.8 cm distance. This further emphasizes the findings previously examined regarding the impact of distribution and how this influences the observed outcomes.

The obtained results validate the analyses conducted by Gomez-Lopez et al. [36] and Kaya et al. [10] regarding the impact of sample positioning relative to the xenon lamp in the horizontal plane. Additionally, Woodling and Morau [37] demonstrated a reduction in the effect on the inactivation of pathogenic microorganisms with an increasing distance between the lamp and the samples. Similarly, Pollock et al. [13] identified varying percentages of microbial inactivation in pulsed light (PL) equipment at different distances (5 to 15 cm), with a peak lethality at 10 cm reaching approximately 60% for Listeria monocytogenes. While they analyzed the distribution of 12 Petri dishes in the PL equipment, they did not determine whether

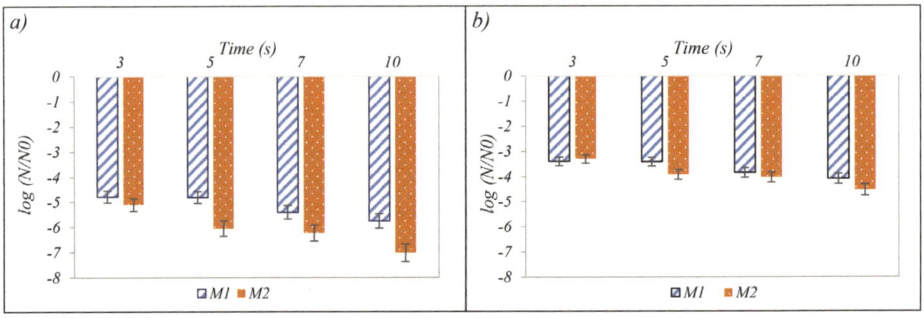

Figure 2.
Inactivation of E. coli *ATTC 25922 at 5.7 cm (a) and 10.8 cm (b).*

this distribution had a significant effect. Despite these analyses, the influence of the fluence emitted by the xenon lamp on the horizontal positioning of the samples at different distances has not been evaluated thus far.

4. Conclusions

The findings in this study not only reveal a significant difference between the two evaluated distances but also demonstrate that at a distance of 10.8 cm, the distribution of pulsed light (PL) in the equipment is more uniform, with respect to color change, absorbance, temperature, and inactivation, compared to a shorter distance. Despite the higher irradiance at 5.7 cm, this phenomenon persists across various activation times. Likewise, it was noted that a photothermal effect is present in the treated samples. This effect not only impacts the inactivation of *E. coli* but is also evident in the changes in color and absorbance, particularly in the M2 samples at 5.7 cm. In conclusion, the importance of analyzing the light dose distribution in pulsed light equipment lies in its potential to ensure result reproducibility, optimize processes, meet regulatory requirements for future industrial scale-up, and identify cost-effective strategies for using energy more efficiently while minimizing resource wastage.

Author details

Christelle Pihen*, Aurelio López Malo and Nelly Ramírez Corona
Departamento de Ingeniería Química, Alimentos y Ambiental, Universidad de las Américas Puebla, ExHda Santa Catarina Mártir s/n, San Andrés Cholula, Puebla, México

*Address all correspondence to: christelle.pihenmz@udlap.mx

References

[1] Bhavya ML, Umesh Hebbar H. Pulsed light processing of foods for microbial safety. Food Quality and Safety. 2017;**1**(3):187-202. DOI: 10.1093/fqsafe/fyx017

[2] Mahendran R, Ramanan KR, Barba FJ, Lorenzo JM, López-Fernández O, Munekata PES, et al. Recent advances in the application of pulsed light processing for improving food safety and increasing shelf life. Trends in Food Science & Technology. 2019;**88**:67-79. DOI: 10.1016/j.tifs.2019.03.010

[3] Profiroiu CM, Bodislav DA, Burlacu S, Rădulescu CV. Challenges of sustainable urban development in the context of population growth. European Journal of Sustainable Development. 2020;**9**(3):51-51. DOI: 10.14207/ejsd.2020.v9n3p51

[4] Bhagat B, Chakraborty S. Potential of pulsed light treatment to pasteurize pomegranate juice: Microbial safety, enzyme inactivation, and phytochemical retention. LWT. 2022;**159**:113215. DOI: 10.1016/j.lwt.2022.113215

[5] Vollmer K, Chakraborty S, Bhalerao PP, Carle R, Frank J, Steingass CB. Effect of pulsed light treatment on natural microbiota, enzyme activity, and phytochemical composition of pineapple (*Ananas Comosus* [L.] Merr.) juice. Food and Bioprocess Technology. 2020;**13**(7):1095-1109. DOI: 10.1007/s11947-020-02460-7

[6] Mandal R, Mohammadi X, Wiktor A, Singh A, Pratap Singh A. Applications of pulsed light decontamination technology in food processing: An overview. Applied Sciences. 2020;**10**(10):3606. DOI: 10.3390/app10103606

[7] Preetha P, Pandiselvam R, Varadharaju N, Kennedy ZJ, Balakrishnan M, Kothakota A. Effect of pulsed light treatment on inactivation kinetics of *Escherichia coli* (MTCC 433) in fruit juices. Food Control. 2021;**121**:107547. DOI: 10.1016/j.foodcont.2020.107547

[8] Santamera A, Escott C, Loira I, Del Fresno JM, González C, Morata A. Pulsed light: Challenges of a non-thermal sanitation technology in the winemaking industry. Beverages. 2020;**6**(3):45. DOI: 10.3390/beverages6030045

[9] Rowan NJ, Valdramidis VP, Gómez-López VM. A review of quantitative methods to describe efficacy of pulsed light generated inactivation data that embraces the occurrence of viable but non culturable state microorganisms. Trends in Food Science & Technology. 2015;**44**(1):79-92. DOI: 10.1016/j.tifs.2015.03.006

[10] Kaya Z, Unluturk S, Martin-Belloso O, Soliva-Fortuny R. Effectiveness of pulsed light treatments assisted by mild heat on saccharomyces cerevisiae inactivation in verjuice and evaluation of its quality during storage. Innovative Food Science & Emerging Technologies. 2020;**66**:102517. DOI: 10.1016/j.ifset.2020.102517

[11] Krishnamurthy K, Tewari JC, Irudayaraj J, Demirci A. Microscopic and spectroscopic evaluation of inactivation of staphylococcus aureus by pulsed UV light and infrared heating. Food and Bioprocess Technology. 2010;**3**(1):93-104. DOI: 10.1007/s11947-008-0084-8

[12] John D, Ramaswamy HS. Comparison of pulsed light inactivation kinetics and Modeling of *Escherichia coli* (ATCC-29055), clostridium Sporogenes (ATCC-7955) and

Geobacillus Stearothermophilus (ATCC-10149). Current Research in Food Science. 2020;**3**:82-91. DOI: 10.1016/j.crfs.2020.03.005

[13] Pollock AM, Pratap Singh A, Ramaswamy HS, Ngadi MO. Pulsed light destruction kinetics of L. Monocytogenes. LWT. 2017;**84**:114-121. DOI: 10.1016/j.lwt.2017.05.040

[14] Palgan I, Caminiti IM, Muñoz A, Noci F, Whyte P, Morgan DJ, et al. Effectiveness of high intensity light pulses (HILP) treatments for the control of *Escherichia coli* and listeria innocua in apple juice, orange juice and milk. Food Microbiology. 2011;**28**(1):14-20. DOI: 10.1016/j.fm.2010.07.023

[15] Wiktor A, Mandal R, Singh A, Pratap Singh A. Pulsed light treatment below a critical fluence (3.82 J/cm^2) minimizes photo-degradation and Browning of a model phenolic (Gallic acid) solution. Food. 2019;**8**(9):380. DOI: 10.3390/foods8090380

[16] Mandal R, Pratap-Singh A. Characterization of continuous-flow pulsed UV light reactors for processing of liquid foods in annular tube and coiled tube configurations using Actinometry and computational fluid dynamics. Journal of Food Engineering. 2021;**304**:110590. DOI: 10.1016/j.jfoodeng.2021.110590

[17] Romero-Martínez L, Moreno-Andrés J, Acevedo-Merino A, Nebot E. Development of a geometrical model for the determination of the average intensity in a flow-through UV-LED reactor and validation with biodosimetry and actinometry. Journal of Water Process Engineering. 2022;**49**:103137. DOI: 10.1016/j.jwpe.2022.103137

[18] Radjagobalou R, Freitas VDDS, Blanco J-F, Gros F, Dauchet J, Cornet J-F, et al. A revised 1D equivalent model for the determination of incident photon flux density in a continuous-flow LED-driven spiral-shaped microreactor using the Actinometry method with Reinecke's salt. Journal of Flow Chemistry. 2021;**11**(3):357-367. DOI: 10.1007/s41981-021-00179-w

[19] Sperle P, Khan MS, Drewes JE, Wurzbacher C. A practical bacterial biodosimetry procedure to assess performance of lab-scale flow-through ultraviolet water disinfection reactors. ACS EST Water. 2023;**3**(8):2130-2139. DOI: 10.1021/acsestwater.2c00648

[20] Garner AL. Pulsed electric field inactivation of microorganisms: From fundamental biophysics to synergistic treatments. Applied Microbiology and Biotechnology. 2019;**103**(19):7917-7929. DOI: 10.1007/s00253-019-10067-y

[21] Ferrario M, Alzamora SM, Guerrero S. Inactivation kinetics of some microorganisms in apple, melon, orange and strawberry juices by high intensity light pulses. Journal of Food Engineering. 2013;**118**(3):302-311. DOI: 10.1016/j.jfoodeng.2013.04.007

[22] Dhar R, Chakraborty S. Influence of voltage and distance on quality attributes of mixed fruit beverage during pulsed light treatment and kinetic Modeling. Journal of Food Process Engineering. 2020;**43**(11):e13517. DOI: 10.1111/jfpe.13517

[23] Fine F, Gervais P. Efficiency of pulsed UV light for microbial decontamination of food powders. Journal of Food Protection. 2004;**67**(4):787-792. DOI: 10.4315/0362-028X-67.4.787

[24] Bialka KL, Demirci A. Efficacy of pulsed UV-light for the decontamination of *Escherichia coli* O157:H7 and

salmonella Spp. on raspberries and strawberries. Journal of Food Science. 2008;**73**(5):M201-M207. DOI: 10.1111/j.1750-3841.2008.00743.x

[25] Chakraborty S, Ghag S, Bhalerao PP, Gokhale JS. The potential of pulsed light treatment to produce enzymatically stable Indian gooseberry (*Emblica Officinalis* Gaertn.) juice with maximal retention in total phenolics and vitamin C. Journal of Food Processing and Preservation. 2020;**44**(12):e14932. DOI: 10.1111/jfpp.14932

[26] Gómez-Sánchez DL, Antonio-Gutiérrez O, López-Díaz AS, Palou E, López-Malo A, Ramírez-Corona N. Performance of combined technologies for the inactivation of Saccharomyces cerevisiae and *Escherichia coli* in pomegranate juice: The effects of a continuous-flow UV-microwave system. Journal of Food Process Engineering. 2020;**43**(12):e13565. DOI: 10.1111/jfpe.13565

[27] Salazar-Zúñiga MN, Lugo-Cervantes E, Rodríguez-Campos J, Sanchez-Vega R, Rodríguez-Roque MJ, Valdivia-Nájar CG. Pulsed light processing in the preservation of juices and fresh-cut fruits: A review. Food and Bioprocess Technology. 2023;**16**(3):510-525. DOI: 10.1007/s11947-022-02891-4

[28] Koutchma T. Advances in ultraviolet light technology for non-thermal processing of liquid foods. Food and Bioprocess Technology. 2009;**2**(2):138-155. DOI: 10.1007/s11947-008-0178-3

[29] Pellicer JA, Gómez-López VM. Pulsed light inactivation of horseradish peroxidase and associated structural changes. Food Chemistry. 2017;**237**:632-637. DOI: 10.1016/j.foodchem.2017.05.151

[30] Siddique MAB, Maresca P, Pataro G, Ferrari G. Influence of pulsed light treatment on the aggregation of whey protein isolate. Food Research International. 2017;**99**:419-425. DOI: 10.1016/j.foodres.2017.06.003

[31] Müller A, Noack L, Greiner R, Stahl MR, Posten C. Effect of UV-C and UV-B treatment on polyphenol oxidase activity and shelf life of apple and grape juices. Innovative Food Science & Emerging Technologies. 2014;**26**:498-504. DOI: 10.1016/j.ifset.2014.05.014

[32] Unluturk S, Atilgan MR. Microbial safety and shelf life of UV-C treated freshly squeezed white grape juice. Journal of Food Science. 2015;**80**(8):M1831-M1841. DOI: 10.1111/1750-3841.12952

[33] Seiji M, Iwashita S. Enzyme inactivation by ultraviolet light and protective effect of melanin. The Journal of Biochemistry. 1965;**57**(3):457-459

[34] Gómez PL, Salvatori DM, García-Loredo A, Alzamora SM. Pulsed light treatment of cut apple: Dose effect on color, structure, and microbiological stability. Food and Bioprocess Technology. 2012;**5**(6):2311-2322. DOI: 10.1007/s11947-011-0598-3

[35] Pataro G, Muñoz A, Palgan I, Noci F, Ferrari G, Lyng JG. Bacterial inactivation in fruit juices using a continuous flow pulsed light (PL) system. Food Research International. 2011;**44**(6):1642-1648. DOI: 10.1016/j.foodres.2011.04.048

[36] Gómez-López VM, Ragaert P, Debevere J, Devlieghere F. Pulsed light for food decontamination: A review. Trends in Food Science & Technology. 2007;**18**(9):464-473. DOI: 10.1016/j.tifs.2007.03.010

[37] Woodling SE, Moraru CI. Effect of spectral range in surface inactivation of listeria Innocua using broad-Spectrum pulsed light. Journal of Food Protection. 2007;**70**(4):909-916. DOI: 10.4315/0362-028X-70.4.909

Chapter

The Wall-Coral: An Autoethnography of a Performative Figure

Cristina Roxana Xaman Mc. Gregor

Abstract

The present chapter is an approximation to the concept of performativity to analyze how it can help us to intervene in reality. Theoretically, we mainly dialog with Judith Butler, Sara Ahmed, Jack Halberstam, and Ana Cristina Aguirre Calleja. The conceptual approaches are in dialog with a case study: the Instagram collaborative project wall-coral, a linguistic play that implies renaming the humidity of the walls and asking for conservation photography to keep this earthly-oceanic being alive. The result was a performative figure that shows possible changes and resistances on how we relate with other beings on a daily basis.

Keywords: performativity, queer theory, autoethnography, coral reef, figure, sediment

1. Introduction

An art project online cannot end the destruction, crack a wall at home, or bring back life, can it? This chapter answers this question using the figure of a wall-coral, a creative approximation of the humidity of the walls. The response has been formulated from a theoretical and a practical approach.

From a theoretical perspective, this research is a study of the concept of performativity. While it is true that performance theory can be a conceptual tool to understand why things happen in specific ways repeatedly, we are only interested in how things remain assembled to be able to intervene in the cycle of reiteration. In other words, our question is how can we, in J. L. Austin's words [1], "inaugurate" a performative? Or in Judith Butler's words: How can we use our agency to inhabit the linguistic *gap* between a saying and its response? [2] What is the performative move we provoke when instead of saying "humidity," we say "wall-coral"?

From a practical perspective, on March 23, 2022, we renamed the wall humidity, an unwanted reality subject to direct annihilation, no questions asked, and instead, we named them wall-coral. The wall-coral was intended as a performative figure, following Aguirre Calleja: one with the capacity to blow concepts up and that allowed us to follow the trajectory of its effects [3]. This performative figure was shared on social media via Instagram @coral_de_pared [4] and replicated on our personal Facebook account, with an open invitation to collaborate with photographs and written narratives as a form of conservation. By the moment this chapter was written, there were 58 wall corals conservated on photographs, and 42 had an oceanic narrative attached to them.

But what is the wall coral? Theoretically, it is a performative figure that allowed us to experiment with what we call an agency of escapism. In practice, it is a collaborative production of a linguistic reality, which is a view on the go of how words gain force. The result has been 58 moments of stop to pay attention to what is normally ignored or instantly annihilated by different people, an unfixed relationship to a well-known reality, on continuous change. Finally, the wall-coral is a collective representation of any other kind of beings we eliminate just because we feel threatened, in danger or we simply do not like their looks. Hopefully, one-stop allows us to stop again and again in a wide diversity of situations in which we relate to other beings with disdain, violence, or indifference.

2. Placing the discussion: The moment of response

"Unlivable lives", says Judith Butler, are those "whose legal and political status is suspended." Sign as the face of evil, and their deaths are not considered as a loss, so there is no mourning around them ([5], pp. XV–XVIII). In this category, fix terrorists, migrants, poor people, and populations in invulnerable positions within a society (e.g., women). I include microorganisms, bacteria, and all animal species we call "food" (what we include here changes from country to country). What Butler wants to acknowledge when calling these lives "unlivable" is that while still living, these lives do not count as any other life. There is the case of how news breaks in tears with a dead soldier from the USA, England, France, Germany... etc. While celebrating bombing and killing the opposition forces that are dismissed as "terrorists", "guerrillas", and so. Sara Ahmed refers to the same population as "bearable and unbearable lives" ([6], p. 205). We prefer Ahmed's terms as they allow us to keep in mind that actually, these lives exist and that while some lives could be livable, they are not necessarily bearable. Communities of human and non-human beings manage to exist in extremely vulnerable situations. To keep them living (breathing, working, and producing) is in the world's best interest, to squeeze labor, bodies, and minds from them. It seems this time in history can be defined as a use-waste system, where the conditions of many are right now unbearable, filled with suffering, scarcity, and exploitation.

Butler affirms – recurring to Althusser's concept of interpellation – that being called a name is what gives our body a social existence; it allows us to be recognizable ([2], p. 5). On this world economy of names, many are pushed to a road of perish. But, says Butler:

...the injurious address may appear to fix or paralyze the one it hails, but it may also produce an unexpected and enabling response. If to be addressed is to be interpellated, then the offensive call runs the risk of inaugurating a subject in speech who comes to use language to counter the offensive call ([2], p. 2).

When being called a terrorist, poor, stupid, queer...? Do we fix what that name intent to address? How do we respond? "The gap between redundancy and repetition is the space of agency." ([2], p. 129), affirms Butler. It is not mandatory to respond accurately to others' calls. By attaching or not to the intention of a call, especially if the call was designed to be hurtful, we can make a difference in our chance to survive.

We find a close relationship between Butler's gap and Ahmed's proposal of the *hap* of happiness. Sara Ahmed [6], a wise critic of the idea of happiness, argues that mandatory happiness implies that whatever does not qualify as a happy life or a happy object must be afraid of and avoided at all costs, while happy lifestyles and objects must be pursued at any cost. Ahmed explores three paradigmatic unhappy figures: the feminist killjoy, the unhappy queers, and the melancholic migrants showing how, even if the call renders these lives unhappy, we can find or create happiness there. Ahmed proposes a way to get out of the trap of a fixed idea of happiness by pointing out that the *hap* of happiness has the same root as *happens* and *perhaps*, implying that happiness can be what happens, instead of an ideal to pursue, and an open possibility to what could happen, hope. That way, a life signed as unhappy by happening, by being what *happens*, could expand the meaning of happiness in the ambit of perhaps [6]. We interpret the *hap* of Ahmed as an example of agency, an appropriation of the linguistic *gap* proposed by Butler. The call could be identified as a cause for unhappiness, but the response to that call by living a queer, migrant, feminist life and exploring what happens could result in happiness.

Following Butler and Ahmed's path, Halberstam presents "illegibility" as "one way of scaping the political manipulation to which all university fields and disciplines are subject" ([7], p. 10). He continues "we may, ultimately, want more undiscipline knowledge more questions and fewer answers" ([7], p. 10). According to the performative theory of Judith Butler, total illegibility is an impossible task. We must remember that legibility is the first stance for social existence ([2], p. 2). Although we can find different ways to respond to that first call, we cannot elude it. We interpret Halberstam's illegibility as placing ourselves on ambiguity, in-between fixed meanings, similar to creating a crack or a hole in the meaning that defines our life, the way the architecture of Matta-Clark works on buildings.

According to Halberstam, Matta-Clark intervened in buildings to place them between equilibrium and collapse [8]. By scaping the categorization of a name, by fixing not in one but in multiple names at the same time, we make ourselves illegible. A trick that reminds us of the analysis of Foucault in Magritte's work *This is Not a Pipe*: "Instead of blending identities, it happens that similitude also has the power to destroy them" ([9], p. 50). Illegibility can also be related to the parody proposed by Butler, "In imitating gender, drag implicitly reveals the imitative structure of gender itself—as well as its contingency [...] the parody is *of* the very notion of an original [...] As imitations which effectively displace the meaning of the original, they imitate the myth of originality itself" ([10], pp. 137–138). As in the work of Magritte, there is not one real pipe that the painter can capture on canvas (there are infinite real pipes), and according to Butler, there is no original gendered body to look back and compare, on Halberstam proposal there are countless real persons, and therefore a dissolution of ideal identities, or a useless call of names that fails on identifying an ideal subject that does not exist.

Ahmed asks us to stop following the path of an ideal happiness that is nowhere, Halberstam asks us to escape from the legibility of a name that cannot define us, and Foucault tells us to realize idealistic world is not anywhere. Finally, Butler, as Halberstam, Ahmed, and Foucault, show us it is possible to break meanings because they have never been completely fixed.

What Butler, Ahmed, and Halberstam's proposals show us is that an interesting place to set the discussion of performativity is at the moment of response.

3. The case of study: Creating the figure of a wall-coral

Here is what happened: In November 2021, the humidity of my wall got the perfect coralline structure similar to the ones deep in the ocean. Bleached as the corals in warm deep salty waters, that humidity was completely white. It was getting bigger and bigger. I felt as if the ocean was invading my house, or perhaps my house was on an immersive trip to the ocean. But, first things first: Is this dangerous? Do I need to rid of it, no matter how interesting, or curious it is? Before destroying that beautiful humidity (I did paint the wall), I took a photo of it (See **Figure 1**) and opened an Instagram account with the name of wall-corals [4]. Renaming humidity and calling it wall-corals was a Butlerian move, I occupy the *gap* between the name and the materiality of humidity. I gave a different response to a common reality. Just because. To see what happens, as in the *hap* of Ahmed, for the sake of making it happen, in the hope that someone else thought it was worth preserving them, at least in pictures.

To inhabit the Butlerian *gap* I created a figure and provided it with a name: wall-coral; the figure got a social existence. It was delivered to the world on social media via Instagram and spread on Facebook too. From then, as Frankenstein, it did not take long to detach from its creator. Why? Because it plays the Butlerian parody breaking the idea of an original; because it scapes legibility by placing itself in-between meanings: the wall-coral is not a wall, is not humidity, is not a coral, but is somehow all of that; because when existence sustain on what *happens* and *hope,* there is no home to come back or origin to attach but a constant building always in transition. This image was no longer mine but ours. It grew, it multiplied, it responded to unsaid questions, it followed its own path. Sometimes I provided the narrative that got attached to the image; others I just let others provide the narrative. Some sent pictures with question marks, asking: Is this a wall-coral? Then, I stopped and needed to ask myself the same question: Is it?

Figure 1.
Coral de pared [@coral_de_pared]. "Departamento. @roxanaxaman San Andrés Cholula, Puebla, México". Instagram, *photographed by Roxana Xaman, 23 March 2022, www.instagram.com/p/Cbb5iUYglqk/?igsh=YjJwcGYweWZiZXRk. Description: Apartment. @roxanaxaman San Andres Cholula, Puebla, Mexico. This is the start of the project: I asked my landlord to bring me the paint to fix the humidity of the walls, but the days passed by and I am still waiting for it. I rejected that piece of wall every passing day until I saw it: It was a coral reef emerging from my bedroom walls. The sea was making a wink; then, my feeling changed from rejection to excitement. I took this photograph to preserve my wall-coral reef.*

3.1 A performative figure

Even if we were able to come up with a creative figure, a new name for the humidity, how can our response have the force not to be dismissed? In other words, how can a response have a performative effect to make things happen?

J. L. Austin proposes that the performative occurs when "the issuing of the utterance is the performing of an action" ([1], p. 6) that requires a known authority enunciating conventional utterances on conventional circumstances ([1], p. 14–15). The authority creates what was intended to. If we play on Austin terms, we would need to first become an authority, and then exercise the power that comes with that authority. But how come some unexpected responses create immediate effects? Butler will note that when a performative occurs – that constitutive call we talked about earlier – there is always an exceeded meaning.

...the performative, the call by the law which seeks to produce a lawful subject, produces a set of consequences that exceed and confound what appears to be the disciplining intention motivating the law. Interpellation thus [...] creates more than it ever meant to, signifying in excess of any intended referent. It is this constitutive failure of the performative, this slippage between discursive command and its appropriated effect, which provides the linguistic occasion and index for a consequential disobedience ([11], p. 122).

Paradoxically, the same call that intended to subjugate a subject is the one that will allow the subject to liberate itself. This will be a different strategy than Austin's formula for the instauration of as performative by betting away with it because of an absence of rules ([1], p. 30). In Butler's proposal, the force comes from what, following the Derridian tradition, she calls *historicity*:

...the history which has become internal to a name, has come to constitute the contemporary meaning of a name the sedimentation of its usages as they have become part of the very name, a sedimentation, a repetition that congeals, that gives the name its force ([2], p. 36).

A force that does not preexist on the subject, that is not a product of sovereign power ([2], p. 69) can work in favor instead of against us if we manage to take advantage of a force that has been wrought out of reiteration.

Has the wall-coral figure earned the force performative? Aguirre Calleja says that a performative figure blows up concepts; creates its own rules and regulations in ambiguity; it cannot be anticipated, although it is located; it is being, it is figuring; it allows us to follow its practices, processes, specific locations and historicity ([3], pp. 18–33). We follow Aguirre Calleja's proposal to explore the elements that make the wall-coral a performative figure.

Blowing up concepts. The wall-coral is not *humidity,* it is all the dynamics, practices, material transformation, eruptions, and links that occur when humidity comes in contact or interaction with the wall, and therefore with the humans living in the house and other living and non-living beings. And it is true also that the wall-coral is not a coral, in-between concepts, the wall-coral plays on Halberstam illegibility. Another consequence of using the figure of the wall-coral, we acknowledged that when we said *humidity,* we were not referring to humidity (H_2O) but that we were talking about the consequences of the humidity in the wall. When saying *humidity* we meant the humus, the black stain on the wall, the peeling paint, etc. So, by saying *wall-coral,* we are referring to the sediment of humidity in connection with the wall and its materials (sand, construction lime, cement, and paint). The previous and the actual name have been blown up by the specific practice of renaming the wall humidity.

Creating its own rules and regulations on ambiguity. The wall-coral indeed started by being white, but then it kept multiplying its colors with every conservated specimen. There is one special wall coral that we love: one of three colors, from a kitchen in Ayala, Morelos (see **Figure 2**). Later, we conservated a red, gray, and white wall coral of Tequisquiapan, Queretaro, and wrote about it: "I found this wall-coral while I was walking on the street. I keep thinking about the layers of paint and its colors. We need layers and layers of paint to cover a red wall. How much does the temperature need to go down, how many degrees to bleach a coral? How many hours of how many warm days can the coral color resist?" (see **Figure 3**). Now, there are some wall corals in blue, white, purple, red, gray, pink, orange, and many more.

Another clue moment about how this figure created its own rules and regulations was when we received a picture of a wall-coral that had grown some leaves, asking: Is this a wall-coral? The question was directed to us as we were the ones who created the image, but we needed to ask the same question too before answering. From the perspective of the interaction, from the perspective of the wall-coral as an interaction, our response was: Of course, yes! And that response opened a new wave of possibilities for the wall corals, and that allowed us to encounter the amazing wall coral that grew a tree from Lerma, Campeche, our grandfather's hometown. We wrote about it: "If the wall-coral was only toxicity, this mixed survival could not fix. The three grew roots, trunk, and branches where I speculate, there lay more than just humidity" (see **Figure 4**).

Regulations were and have been ambiguous, not just in the senses that Victor Turner "The very word 'ambiguity' is derived from the Latin *agere* to 'act' for it comes from the verb *ambigere,* to 'wander', *ambi-,* 'about, around' + *agree,* 'to do,' resulting inthe sense of having two or more possible meanings, 'moving from side to side,' 'of

Figure 2.
Coral de pared [@coral_de_pared]. "Coral de pared de tres colores en la cocina de una casa en Ayala, Morelos". Instagram, *photographed by Elena Andrade, 6 April 2022, www.instagram.com/p/CcBCIdzuolL7/?igsh=MXVvZHYoMm5zbmo1Ng==*

DOI: http://dx.doi.org/10.5772/intechopen.115472

Figure 3.
Coral de pared [@coral_de_pared]. "Coral de pared en un muro de calle en Tequisquiapan, Queretaro". Instagram, *photographed by Roxana Xaman, 25 April 2022, www.instagram.com/p/CcyRrw6uH8n/?igsh=MWw5d3U1amQ3bXNvYw==. Description: Wall-coral on a street wall in Tequisquiapan, Queretaro. I found this wall-coral while I was walking on the street. I keep thinking about the layers of paint and its colors. We need layers and layers of paint to cover a red wall. How much does the temperature need to go down, how many degrees to bleach a coral? How many hours of how many warm days can the coral color resist?*

Figure 4.
Coral de pared [@coral_de_pared]. "Coral de pared y otras especies en Lerma, Campeche". Instagram, *photographed by Roxana Xaman, 25 April 2023, www.instagram.com/p/CrePreTuetR/?igsh=Z2hqcGk5bzVjYnB2. Description: Wall-coral and other species in Lerma, Campeche. If the wall-coral was only toxic, this mixed survival could not be fixed. The three grew roots, trunk, and branches where I speculate, there lay more than just humidity.*

doubtful nature" ([12], p. 102). Apart from movement and indecision from the Latin roots provided by Turner, it is possible to make out a third meaning: exploration; as the result of moving and oscillation.

It cannot be anticipated. It is being, it is figuring. The wall-coral has been and will keep changing, as long as we keep looking. Other places, other wall structures, other materials, and other ways of life will entail different or unheard wall-corals with different looks, forms, and ways of interacting. While we can anticipate certain elements or aspects, as we have a comfortable tendency to reproduce, repeat, and constrain, while we keep the project open to anyone's collaboration, it will imply a certain level of surprise.

Another aspect we have to take into account is that the conservation of the wall-coral is just the conservation of a moment; even the same wall-coral will be different on a week, on a month. We can follow the track of one specimen, and its ways will always lead us somewhere else. The wall coral is always being, always figuring. There is no ideal figure of the wall-coral. The oceanic coral allowed us to outline a phenomenon on the surface, now, we are letting real events define what that phenomenon is.

The photographic conservation on Instagram has *allowed us to follow its practices, processes, specific locations, and historicity*. The Instagram project has wall-corals from Mexico, Spain, Portugal, Argelia, and Palestina; in bathrooms, bedrooms, living rooms, bars, and patios. Some narratives have linked wall-corals to sadness, Gallardo says about her wall-coral: "...my therapist told me that when the wall-coral emerges in a house (due to humidity) is because there are sad energies inhabiting that place" (see **Figure 5**). Others have put their interest in the possibility of new species, Andrade says about her wall-coral: "While I was documenting this wall-coral some neighbors approached me and asked me what was going on. I explained to them about the wall-corals and they were surprised, as they told me: They did not know corals could appear on walls. They asked me if this was bad or dangerous, and I responded there was nothing bad about them" (See **Figure 6**). Some have even defined with them the character of a city: "Certainly, this is a very coral city" said Dinorin in her Casablanca

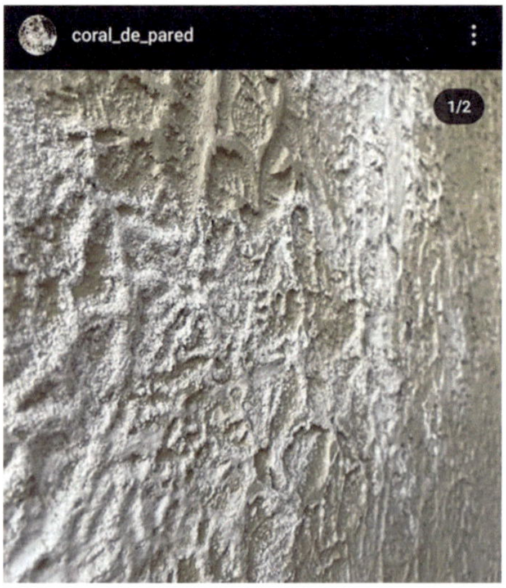

Figure 5.
Coral de pared [@coral_de_pared]. "Coral de pared de @airygallardo en departamento de la ciudad de Puebla, Pue., México". Instagram, *photographed by Airy Gallardo, 28 March 2022, www.instagram.com/p/Cbp47DoudZE/?igsh=c2ZpMGszYnA1Njg2. Description: Wall-coral @airygallardo in an apartment in Puebla, Pue., Mexico. I went to therapy one day and my therapist told me that when the wall-coral emerges in a house (due to humidity) is because there are sad energies inhabiting that place."*

Figure 6.
Coral de pared [@coral_de_pared]. "Coral de pared en el muro exterior de una casa en Tlayecac, Morelos". Instagram, *photographed by Elena Andrade, 22 April 2022, www.instagram.com/p/CcqxujrOmcE/?igsh=NnR6bGF3cnh4ZDJm. Description: Wall-coral on a street wall in a house of Trayecac, Morelos. While I was documenting this wall-coral some neighbors approached me and asked me what was going on. I explained to them about the wall-corals and they were surprised, as they told me: They did not know corals could appear on walls. They asked me if this was bad or dangerous, and I responded there was nothing bad about them. They seem interested in the wall-corals.*

collaboration (see **Figure 7**). Apart from these, many other linkages have been made and are now part of what a wall-coral is.

By collectively conserving the wall-corals, we have asked ourselves: Is this one? How? Why? The photographs preserve our answers, showing a diversity of textures,

Figure 7.
Coral de pared [@coral_de_pared]. "Corales de pared en la ciudad de Casablanca, Marruecos conservados por @Melgosh". Instagram, *photographed by Vianny Dinorin, 29 July 2023, www.instagram.com/p/CvTGZ5MOA7W/?igsh=MWlvbjFtdHBjZ3Rxaw==. Description: Wall-corals on the city of Casablanca, Marruecos conservated by @Melgosh. Certainly, this is a very coral city.*

forms, spaces, and territories of where and how the wall-coral live. By creating narratives around them, we have disclosed, created, and recreated our relationship with them. Since the 23rd of March, 2022 the name of *wall-coral* has been gaining a force.

If historicity means sedimentation of the name usages, as previously said with Butler ([2], p. 36), then we can think of the wall-coral as sediment (for the usages of the sediments of previous meanings of "wall", "coral", "humidity") sedimented (for the usages of the name "wall-coral" and the meanings surrounded those usages) sediment*ing* (as while writing these chapter the name keeps earning meanings and changing over its usages).

4. Conclusions

The word *humidity* hides a complex reality, that as far as the photographs from the wall-coral conservation project have shown: it implies from humus to an entire tree growing in the middle of a wall. What we want to discard when saying *humidity,* as if it were nothing, is a whole ecosystem of microorganisms (bacteria, humus, plants, etc.). By inhabiting the gap present between the common name *humidity* and the complex interaction this word intended to refer to, we uncover those interactions. To do it, we rename that reality and call it *wall-coral*. By not being a coral nor a humidity, the wall-coral became illegible in Halberstam terms.

This illegibility allowed us to pay attention to the elements that interact with human constructions. We wanted to promote a look that could make us acknowledge the dynamics we are about to eliminate just because we believe they are inconvenient to us: it looks awful, it is dangerous, and it weakens the walls of our home. We wanted to focus on an unbearable life/reality, to acknowledge a routinary annihilation.

The wall-coral conservation project on Instagram encountered collaborators who sent pictures of their wall-corals and/or wrote a narrative around their specimens. These collaborations allowed us to keep constructing this performative figure of the wall-coral, in Aguirre Calleja's term: a figure that blows up concepts, that cannot be anticipated, that creates its own rules and regulations on ambiguity, that it is being, figuring, that allow us to follow its practices, processes, specific locations and historicity.

This project also found detractors, questions, and critics. There were some worries expressed about the conservation project. The open call for collaborations says: "Have you seen a coral growing on your wall? Ocean coral and wall-coral live surrounded by humidity. Human lives kill both kinds of coral. Human beings are destructive! How can we conserve them? Link the space where you live with the ecosystem. Took a photograph of your coral...". It was dangerous, one said on Facebook [13], and it can damage our respiratory system, both the oceanic corals and the wall-corals were incomparable, said another on Instagram [14]. While we understand the concerns, we continue under two main arguments:

(1) We were not telling people to keep their walls humidity, but asking them to preserve them by taking a photograph. Before painting the wall, take a photograph, and realize that you are eliminating something that was there. A life! Apparently, it was a dangerous thing to say, it sounded almost insulting. A life? What are you talking about? This was the reaction. Just because our human respiratory system does not support humidity, that does not mean there are no microorganisms there. Humus, as far as we know is alive. Let us put another example: just because some heterosexual feels uncomfortable when imagining themself having a sexual encounter with

DOI: http://dx.doi.org/10.5772/intechopen.115472

someone of their same gender, that does not mean same-sex encounters are unpleasant. (2) We did not open a laboratory to grow wall humidity, the humidity is, has, and will be there. We are just asking people to pay attention to something that is already there. The one who does not have humidity at home, raise your hand. If you raised your hand: Are you sure? Really? Did you check on all of your bathrooms? How about the laundry room? In other words: We did not crack the building; we did not create the imbalance as Matta-Clark did (that would have been interesting). The imbalance was already there, we just showed it. The wall-coral project has the function that Halberstam notes for trans* bodies: "Trans* embodiment [...] is the visual confirmation that all bodies are uncomfortable and wrong-ish, situated as they are within confining grammars\ of sense and security" (8). Wall coral shows that your house our not *ours,* that no matter how hard we try, nature will find a way to claim its belonging to that part of the world.

At home, these bearable lives previously condemned to instant destruction now live freely spread on the walls of the room, bathroom, etc. This time, we have not been able to get rid of them. It seems that, according to linguistic theorists, how we name a reality takes part in how we interpret reality, changing how we treat it, and what kind of linkage we create with it.

It is our concern that the desire behind the act of keeping our walls smooth and clean, apart from health (we do many unhealthy things such as eating chips, smoking, drinking alcohol... to think this destruction comes from our health concern) and insisting on eradicating that wall-coral is fixed with complex ideas on property, housing, security, and more. We are interested in making you think of the wall-corals as a representation of any other kind of beings we eliminate just because. We are not as ambitious to pretend that this kind, or any kind of annihilation, stops, but at least we want to make sure this common and constant annihilation is recognized as such.

Acknowledgements

This chapter takes part in the PhD thesis "Sea Immersions. Autoethnographic Performance: A Performative Queer Practice". This work has been funded by the National Council of Humanities, Science, and Technology (CONAHCYT, in its Spanish Acronym) and the University of the Americas Puebla (UDLAP, in its Spanish Acronym).

Conflict of interest

We have no conflict of interest to disclose.

Author details

Cristina Roxana Xaman Mc. Gregor
University of the Americas Puebla (UDLAP, in its Spanish Acronym),
San Andres Cholula, Puebla, Mexico

*Address all correspondence to: cristina.xamanmr@udlap.mx

References

[1] Austin JL. How to Do Things with Words. Oxford: Oxford University Press; 1962

[2] Butler J. Excitable Speech. A Politics of the Performative. New York: Routledge; 1997

[3] Calleja A, Capítulo AC. Figuraciones: Una apuesta epistémico-política desde el feminismo. In: Calleja A, Cristina A, editors. Figuras performativas de la acción colectiva: Una trayectoria con la Comisión Civil Internacional de Observación por los Derechos Humanos, desde las políticas de conocimiento feminista y la etnografía crítica [thesis]. Spain: Universitat Autónoma de Barcelona (UAB), Departamento de Psicología Social; 2012. pp. 11-37. Available from: http://hdl.handle.net/10803/97343

[4] Coral de pared. Instagram. Available from: https://www.instagram.com/coral_de_pared?igsh=OXFxaXcxdHp3ejJm

[5] Butler precarious. Precarious Life. The Powers of Mourning and Violence. New York: Verso; 2004

[6] Ahmed S. The Promise of Happiness. Durham: Duke University Press; 2010

[7] Halberstam J. The Queer Art of Failure. Durham: Duke University Press; 2011

[8] Halberstam J. Unbuilding gender. Places Journal. 2018. DOI: 10.22269/181003

[9] Foucault M. This Is Not a Pipe. Trad. James Harkness. Berkeley: University of California Press; 1983

[10] Butler J. Gender Trouble: Feminism and the Subversion of Identity (Thinking Gender). New York: Routledge; 1990

[11] Butler J. Bodies that matter. In: On the Discursive Limits of "Sex". Nueva York: Routledge; 1993

[12] Turner V. Acting in everyday life and everyday life in acting. In: Victor T, editor. From Ritual to Theatre. The Human Seriousness of Play. New York: PAJ Publications; 1992. pp. 102-123

[13] Xaman R. Fieldnote. Facebook post; 2022

[14] Xaman R. Fieldnote. Instagram Private Message; 2022

Chapter

Flood Risk Management as a Water Security Strategy

Edith Bonilla-López and Carlos Patiño-Gómez

Abstract

The concept of water security must be approached from a multidimensional and dynamic viewpoint, including water quality and quantity, considering equitable and affordable access, as well as protection from water-related disasters. Because of the challenges that have been identified for its fulfillment, it is necessary to update regulations and transit toward multisectoral instruments and public policies that allow addressing the obstacles from a comprehensive perspective. In order to measure this condition, the implementation of indicators is a recurrent tool, yet their components hardly cover the totality of its dimensions, especially the ones related to hydrometeorological events, such as floods. Flooding is one of the most relevant phenomena due to its social, environmental, and economic implications, jeopardizing water security, and can be assessed at the basin level to guide actions according to its physiographic and climatologic characteristics, without leading to an under- or overestimation of management actions. The aim of this chapter is to describe general aspects of these terms, their measurement, challenges, expected future, Mexico's water security situation, the relation with floods, and to propose a strategy for the achievement and support of water security through flood risk management.

Keywords: water security, strategy, hydrometeorological events, floods, flood management, Mexico

1. Introduction

As the negative impacts on water resources increase globally, the term water security (WS) has gained popularity. Although it has different definitions [1–6], some more comprehensive than others, it generally consists of guaranteeing the quantity and quality of the volume of water to satisfy the needs of humans and the environment, considering the protection from water-related disasters, with a vision of human rights, viable, and equitable access. This concept is complex, dynamic, and multidimensional [6], as its dimensions behave differently at a spatial, political, hydrological, or social scale [7]. It is for this reason that public policies must consider the particular conditions of the basin or site of interest. Likewise, the Intergovernmental Hydrological Program states that WS is "the capacity of a population to safeguard at the basin level access to water in adequate quantities and with the appropriate quality to sustain the health of people and ecosystems, just as to ensure the effective protection of lives and property during water disasters (floods, droughts, landslides, and land subsidence)" [3].

As indicated by Martinez-Austria, anthropogenic actions such as population growth, urbanization, growing demand for energy and food, climate change (CC), and poor water management have affected water security (shortages, water pollution, conflicts over water, floods, droughts, and environmental deterioration) [2]. It can be seen that the factors that generate the problem have been present for several decades, making it possible to foresee their consequences without carrying out any preventive action. Moreover, this provides an opportunity to generate public policies of a mandatory nature, for the different actors responsible for the components of WS.

Regarding the monitoring of water security trends, it has been carried out with the use of indexes, whose design contemplates its different components, but not always in their entirety, and depending on the domain of interest, the individual indicators are selected. This view could bias the outcome of the assessments and provide a misleading picture of the situation. Therefore, it is crucial to analyze all of the multiple related components, such as human development, ecological sustainability, geopolitics, international relations, vulnerability, and risk [6], and find a way to characterize the situation of a certain region or country, to take it as a reference base for decision-making.

Among the most frequent challenges encountered when assessing WS is the lack of water availability, pollution, food production, real or virtual cross-border exchange of water resources, hydrometeorological phenomena, alteration of currents and water bodies, and the effects of CC. Care actions must take into consideration mechanisms to reduce these obstacles. Some of the hydrometeorological phenomena that cause the greatest structural, social, and economic damage are floods, and within time, the economic losses have been increasing substantially and more than 1.5 million people have been affected by floods in the last decades [8, 9]. Specifically in Mexico, according to the geoprocessing developed by the authors of the global high-resolution dataset of earth's floodplains [10], approximately 7% of Mexican territory serves both hydrologically and hydraulically as a floodplain (refer to **Figure 1**). This aspect gains relevance for the human settlements located

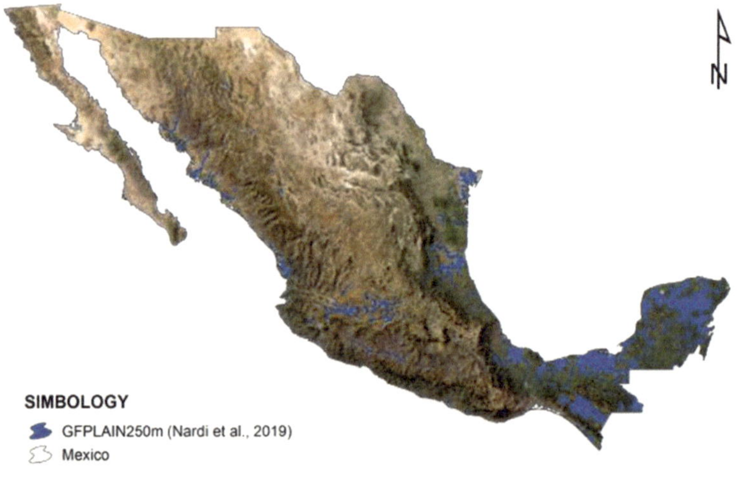

Figure 1.
Mexican floodplains. Processed from: Ref. [10].

in flood-prone areas and the growing trend of vulnerability to flooding owing to climate change. As determined by Mexico's National Center for Disaster Prevention (CENAPRED) [11], the greatest amount of damage and losses corresponded to disasters of hydrometeorological origin, mainly floods, in the period from 2000 to 2020. This is why, it is necessary to consider these events within studies aimed at preserving water security.

A way to guide the adaptation and mitigation actions for flood-risk reduction is through the three components of risk – hazard, related to the atmosphere-land processes – exposure and vulnerability, associated with socioeconomic changes, including population growth, economic development, and urbanization [8]. As for flood risk management (FRM), the University of Technology Hamburg, Harburg, defines it as the actions that are implemented before, during, and after the occurrence of a flood event, and these can be related to the prevention, attention, and recovery dimensions of integral risk management [12]. Public policy and decision-making contemplate FRM actions around the topics of land use, water resources, environment, coastal management, and the consideration of other possible hazards [13]. Moreover, the Sustainability Development Goal 11 and its target 11.5, look for the reduction of deaths, people affected, and economic losses generated by water-related disasters, especially floods, as an imperative aspect [14].

This book chapter describes relevant aspects of water security, the existing metrics, challenges, expected future, Mexico's WS situation, its relationship with floods, and a proposed strategy for the achievement and support of water security, through the knowledge of the basin's behavior, the identification of meteorological hazards, flood exposure analysis, application of metrics, and policy proposal or update, corresponding to the site's characteristics. Some benefits that can be recognized from this approach are the development of a comprehensive approach that contributes to water security, not only from the quality and quantity axes but also from a flood perspective; different temporal and spatial scales can be studied; a technical methodology including data processing, hydrological and hydraulic modeling, use of geographic information systems; based on particular physiographic and climatological basin conditions; oriented to flood risk reduction, policy, and decision-making.

2. Water security

The literature on water security shows different definitions by both individual authors and international institutions. However, the majority revolves around guaranteeing the quantity and quality of the volume of water to meet the needs of human beings and the environment, considering the attention to water-related disasters and with a vision of human rights. One of these definitions is the one proposed by Ganoulis, who states that it refers to the preservation of sufficient quantity and quality of water for human and ecosystem services at all times, including climate change, floods, and droughts [5]. For this context, water is considered as the element that allows the improvement of socioeconomic activities, maintains life on the planet, achieves socioeconomic growth, and guarantees a good environmental and ecological status. It is also important to reiterate that water management will depend on the natural variability of the environment.

In addition to these previous considerations, it is necessary to assure that prices for access to water in quantity and quality must be affordable and equitable, seeking to protect the population and their assets from extreme hydrometeorological phenomena, such as droughts and floods [2]. By taking into account not only the fulfillment of water for human consumption but also the destination for ecosystems and food production, as well as the risks to which the population is exposed and the consequences on the economy, it coincides with what Grey and Sadoff (2007) consider [2].

Water security involves various domains, or components, such as human development, ecological sustainability, geopolitics, international relations, vulnerability, and risk [4]. As can be seen, this concept has a multidimensional character [6] because its dynamic properties occur in space and time; this implies that its analysis and evaluation are carried out at different scales to serve as a basis for sustainable management of water resources, secure human health and well-being, fight poverty, and achieve economic development [2]. According to Su et al., WS generally represents the health of aquatic ecosystems, which must take many factors into account, such as resource reuse, environmental protection, and disaster prevention [15].

2.1 Water security metrics

It is possible to find different ways to measure water security, both individually through water availability, vulnerability, and sustainability, or by using indexes that involve physical, socioeconomic, political, and ecological aspects [16]. As mentioned before, it is important to analyze WS at different scales due to its spatial variation, in order to better guide the corresponding actions. Octavianti and Staddon pointed out that these actions can be addressed from a political scale (municipality-state-country), social (house-community-rural-urban), or hydrological (basin) [7]. A review at the regional or national level could conceal some particularities or changes that occurred at the local scale, as pointed out in [17]. For resource-based measurement, Li et al. established that some approaches are followed, such as pressure-response state, dynamic systems, green and blue water footprint, risk-based water management, integrated assessment, carrying capacity, and vulnerability [18].

In order to express the characteristics of water security, it is preferable to use a composite index [6]. In this regard, some of the most popular indices are Water Disaster Risk (WDR), Water Security Index (WSI), Water Pressure (WP), Water Poverty (WP), and the Water Carrying Capacity Index (WCCI). In future work, the literature proposes to consider the lack of WS in other sectors and revise its significance in the viability and feasibility of investments in cities [7]. From a bibliographic analysis carried out in [4] on WS indicators, it was concluded that issues of water scarcity are usually addressed, while the risk of flooding and rising sea levels are rarely touched on.

In the 2023 Global Water Security Assessment, where 10 components of WS were evaluated (drinking water, sanitation, good health, water quality, water availability, water value, water governance, human safety, economic safety, and water resource stability), it was found that most of the world's population lives in water-insecure countries [19]. **Figure 2** presents a water security estimation, given a maximum value of 10 for each component, obtained from Sustainable Development Goals available data per country, and aggregated by summation. An insecure status is considered when the water security component receives a score below a threshold value of 64. In this comprehensive metric, water-related disasters are considered within human and economic safety.

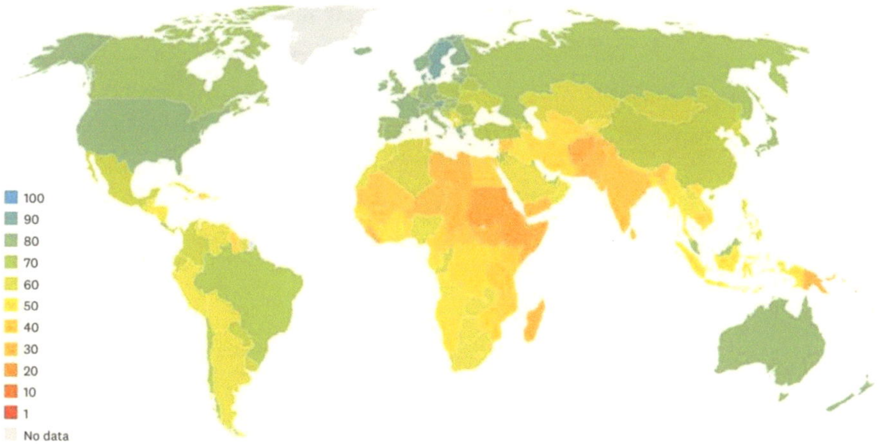

Figure 2.
National water security mapped globally, based on a score of 1–100. Reproduced from: Ref. [19].

2.2 Challenges and future of water security

The global conditions of the different components of water security are exposed to both natural and anthropogenic variations, in addition to expectations of population growth and the consequent overexploitation of natural resources, specifically water. Ganoulis mentions that WS must consider innovative attempts to improve water resource management, public policies, and governance [5]. **Figure 3** exhibits a review of the main challenges in terms of water security and their triggering factors identified by Ref. [2], where the phenomena of scarcity, contamination of water bodies, conflicts over water and environmental deterioration, generated by demographic growth, urbanization, growing demand for energy and food, CC, and poor water management.

One aspect that has resulted in national security problems refers to the import and export of water in transboundary basins, which have been affected by the decrease in surface and groundwater availability, which sometimes prevents compliance with international treaties. The effects of climate change will be reflected in the water supply, as well as demographic mobilizations, among other problems, and to examine these, some methodologies have been proposed, such as the one used by Sone et al., in which hydrological models are carried out in the face of climate projections and shared socioeconomic trajectories and various temporal scenarios [20]. Subsequently, different measures of water security were analyzed, such as the water scarcity and vulnerability index, as proposed in [21], in addition to using two drought indexes for the baseline and future, consecutive dry days, and the standardized precipitation-evapotranspiration index.

The projected population increment will be associated with the need to have adequate food production. Because agricultural activities worldwide use approximately 70% of the available water resource, it is expected that hydraulic infrastructure will be needed for storage and distribution, which minimizes losses and implies an investment by users. On the other hand, water shortage is related to both natural and anthropogenic factors, including political aspects [3], and it has been established that it will increase by 35% if water quality is considered [22]. Among the actions carried

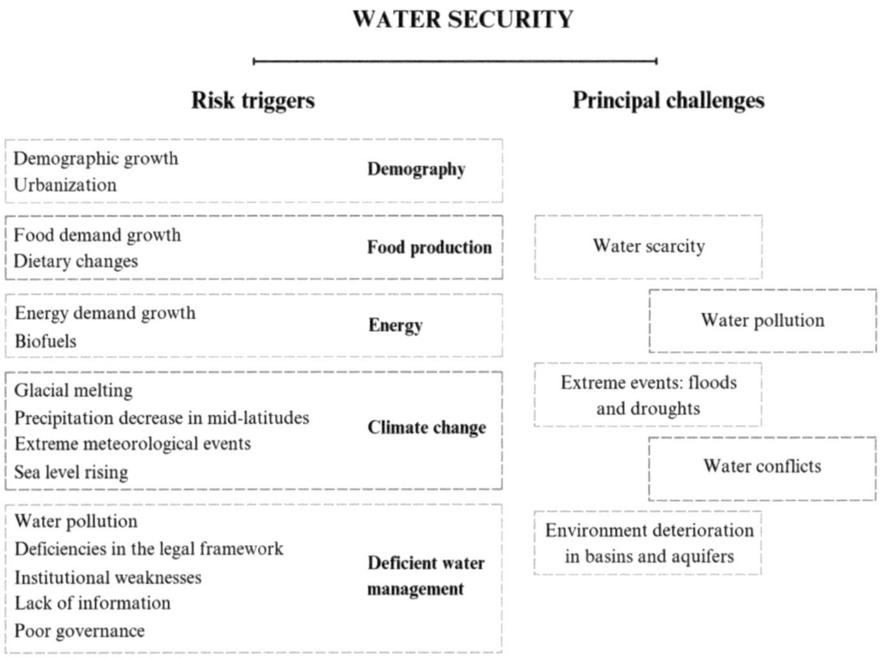

Figure 3.
Water security challenges and its generating factors. Adapted from: Ref. [2].

out by international institutions, there is the UNESCO International Hydrological Program, whose objective is to contribute to Water Security [3], promoting science, innovation, and the development of institutional capacities and humans in hydrology.

It is clear that prevention actions are essential due to the projected impacts on water security in the world; for example, by 2030, there will be a water deficit of 40% for all uses [23]. Sone et al. analyzed that a better understanding of the outcomes of climate change on WS is required to address water vulnerabilities with the increase in extreme weather events [20]. They also acknowledged that the lack of public policies to adapt demands to changing availability exacerbates hydrological droughts. Therefore, a balance between water availability and demand must be sought to improve water security in the uncertain future.

3. Water security in Mexico

By taking a closer look at the region occupied by Mexico, 70% of the territory is arid or semi-arid, and the rest of the territory experiences the risk of recurrent flooding because of tropical weather systems or cold fronts [2]. Regarding the habitant's projection for the period 2016–2050 [24], based on 2015 surveys, for 2030, the Mexican population will be around 138 million and for 2050, 148 million, which goes hand in hand with the supply and sanitation of water required for this growing population. It is expected that population density in the so-called megacities will be affected by conditions of water scarcity. Mexico's National Water Commission (CONAGUA) estimates that by 2030, the deficit will correspond to 25% of the demand for water resources [25]. Considering climate change, precipitation will be reduced by approximately 15%, as stated in [26], affecting runoff, aquifer recharge,

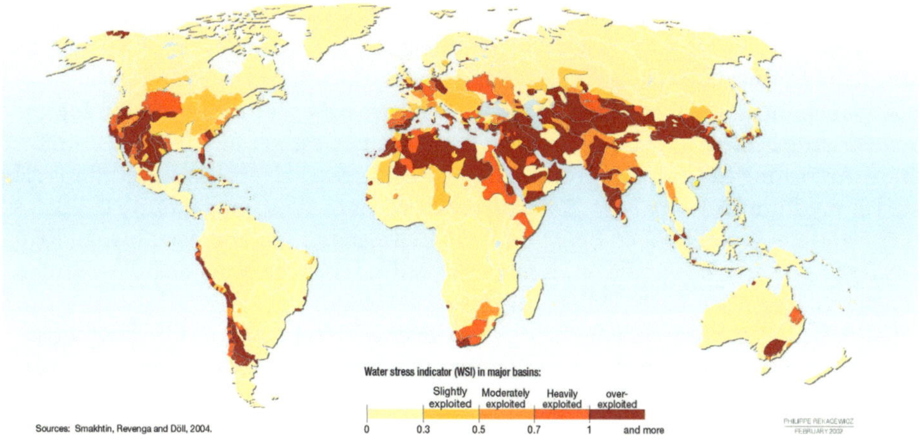

Figure 4.
Water stress indicator in major basins. Reproduced from: Ref. [27].

and availability. In the map presented in **Figure 4** in Ref. [27], Mexico's central and north basins, the Lerma River and the Valley of Mexico, are under a condition of high exploitation or overexploitation.

From the country's total groundwater consumption, 38.9% comes from groundwater, mainly for consumptive uses, irrigation, for urban and industrial public use. However, out of the 653 total aquifers in the country as for 2023, 286 of them are classified as without water availability [28]. In addition, 18 aquifers have saline intrusion, and 32 are subject to soil salinization. On the topic of drinking water coverage, although 90.94% has been covered, it is not reflected in real access to water and sanitation; in the states with the highest degree of poverty, coverage has been less than 60%, as is the situation of Chiapas, Tabasco, Campeche, Veracruz, Oaxaca, and Guerrero [29].

In Mexico, among the most relevant challenges is the contamination of water bodies and rivers, which has effects on human health and the environment, and its attention implies investments. When considering the total water bodies in the country, only 23.4% that are monitored coincide with excellent quality using Chemical Oxygen Demand as metric [30]. While as commented by Jiménez-Cisneros et al. [31], existing wastewater treatment plants (WTP) only treat 16% of the water coming from the industry. At the municipality level, 2786 WTP operated and treated 144710.0 L/s at the end of 2020 [32]. It is necessary to increase the quality and quantity monitoring network, to conserve hydrological systems (basins and aquifers), for decision-making. A review of the discharge regulations for treated wastewater is also needed [29].

For water consumption in agriculture, this corresponds to 76.04% on average for the country, of which 67% is extracted from surface water, and the remaining 9% corresponds to groundwater [33]. At this point, the search for greater efficiency in irrigation systems is desirable, as it is reported as approximately 64.7% in conduction within the irrigation area, according to Peña-Peña (2007) referenced in [29]. Mexico is one of the most relevant importers of meat and cereals worldwide, such as corn, while it exports vegetables and fruits. Although Mexico decreased the export of oil, the export of virtual water has increased through its agricultural and industrial products and services, which raises the vulnerability of the regions where natural resources are consumed [29]. Another aspect is the import and export of water between countries, in the case of the northern and southern borders of Mexico, importing 3.4 × 104 hm^3/year on the southern border and 5900 hm^3/year on the northern border [29].

About changes in stream flows, human activities have altered their hydrological and hydraulic behavior with the construction of dams, canals, and pipelines, among others. According to Garrido Pérez et al., 73% of Mexico's rivers have alterations classified as medium and very high [34]. Some examples of these currents are the Balsas River, Santiago River, Panuco River, Grijalva-Usumacinta, and the Papaloapan River. In order to preserve its operation, it must be verified that the ecological flow is met; however, the regulations in this regard are not mandatory.

In the face of climate change, increases in average temperature will affect water resources, varying the way in which this resource is required for different uses, such as agricultural and urban [29]. Precipitation in northern Mexico is expected to be reduced between 16 and 25% for RCP 6.0, which will be reflected in the volume of surface and groundwater available. The conditions generated in the atmosphere and ocean will cause more extreme hydrometeorological phenomena, such as heat waves, extreme precipitation, or recurrent droughts, which put the population, their property and human right to water, health, and a healthy environment, at risk.

Considering hydrometeorological phenomena, these mainly refer to tropical cyclones, intense rainfall, hail, droughts, and floods. CENAPRED determined that historically, some of the costliest disasters related to hydrometeorological phenomena have been hurricanes Alex, Karl, and Matthew in 2010, with 92.3 billion Mexican pesos [35]. In 2022, damage due to these phenomena represented 81% of the total, with 123 deaths, 3509.08 million Mexican pesos, 9524 homes, 387 schools, 99 hospitals damaged, and 207,754 inhabitants affected. After the landfall of category 5 hurricane Otis, in October 2023 in Mexican Pacific, economic losses of at least 16,000 million dollars were estimated [36].

Particularly, out of the total official statements issued by CENAPRED [35], 77% were related to impacts of hydrometeorological phenomena that caused emergency situations and disasters (see **Figure 5**). In general, referring in a higher percentage to rains, floods, extreme temperatures, mass removal processes, earthquakes, and tropical cyclones.

Defining what a flood is, this event refers to the overflowing of the normal limits of a stream or body of water, or the accumulation of water over areas that are not normally submerged, according to Field et al. (2012) in Ref. [4], and it can be generated by different causes (**Table 1**). Also, this phenomenon can be characterized through precipitation variables (duration, intensity, temporality), temperature patterns, as well as the previous conditions of the basins, such as river levels, soil infiltration capacity, and the presence of protection structures.

Water disasters are the outcome of very complex interactions among the ocean, atmosphere, and land, but also of the exposure and vulnerability of human and natural systems. In accordance with IPCC (2014), there is not yet enough data to demonstrate that the magnitude and frequency of floods have changed attributable to anthropogenic climate change, although it is clear that flooding socioeconomic losses increased during the second half of the twentieth century as a result of greater exposure and vulnerability of the population [3].

Increasing urbanization aggravates the negative socioeconomic impacts derived from climate disasters, such as droughts and floods, as noted by the authors in Refs. [37–39], which is reflected in the increase in the impervious area and the decrease in infiltration,

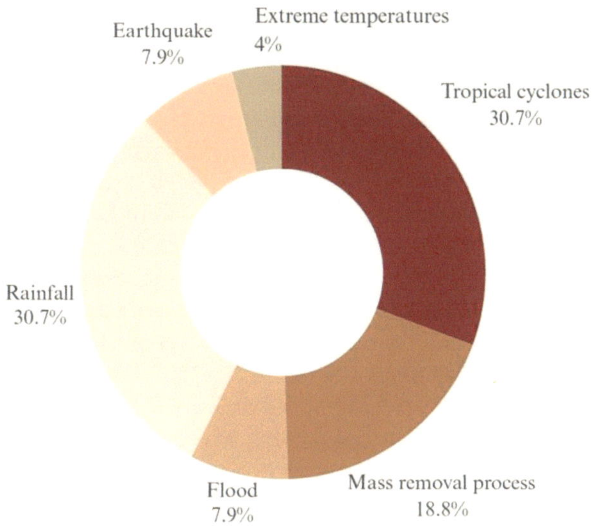

Figure 5.
Official disasters and emergency statements by type of phenomena in Mexico in 2022. Adapted from: Ref. [35].

Type	Characteristics
Convective storms and flash floods	Small, affected areas, rainfall rate exceeds infiltration and drainage capacity
Fluvial	Long-lasting rainfall over large areas
Coastal	Winter storms and tide
Cold season	Tides and snow cover
Dam breakage, tsunamis, waterlogging, glacier outburst	Local effects, however, sometimes extensive, generated instantaneously

Table 1.
Flood types and characteristics. Adapted from: Ref. [5].

which can result in frequent and severe floods. In Mexico, one of the causes of flooding is rainfall generated by tropical cyclones [40]. For loss of life associated with disasters in 2022, 25% is attributed to hydrometeorological phenomena [35].

Another aspect to consider is that the devastating consequences of this type of event in food production have been frequently recorded [41]. The aforementioned becomes important since, on the banks of the rivers and floodplains, human settlements are located, agricultural, industrial, or infrastructure districts, which can place them in a situation of flood risk when an extraordinary flood occurs and in the scenario that, for the characteristics of the channels, it is not possible to contain the volume of water and overflows occur.

In Mexico, 65% of the population is found in areas susceptible to flooding (i.e., in the lower parts of the basins) [42]. For example, in those regions with floodplains, such as the states of Veracruz, Tabasco, Nayarit, Jalisco, Baja California or Chiapas, the risk of flooding is related to the phenomenon hazard, in addition to the vulnerability given by the topographic configuration of the site, existing control infrastructure, social and economic conditions, among other factors.

4. Flood risk management and water security

Within flood management, the variability of climatological and hydrological processes, the hydraulic response, and the dynamics of the phenomenon, must be considered [43, 44]. This, through a proposal that provides technical elements for decision-making and the formulation of public policy, aimed at avoiding the development of new urban settlements in areas at risk of flooding and reducing the vulnerability of existing settlements in these zones. The said management refers to those actions performed before, during, and after flooding phenomena, as well as the policies and administrative decisions that seek to prevent or mitigate the damage provoked by this phenomenon [12], which depends on the characteristics of the site [45].

The complexity of flood events is due to the fact that they involve economic activities and settlements that are located in flood-prone areas [4]. Changes in flood hazard can be generated by human and natural factors, while exposure and vulnerability are caused by anthropogenic activities. The effects of floods not only imply the loss of human life or material property but are also accompanied by the interruption of communication services, electricity, drinking water, food production, and transportation, among others. In addition to this, groundwater is also impacted by this phenomenon, leading to the contamination of stagnant water [4].

Another relevant aspect of floods is the perception of fear of being flooded, which is why, by having protection infrastructure, a false sense of security is given, forgetting vulnerability. In turn, as these structures are designed for large but unlikely events, they could fail for events with less severity but greater extension [4]. Floods affect social security and structural stability, threatening water security, which is a reason for necessary update of water management policies and prioritize WS, by considering the influence of urbanization on the floods [46]. Furthermore, throughout an adequate FRM, the use of the floodplain can support economic activities and improve ecosystems, contributing to regulating cultural services. On the other hand, some obstacles to flood risk management are the lack of clear roles and responsibilities for policy, planning, and implementation [13].

4.1 Strategy

Once the relation between flood and its management, and water security are stated, **Figure 6** presents a strategy to strengthen water security through flood risk management. As indicated by Valencia Vargas et al., it is recommended that integrated water resources management (IWRM) is addressed at basin scale, and since WS is supported by a correct IWRM, the use of the same spatial scale is selected [47].

The hydrologic response of a basin is related to physiographic conditions given by the topographic configuration, such as the area, shape, length, and slope of the main river and the tributary streams. Another factor that determines runoff potential is the type of soil and the land use cover existing in the basin. These properties are integrated in hydrologic modeling, from the application of simple mathematical equations to the use of specialized software. The latter requires climatological data related to extreme precipitation events to be studied for flood studies.

After knowing the hydrological behavior of the basin of interest in the face of certain rainfall scenarios, past and future meteorological phenomena with a flood hazard development possibility need to be identified. Subsequently, through forecasting, runoff can be quantified in order to compare it with the hydraulic capacity of the streams and estimate the flood risk situation. These previous steps, accompanied

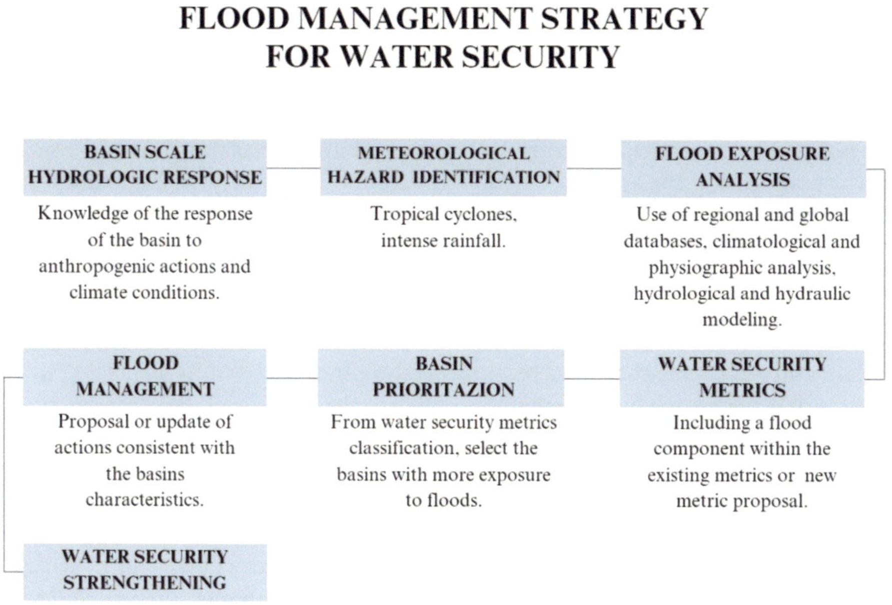

Figure 6.
Strategy for water security strengthening through flood management.

by the verification of regional and global databases that help to define areas prone to flooding, constitute a flood exposure analysis.

From the mentioned approach, the water security metrics available in literature that contemplate a flood dimension can be integrated and improved [4], or in the case that this component is not contemplated, further work on the development of an individual flood indicator can be addressed. By having a classification of the basin around its flood condition, a better prioritization of investment and risk mitigation proposals is possible.

Within a typical flood risk management scheme, structural measures are considered to increase resilience; some examples of these are levees, dikes, streams channelization, and dams, among others. Nevertheless, their construction is not always considered fair at environmental or societal level, and decision-makers should bear in mind the probability of failure and uncertainty. Complementary to physical options, the non-structural measures or actions have the goal of reducing the exposure and vulnerability of people, economy, and ecosystems [13], through planning control of flood-prone areas, warning systems, evacuation planning, forecast, building codes, and insurance, as well as fair flood protection. From the definition of Sayers et al. (2013), for a good FRM, it must include an uncertainty analysis, spatial planning, infrastructure management, emergency planning and management, flood hazard and risk mapping, early warning systems, effective land controls and building codes and insurance [13].

By integrating the described steps, the proposed strategy serves as a technical-based one that supports water security from its flood dimension and can be employed by its own or be included in existing WS metrics. The methodology has the advantage that can be applied for different spatiotemporal scales and is adapted to the specific physiographic and climatological conditions of basins. The information given by this methodology is oriented toward flood risk reduction, policy, and decision-making.

5. Conclusions

Through the analysis of the general aspects of water security, it is possible to have an overview of the complexity and challenges faced both at a global and regional level, noting the need to update various regulations around its domains and components by considering the approach of multisectoral regulations and instruments that could better resolve the improvement gap.

In addition to the challenges presented in this chapter, one is visualized where, for the evaluation of the WS situation of a certain region or at a global level, there is not sufficient basic information available for the application of the different existing metrics at the moment. Thus, a prioritization of actions could not be established to maintain and guarantee the water resource as the concept seeks. As suggested in the literature, it is necessary to study water security at the local level. For Mexico, derived from the fact that the integral water resource management unit is the basin, the allocation of resources could be based on prioritizing attention to those that result in critical values of the water security indicators, including the flood dimension.

All the challenges that prevent water security from being guaranteed are important, have significant repercussions, and require to be addressed in order to achieve social and economic development. However, in this WS strategy proposal, main attention was given to the hydrometeorological phenomenon of floods, as it has not been commonly contemplated in metrics and is a dynamic process derived from climatological, physiographic, hydrological, and hydraulic conditions, and in turn, involves multisectoral aspects for its management. Moreover, it allows its analysis at the hydrological basin scale, as recommended in the literature, so that decision-making and public policy proposals are carried out according to the requirements and particularities of each interest basin and avoids a generalized flood risk management, which could turn out to be under or overestimated.

With constant changing conditions of climate, urbanization, economic development, and environmental damage, FRM needs to be continuously updated and adapted. In this sense, the implementation of international programs must also seek mechanisms that are binding or obligatory so that the countries or actors involved comply with the global sustainable development objectives. Collective awareness must also be promoted to allow directives to permeate with a global vision at local levels.

Acknowledgements

The authors would like to thank the Universidad de las Américas Puebla and CONAHCYT for the support they provided in carrying out this work. Special thanks to Professor Carlos Patiño- Gómez (†)for his invaluable mentoring and participation in this chapter, as well as for the leadership of the UNESCO Chair in Hydrometeorological Risks at Universidad de las Américas Puebla. The authors also thank Dr. Benito Corona-Vasquez for his contribution to the preparation of this chapter.

Conflict of interest

The authors declare that they have no competing interests.

DOI: http://dx.doi.org/10.5772/intechopen.115472

Author details

Edith Bonilla-López* and Carlos Patiño-Gómez
Department of Civil and Environmental Engineering, University of the Americas Puebla, Puebla, Mexico

*Address all correspondence to: edith.bonillalz@udlap.mx

References

[1] UNESCO. International Hydrological Programme (IHP). Eight Phase "Water Security: Responses to Local, Regional, and Global Challenges" Strategic Plan IHP-VII (2014-2021) [Internet]. 2012. Available from: https://unesdoc.unesco.org/ark:/48223/pf0000218061 [Accessed: December 14, 2023]

[2] Martínez-Austria PF. Los retos de la seguridad hídrica. Tecnología y Ciencias del Agua. 2013;**IV**(5):165-180

[3] Jiménez-Cisneros B. Seguridad hídrica: Retos y respuestas, la fase viii del programa hidrológico internacional de la unesco (2014-2021). Aqua-LAC. 2015;7(1):20-27

[4] Octavianti T. Rethinking water security: How does flooding fit into the concept? Environmental Science & Policy. 2020;**106**:145-156. DOI: 10.1016/j.envsci.2020.01.010

[5] Ganoulis J. Water security in the anthropocene: A dialectical water-man interaction model. Sustainability. 2022;**14**(12):6955. DOI: 10.3390/su14126955

[6] Qiao Y, Chen Y, Lu H, Zhang J. Integrating water-related disaster and environment risks for evaluating spatial-temporal dynamics of water security in urban agglomeration. Environmental Science and Pollution Research. 2022;**29**:58240-58262. DOI: 10.1007/s11356-022-19832-5

[7] Octavianti T, Staddon C. A review of 80 assessment tools measuring water security. Wiley interdisciplinary reviews. Water. 2021;**8**(3). DOI: 10.1002/wat2.1516

[8] Swiss Re Institute. Sigma research. In: 5 Charts: What you Need to Know about Growing Flood Risk [Internet]. Swiss Re Institute; 2022. Available from: https://www.swissre.com/institute/research/sigma-research/sigma-2022-01/five-charts.html [Accessed: October 23, 2023]

[9] UNDRR. Tendencias y estadísticas [Internet]. UNDRR; 2021. Available from: https://www.undrr.org/sites/default/files/inline-files/7-Tendencias%20y%20estadi%CC%81sticas_2.pdf [Accessed: October 23, 2023]

[10] Nardi F, Annis A, Di BG, Vivoni ER, Grimaldi S. GFPLAIN250m, a global high-resolution dataset of earth's floodplains. Scientific Data. 2019;**6**. DOI: 10.1038/sdata.2018.309

[11] CENAPRED. Impacto Socioeconómico de los Principales Desastres Ocurridos en México en el Año 2019 (RESUMEN EJECUTIVO) [Internet]. CENAPRED; 2019. Available from: http://www.cenapred.unam.mx/es/Publicaciones/archivos/429-RESUMENEJECUTIVOIMPACTO2019.PDF [Accessed: May 17, 2023]

[12] University of Technology Hamburg - Harburg. Flood Manager E-Learning [Internet]. 2010. Available from: http://daad.wb.tu-harburg.de/ [Accessed: July 9, 2021]

[13] Sayers P, Yuanyuan L, Galloway G, Penning-Rowsell E, Fuxin S, Kang W, et al. Flood risks management: A strategic approach. Asian Development Bank. 2013. 206 p. ISBN: 9789230011598

[14] United Nations. Sustainable Development Goals. Goal 11: Make Cities Inclusive, Safe, Resilient and Sustainable [Internet]. United Nations. Available from: https://www.un.org/

sustainabledevelopment/cities/; [Accessed: December 14, 2023]

[15] Su Y, Gao W, Guan D. Integrated assessment and scenarios simulation of water security system in Japan. Science of the Total Environment. 2019;**671**:1269-1281. DOI: 10.1016/j.scitotenv.2019.03.373

[16] Nkiaka E. Exploring the socioeconomic determinants of water security in developing regions. Water Policy. 2022;**24**(4):608-625. DOI: 10.2166/wp.2022.149

[17] Babel MS, Shinde VR, Sharma D, Dang NM. Measuring water security: A vital step for climate change adaptation. Environmental Research. 2020;**185**. DOI: 10.1016/j.envres.2020.109400

[18] Li X, Su X, Wei Y. Multistage integrated water security assessment in a typical region of northwestern China. Journal of Cleaner Production. 2019;**220**:732-744. DOI: 10.1016/j.jclepro.2019.02.033

[19] MacAlister C, Baggio C, Perera D, Qadir M, Taing L, Smakhtin V. Global Water Security 2023 Assessment [Internet]. Hamilton, Canada: United Nations; 2023. Available from: http://inweh.unu.edu/publications/ [Accessed: December 5, 2023]

[20] Sone JS, Araujo TF, Gesualdo GC, Ballarin AS, Carvalho GA, Oliveira PTS, et al. Water security in an uncertain future: Contrasting realities from an availability-demand perspective. Water Resources Management. 2022;**36**(8):2571-2587. DOI: 10.1007/s11269-022-03160-x

[21] Rodrigues DBB, Gupta HV, Mendiondo EM. A blue/green water-based accounting framework for assessment of water security. Water Resources Research. 2014;**50**(9):7187-7205. DOI: 10.1002/2013WR014274

[22] van Vliet MTH, Jones ER, Flörke M, Franssen WHP, Hanasaki N, Wada Y, et al. Global water scarcity including surface water quality and expansions of clean water technologies. Environmental Research Letters. 2021;**16**(2). DOI: 10.1088/1748-9326/abbfc3

[23] WRG. Charting our Water Future Economic Frameworks to Inform Decision-Making [Internet]. WRG; 2009. Available from: https://www.mckinsey.com/~/media/mckinsey/dotcom/client_service/sustainability/pdfs/charting%20our%20water%20future/charting_our_water_future_full_report_ashx [Accessed: December 14, 2023]

[24] CONAPO. Colección Proyecciones de la población de México y de las entidades federativas 2016-2050 [Internet]. CONAPO. Available from: http://www.conapo.gob.mx/work/models/CONAPO/Cuadernillos/33_Republica_Mexicana/33_RMEX.pdf; 2018 [Accessed: October 14, 2023]

[25] CONAGUA. Agenda del Agua 2030. Avances y logros 2012 [Internet]. Ciudad de México: CONAGUA; 2012. Available from: www.conagua.gob.mx [Accessed: November 21, 2023]

[26] Martínez-Austria PF, Patiño-Gómez C. Atlas de vulnerabilidad hídrica en México ante el cambio climático [Internet]. Jiutepec; 2010. Available from: https://www.imta.gob.mx/biblioteca/libros/atlas.pdf Accessed: December 14, 2023

[27] Rekacewicz P. Water Scarcity Index [Internet]. 2006. Available from: https://www.grida.no/resources/5586 [Accessed: December 14, 2023]

[28] Diario Oficial de la Federación. Acuerdo por el que se actualiza la

disponibilidad media anual de agua subterránea de los 653 acuíferos de los Estados Unidos Mexicanos, mismos que forman parte de las regiones hidrológico-administrativas que se indican [Internet]. México: Diario Oficial de la Federación; 2023. Available from: https://dof.gob.mx/nota_detalle.php?codigo=5708074&fecha=09/11/2023#gsc.tab=0

[29] Martínez Austria PF, Díaz-Delgado C, Moeller-Chavez G. Seguridad hídrica en México: diagnóstico general y desafíos principales. Ingeniería del agua. 2019;**23**(2):107. DOI: 10.4995/ia.2019.10502

[30] CONAGUA. Calidad del agua en México [Internet]. CONAGUA; 2023. Available from: https://www.gob.mx/conagua/articulos/calidad-del-agua [Accessed: December 14, 2023]

[31] Jiménez-Cisneros B, Torregrosa Y, Armenta ML, Aboites-Aguilar L. El agua en México: cauces y encauces. Ciudad de México: Academia Mexicana de Ciencias; 2010

[32] CONAGUA. Inventario de Plantas Municipales de Potabilización y de Tratamiento de Aguas Residuales en Operación [Internet]. CONAGUA; 2020. Available from: https://www.gob.mx/conagua/documentos/inventario-de-plantas-municipales-de-potabilizacion-y-de-tratamiento-de-aguas-residuales-en-operacion [Accessed: October 14, 2023]

[33] CONAGUA. Atlas del Agua en México 2018 [Internet]. Ciudad de México: CONAGUA; 2018. Available from: www.gob.mx/conagua [Accessed: September 24, 2023]

[34] Garrido Pérez A, Cuevas ML, Cotler H, González DI, Tharme R. Evaluación del grado de alteración ecohidrológica de los ríos y corrientes superficiales de México. Investigación ambiental Ciencia y política pública. 2010;**2**(1):25-46. ISSN: 2007-4492

[35] CENAPRED. Impacto Socioeconómico De Los Principales Desastres Ocurridos En México Resumen Ejecutivo 2022 [Internet]. Ciudad de México: CENAPRED; 2023. Available from: www.gob.mx/cenapred [Accessed: November 3, 2023]

[36] Forbes Staff. Otis dejó pérdidas por al menos 16,000 millones de dólares: Fitch Ratings [Internet]. Forbes México: Forbes Staff; 2023. Available from: https://www.forbes.com.mx/otis-dejo-perdidas-por-al-menos-16000-millones-de-dolares-fitch-ratings/#:~:text=%E2%80%9CLas%20estimaciones%20preliminares%20de%20p%C3%A9rdidas,dijo%20Fitch%20en%20un%20comunicado [Accessed: December 14, 2023]

[37] United Nations. Global Assessment Report on Disaster Risk Reduction. Risk and Poverty in a Changing Climate. Geneva: UN-HABITAT; 2009. ISBN: 9789211320282

[38] United Nations. Global Assessment Report on Disaster Risk Reduction: Revealing Risk, Redefining Development. Geneva: United Nations International Strategy for Disaster Reduction; 2011. ISBN: 9789211320305

[39] Jha AK, Bloch R, Lamond J. Cities and Flooding. A Guide to Integrated Urban Flood Risk Management for the 21st Century. World Bank Publications, Washington; 2012. DOI: 10.1596/978-0-8213-8866-2

[40] CENAPRED. Inundaciones. Ciudad de México: Secretaría de Seguridad y Protección Ciudadana; 2021. 51 p

[41] Atanga RA, Tankpa V. Climate change, flood disaster risk and food

security nexus in northern Ghana. Frontiers in Sustainable Food Systems. 2021;**5**. DOI: 10.3389/fsufs.2021.706721

[42] Helena C-A. Las cuencas hidrográficas de México: diagnóstico y priorización. Pluralia Ediciones e Impresiones; 2010. 231 p

[43] Visessri S, Ekkawatpanit C. Flood management in the context of climate and land-use changes and adaptation within the Chao Phraya river basin. Journal of Disaster Research. 2020;**15**:579-587. DOI: 10.20965/jdr.2020.p0579

[44] Kundzewicz Z, Plate EJ, Rodda HJE. Changes in flood risk in Europe-setting the stage. In: Kundzewicz ZW, editor. Changes in Flood Risk in Europe IAHS Special Publication 10. Poznan, Poland: International Association of Hydrological Sciences; 2012. pp. 0-526. ISBN: 9781907161285

[45] Comisión Nacional del Agua. In: Secretaría de Medio Ambiente y Recursos Naturales, editor. Manual para el control de inundaciones [Internet]. 2011th ed. Ciudad de México: Comisión Nacional del Agua; 2011. Available from: www.conagua.gob.mx [Accessed: July 16, 2019]

[46] Shao M, Zhao G, Kao SC, Cuo L, Rankin C, Gao H. Quantifying the effects of urbanization on floods in a changing environment to promote water security — A case study of two adjacent basins in Texas. Journal of Hydrology (Amst). 2020;**589**. DOI: 10.1016/j.jhydrol.2020.125154

[47] Valencia Vargas JC, Díaz Nigenda JJ, Vargas Martínez L. La gestión integrada de los recursos hídricos en México. Available from: http://www2.inecc.gob.mx/publicaciones2/libros/528/gestion.pdf [Accessed: September 11, 2022]

Chapter

Load Forecasting Based on Machine Learning

Erick Rodriguez-Orduna and Jorge Rodriguez-Asomoza

Abstract

Load forecasting is a critical part of electrical systems, and it allows operators, utilities, and distribution companies to be aware of what to expect from their user's demand; hence, they can make decisions to meet the expected demand of users. Forecasting is a complex task that nowadays relies on machine learning because of its potential and customization. Machine Learning (ML) in research has demonstrated to be an efficient approach to achieve predictions for multistep applications. Since there are multiple ML models that can perform forecast in this work, an introduction to such methods is presented. This paper describes techniques such as Support Vector Regressor (SVR), Artificial Neural Networks (ANN), and Gradient Boosting. Researchers have found that there is a gap in the application and implementation of forecasts in continuous environments; hence, in this work, we also mention some challenges that ML forecasts face in this production environment and what problems could be faced by researchers in the area. This work aims to contribute with a general insight into load forecast based on machine learning and the current challenges.

Keywords: load forecasting, machine learning, neural networks, IoT, gradient boosting

1. Introduction

The usage of monitoring systems to supervise and control industrial processes eventually creates an operational necessity to anticipate and be prepared for events. In the case of electrical systems, the forecast, which is the process of anticipating future demand for power, is necessary in all parts of the system (Generation, Transmission, Distribution, and Consumption). The application of load forecasting becomes relevant as there is increasing digitalization enabling technologies in industry such as Industrial Internet-of-Thing (IIoT) or Operation Technology (OT) systems.

Demand Load, which is commonly the forecasted variable, is the instantaneous power consumed by users and in these terms, Load Forecasting is the process from which we can obtain by prediction algorithms the future expected demand of a load. The load can be a Home, Factory, a Cluster, a State, or a Country. Hence, the demand varies depending on social, productive, and environmental variables. Due to the large number of elements that can affect the load forecast, there are many ways to perform this task.

To perform the load forecast, researchers and companies rely on power quality (PQ) measurements obtained from specialized devices known as PQ-Meters, these

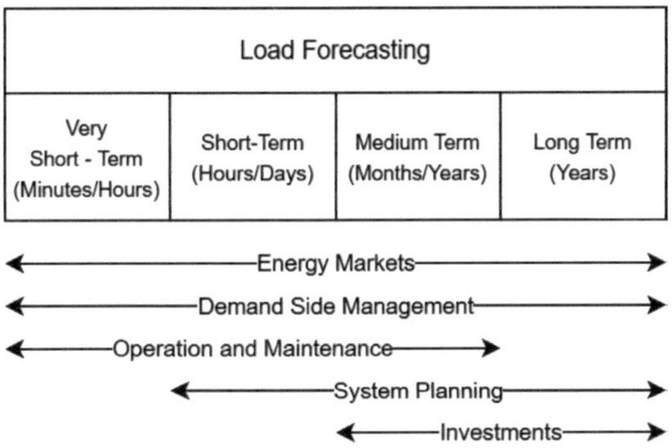

Figure 1.
Load forecast time and applications based on [2].

meters comply with several accuracy and manufacture standards such as [1]. An advantage of using PQ-Meters relies on the quality of the measurement and are commonly used in industrial environments.

Even though the PQ-Meters ensure the quality of the measurements, retrieving it in a remote way is performed commonly by automation processes in other devices such as loggers, Programmable Logic Controllers or IoT gateways requiring communication over long distances. Hence there is a possibility that data may contain missing values or can be unavailable at certain times. These problems can impact negatively on the forecast results.

1.1 Types of load forecasting

Given the diverse stages at which load forecast may be used in the electrical cycle, there are several categories and applications. **Figure 1** illustrates a classification of load forecast times and how those are relevant in major operations of electrical industry. It is possible to note that applications vary from engineering operations to system planning, and economic projections.

According to Forecasting books and authors such as [3], before the development of any time series forecasting modeling, it is necessary to answer some questions such as How the forecast will be used, What type of time-series is used, and its properties (Trends, Seasonality, Stationarity), How many Observations are, The length of the Forecast (One-Step or Multi-Step), Which is the cost of forecast per series, Skills and experience of the analyst.

2. Methods and models

Load forecasting may use several methodologies, and each of them has benefits and disadvantages. It is also able to capture or empathize with different patterns in the demand data. Traditional approaches include Linear Regression, which establishes linear relations in the data and the predictors; alternatively, Time Series Analysis, which involves techniques such as Autoregressive Moving Averages family or Exponential Smoothing are used to capture temporal dependencies and seasonal patterns, at the same time, Series Analysis also consider nonlinear methods such as Nonlinear autoregressive moving average model with exogenous inputs defines a system in terms of nonlinearity and achieved efficient results in the Load Forecast Problem.

In addition to Time Series Analysis methods, there are other techniques such as Machine Learning (ML)-based systems. Some of the techniques used are artificial neural networks (ANN) such as Long Short Term Memory (LSTM) or Gate Recurrent Unit (GRU) networks, Support Vector Regressor (SVR), Decision Trees (DT), Gradient Boosters, and ensemble architectures [4–7].

The advantage or reason behind the usage of ML methods is that these methods can capture patterns in non-stationarity and can learn trends and shapes within the data; also, these methods allow to increase the explanatory variables to add such as Weather, Social or Timestamp features easily to reduce the forecast error. **Figure 2** presents a brief classification of Load Forecast techniques found in literature and commercially.

The ML models have proved to be efficient in performing a multistep forecast [4]. This capability is relevant since many applications related to decision-making require a larger forecast timeframe than one step in the future.

One way to achieve the multistep forecast based on ML is by using the sliding window approach, which is a common technique in which the features are several timesteps before and the output or target are the following timesteps. This preprocessing allows us to generate many samples for the batches and, depending on the sizes, can impact the efficiency of the learning process. The sliding window approach allows models to adapt to new and dynamic environments, it continually can be used to keep learning in future steps and allows to have a better model. **Figure 3** shows how the sliding window works.

2.1 Support vector regressor

The core idea behind an SVR is Support Vector Machine (SVM), where the primary objective is to find a hyperplane that, by optimizing a risk function it, can achieve the maximum separation of classes in a multiclass problem. However, SVR extends this knowledge to a regression problem; hence, it can be used to fit a regression hyperplane to produce forecasts (**Figure 4**).

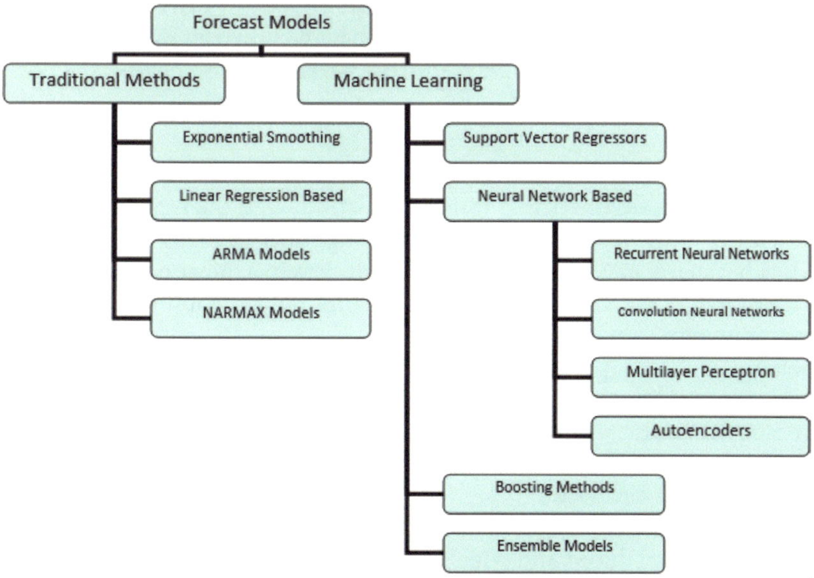

Figure 2.
Load forecast models.

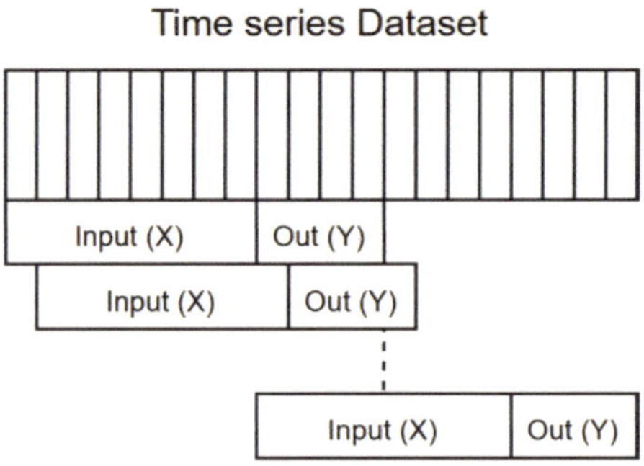

Figure 3.
Sliding window for time series learning.

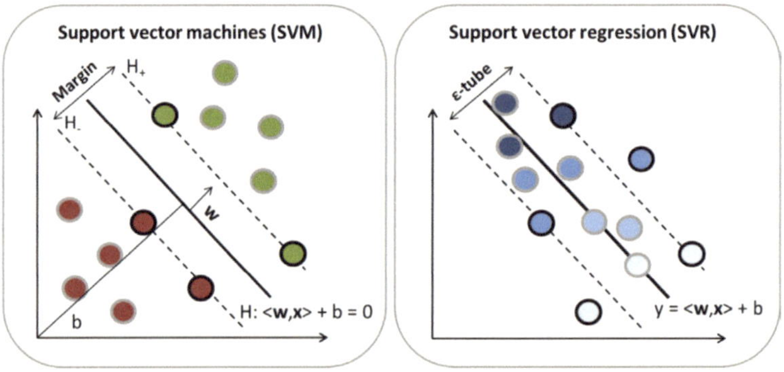

Figure 4.
SVM and SVR [9].

According to Ref. [8], SVR states that a Linear Function f can be used to map a nonlinear relationship on input/output data in a high-dimensional features space. The basis function that describes a Hyperplane that fits the data is known as

$$f(x) = \omega^T \varphi(x) + b \tag{1}$$

where $f(x)$ denotes the fitting forecast function, x is the input vector, w is the weight vector, b is the bias, and φ is the mapping function. This function is the core of the SVM because it can map a nonlinear set of data to a linear pattern in another feature space. There are several mapping functions, but the common for regression are linear, polynomial and radial basis.

Depending on the mapping function, one function can yield better results over others, this is due to the presence of different distributions and similarities to data and mapping functions.

2.2 Artificial neural networks

The ANN was introduced and widely used in the 1990s [10] and is intended to resemble the human process of knowledge. It is commonly used due to its flexibility to change and manipulate the shapes of the inputs and outputs.

The ANN has evolved into a new type of network, the Recurrent Neural Networks, which works better for regression tasks such as load forecast. RNN works by the same principle of ANN in which the input is propagated along the units of a network based on their weights then adjusted by a loss function. Because of propagation in time, the RNN can resemble a memory; nonetheless, the RNN has a problem known as the gradient problem. In the gradient problem, it is stated that in the long run of a time series regressor based on RNN, the gradient and computations based on it will then reach zero (Vanishing Gradient) or reach a very large value (Exploding Gradient) [11].

To solve problems related to gradients, another type of cells was introduced: the Long Short Term Memory (LSTM) and Gate Recurrent Unit (GRU); later, these cells can store memory for long trends and small trends, hence allowing an effective convergence of the gradient.

LSTM is a type of neuron unit that is used for Regression tasks such as Natural Language Processing and Forecasting. The building of the LSTM network allows for faster convergence while training and can detect long-term trends in the data. LSTM has two vectors for state $\boldsymbol{h}_t$ and $\boldsymbol{c}_t$ which can be considered as short-term state and long-term state, respectively. Similarly, to LSTM cells, the GRU was developed to tackle the gradient problems of RNN. GRU is a simplified version of the LSTM cell. The GRU cell works by performing some simplifications of LSTM cell having only one memory state.

Figure 5 from Ref. [12] presents how the computations are made for each of the common cells based on RNN and the updated cells like LSTM and GRU.

The cells are more complex but eventually, they can preserve relevant information of the data.

2.3 Gradient boosting

The gradient Boosting algorithms are based on the principle of Boosting, which is a strategy that increases the efficiency of a "strong" predictor by combining multiple predictors that are "weak" [13]. The typical weak learner for many Gradient Boosting libraries is the decision tree; this model, according to [14], works by partitioning the input parameter space into distinct and non-overlapping regions following a set of if-then rules. The partitioning must be done so a loss function can be minimized

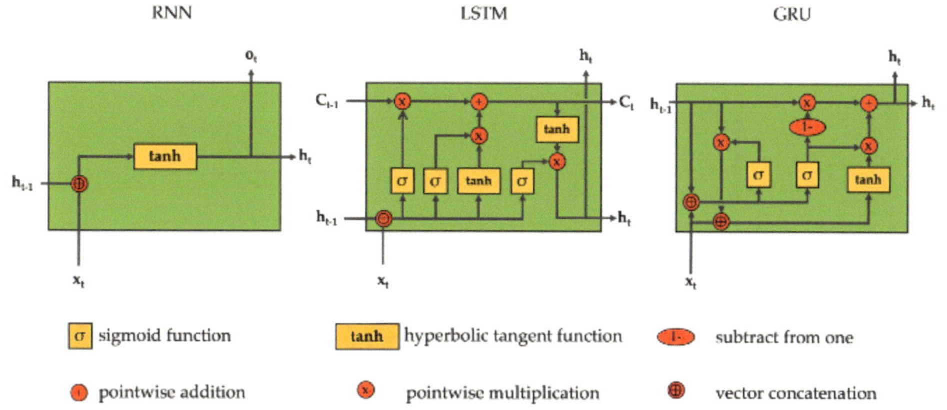

Figure 5.
Cell computations of RNN, LSTM and GRU [12].

or maximized. For typical classification trees, one of the loss functions is the Gini impurity index, whereas for regression problems, it can be the mean squared error.

According to Ref. [15], in a gradient boosting each step at each step, a new "weak model" is trained to predict the error of the strong model (so-called pseudo-response). The error can be a performance metrics and the weak model is added to the strong model to reduce the error of the strong model. Gradient Boosting has iterative training following the formula:

$$F_{i+1} = F_i - f_i \tag{2}$$

where F_i is the strong model at iteration i and f is the weak model. The operation is repeated until a criterion is found. Currently, there are diverse ways to perform the gradient boosting algorithm such as XGBoost, SKlearn Gradient Boosting and LightGBM among others. **Figure 6** shows how the learning process is made on the gradient boosting method, where it can be seen the weak and strong predictors.

2.4 Performance metrics

The typical functions used to evaluate the model or to use as a loss function are the Mean Squared Error (MSE), Root Mean Squared Error (RMSE) and Mean Absolute Error (MAE):

$$MAE = \frac{1}{N}\sum_{t=1}^{n}\left|F_t - A_t\right| \tag{3}$$

$$MSE = \frac{1}{N}\sum_{t=1}^{n}\left(F_t - A_t\right)^2 \tag{4}$$

$$RMSE = \sqrt{\frac{1}{N}\sum_{t=1}^{n}\left(F_t - A_t\right)^2} \tag{5}$$

where F_t is the Forecast, A_t is the actual value, N is the series length. The lower the value resulting from the formulas above means that the forecast is better. These formulas are commonly used by many researchers to demonstrate the effectiveness of their models.

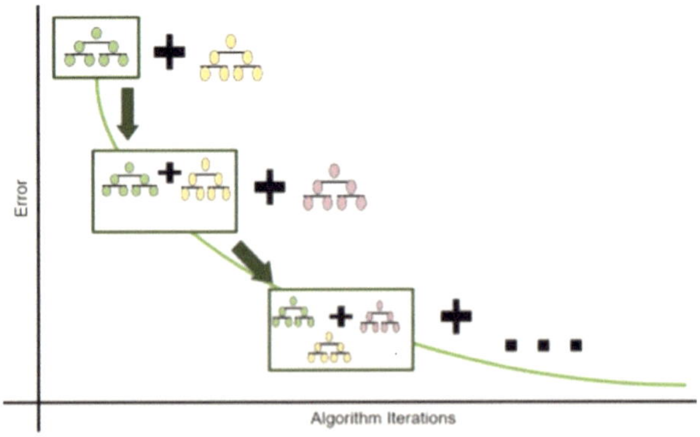

Figure 6.
Gradient boosting method [16].

3. Current trends and challenges of load forecast

The load forecasting research, as stated by Somu and Ramamritham [7] and Ahmad [17], presents gaps such as limitations in evaluations of the forecast as a continuous system; this is a production environment with new addition of samples and retraining models, yet there are research works such as [18] that consider the forecast as a system and present the retraining and evaluation of forecast.

In Ref. [19], these production problems are presented by researchers and engineers of Comision Federal de Electricidad (CFE) in which they implemented an ML Pipeline for addressing some of the problems such as continuous forecasting, Big Data Storage issues and communication problems such as missing data; however, these problems present an opportunity area in software development for load forecast problem.

Despite having little information about the production environments in research, Cloud solution providers such as Ref. [20] or Ref. [21] present architectures considering Developer Operations (DevOps) and ML Operations (MLOps) for Load or Energy Forecasting; in this system, they consider pipelines for the data flow but do not provide insights on how to perform effective retraining and other ML techniques such as Continual learning or Task Incremental Learning, some of these techniques can be seen in Ref. [22].

Authors like [17] also point out that research challenges are related to choosing a simple or an ensemble model. This is because the demand varies on other variables such as weather and unexpected disturbances, and distribution companies generally do not consider the necessity of a forecast.

In Ref. [23], a forecast pipeline is presented, incorporating elements relevant to any ML time series forecast, such as a comprehensive understanding of the data provided. It also considers preprocessing steps such as correcting outliers, missing values, and transformations; feature engineering to highlight relations or add more features; model selection and hyperparameter optimization and lastly ensembling the forecast.

4. Conclusions and future work

Finally, this work reviews some of the common techniques used for load forecasting based on ML, with a particular focus on SVR, RNN and Gradient Booster. The application of ML in load forecasting emerges as a viable solution but requires further analysis to ensure that the information added to these models is relevant, yielding an effective forecast application.

This work also points out that even though in research there are many remarkably effective models, there is still a gap in a continuous environment for these forecasts, considering the data lifecycle and the fine tuning and retraining of models is still a challenge for research to validate the effectiveness of a model. One of the issues still pointed out by authors is that demand depends on other variables, such as climatic or social, and therefore, it is required to map these features into the training data for the model; in this way, more robust systems can be made.

Even though there are notable advances in load forecasting models, many researchers have pointed out the gap in integration into production environments, and many authors still consider this as a major issue because the forecast will tend to increase the error if the model does not have a retraining process which is required to provide a product or service based on the load forecast result.

Author details

Erick Rodriguez-Orduna* and Jorge Rodriguez-Asomoza
Computing and Electronic Department, Universidad de las Americas Puebla, Cholula, Puebla, Mexico

*Address all correspondence to: erick.rodriguezoa@udlap.mx

References

[1] ANSI C12.1-2022 - Electric Meters - Code for Electricity Metering (incorporates ANSI C12.20-2015). [Online]. Available from: https://webstore.ansi.org/standards/nema/ansic122022?source=blog&_ga=2.132303885.2885614.1679667766-1449039159.1679667765 [Accessed: March 24, 2023]

[2] Zor K, Timur O, Teke A. A state-of-the-art review of artificial intelligence techniques for short-term electric load forecasting. In: 2017 6th International Youth Conference on Energy (IYCE). IEEE; 2017. pp. 1-7. DOI: 10.1109/IYCE.2017.8003734. Available from: https://ieeexplore.ieee.org/document/8003734

[3] Chatfield C. The analysis of time series - an introduction. In: Chapman & Hall/CRC Texts in Statistical Science. 6th ed. Chapman and Hall/CRC; 2003. Available from: https://www.taylorfrancis.com/books/mono/10.4324/9780203491683/analysis-time-series-chris-chatfield

[4] Guo W, Che L, Shahidehpour M, Wan X. Machine-learning based methods in short-term load forecasting. The Electricity Journal. 2021;**34**(1):106884. DOI: 10.1016/j.tej.2020.106884

[5] Inteha A, Nahid-Al-Masood FH, Khan IA. A data driven approach for day ahead short term load forecasting. IEEE Access. 2022;**10**:84227-84243. DOI: 10.1109/ACCESS.2022.3197609

[6] Li H, Liu H, Ji H, Zhang S, Li P. Ultra-short-term load demand forecast model framework based on deep learning. Energies. 2020;**13**(18):18. DOI: 10.3390/en13184900

[7] Somu N, Gauthama Raman MR, Ramamritham K. A deep learning framework for building energy consumption forecast. Renewable and Sustainable Energy Reviews. 2021;**137**:110591. DOI: 10.1016/j.rser.2020.110591

[8] Kao Y-S, Nawata K, Huang C-Y. Predicting primary energy consumption using hybrid ARIMA and GA-SVR based on EEMD decomposition. Mathematics. 2020;**8**(10):10. DOI: 10.3390/math8101722

[9] Rodríguez-Pérez R, Bajorath J. Evolution of support vector machine and regression modeling in chemoinformatics and drug discovery. Journal of Computer-Aided Molecular Design. 2022;**36**(5):355-362. DOI: 10.1007/s10822-022-00442-9

[10] Kermanshahi BS et al. Artificial neural network for forecasting daily loads of a Canadian electric utility. In: 1993 Proceedings of the Second International Forum on Applications of Neural Networks to Power Systems. IEEE; 1993. pp. 302-307. DOI: 10.1109/ANN.1993.264330. Available from: https://ieeexplore.ieee.org/document/264330

[11] Géron A. Hands-on Machine Learning with Scikit-Learn, Keras and TensorFlow: Concepts, Tools, and Techniques to Build Intelligent Systems. 3rd ed. Sebastapol, CA: O'Reilly; 2023

[12] Sit M, Demiray BZ, Xiang Z, Ewing GJ, Sermet Y, Demir I. A comprehensive review of deep learning applications in hydrology and water resources [online]. Water Science and Technology. 2020;**82**(12):2635-2670. Available from: https://eartharxiv.org/repository/view/149/ [Accessed: January 26, 2024]

[13] Friedman J, Hastie T, Tibshirani R. Additive logistic regression: A statistical

view of boosting. The Annals of Statistics. 2000;**28**(2):337-407. DOI: 10.1214/aos/1016218223

[14] Touzani S, Granderson J, Fernandes S. Gradient boosting machine for modeling the energy consumption of commercial buildings. Energy and Buildings. 2018;**158**:1533-1543. DOI: 10.1016/j.enbuild.2017.11.039

[15] Google for Developers, Gradient Boosted Decision Trees. Machine Learning [Online]. Available from: https://developers.google.com/machine-learning/decision-forests/intro-to-gbdt [Accessed: January 26, 2024]

[16] Rufaida SI, Leu J-S, Su K-W, Haniz A, Takada J-I. Construction of an indoor radio environment map using gradient boosting decision tree. Wireless Networks. 2020;**26**(8):6215-6236. DOI: 10.1007/s11276-020-02428-7

[17] Ahmad N, Ghadi Y, Adnan M, Ali M. Load forecasting techniques for power system: Research challenges and survey. IEEE Access. 2022;**10**:71054-71090. DOI: 10.1109/ACCESS.2022.3187839

[18] Oprea S-V, Bâra A. Machine learning algorithms for short-term load forecast in residential buildings using smart meters, sensors and big data solutions. IEEE Access. 2019;**7**:177874-177889. DOI: 10.1109/ACCESS.2019.2958383

[19] Flores JJ, Garcia-Nava JL, Cedeno Gonzalez JR, Tellez VM, Calderon F, Medrano A. A machine-learning pipeline for large-scale power-quality forecasting in the Mexican distribution grid. Applied Sciences. 2022;**12**(17):17. DOI: 10.3390/app12178423

[20] AWS. Building AI-Powered Forecasting Automation with Amazon Forecast by Applying MLOps. AWS Machine Learning Blog; 2023. [Online]. Available from: https://aws.amazon.com/blogs/machine-learning/building-ai-powered-forecasting-automation-with-amazon-forecast-by-applying-mlops/ [Accessed: October 25, 2023]

[21] Azure, Forecast Energy and Power Demand - Azure Solution Ideas [Online]. Available from: https://learn.microsoft.com/en-us/azure/architecture/solution-ideas/articles/forecast-energy-power-demand [Accessed: October 25, 2023]

[22] Yin S-Y, Huang Y, Chang T-Y, Chang S-F, Tseng VS. Continual learning with attentive recurrent neural networks for temporal data classification. Neural Networks. 2023;**158**:171-187. DOI: 10.1016/j.neunet.2022.10.031

[23] Meisenbacher S et al. Review of automated time series forecasting pipelines. Wiley Interdisciplinary Reviews Data Mining and Knowledge Discovery. 2022;**12**(6):e1475. DOI: 10.1002/widm.1475

Chapter

Review of Control Methods for DC-DC Power Converters from a Traditional and Artificial Intelligence Perspective

Erik Martínez-Vera and Pedro Bañuelos-Sánchez

Abstract

In this work, a review of control methods for DC-DC power converters is performed. The review is divided into traditional control (TC) methods and Artificial Intelligence (AI) methods. Brief descriptions of the major methods for both categories are presented. Then, a review of the state-of-the-art is developed based on the proposed classification. A discussion about the benefits and drawbacks of each method is summarized with emphasis on the intended converter topology. In order to provide a comprehensive analysis and comparison of the main control methods, simulations at 1 MHz switching frequency of four basic of-the-shelve algorithms are implemented.

Keywords: DC-DC power converters, control systems, neural networks, fuzzy logic, artificial intelligence

1. Introduction

Power converters are key components of modern electronic devices. They work as interfaces between energy sources and loads. Telecommunications, personal computers, aircraft, and electric vehicles are some examples of applications. Converters are classified according to the type of input and output energy: AC-DC (rectifiers), DC-DC, DC-AC (inverters), and AC-AC. Research on the topic is focused on developing efficient devices with high power density and low cost. Different circuit topologies allow to reduce the number of components while increasing the output voltage range. Emerging semiconductor materials, such as silicon carbide (SiC) and gallium nitride (GaN), permit higher switching frequencies than traditional silicon (Si) power switches. Thus, a reduction in the size of the passive components can be achieved. Their application in dynamic systems requires precise regulation of the converter behavior during power transients. Large current and voltage overshoots could damage not only the converter components but also more importantly, the loads. Then, reliable and accurate control methods are needed for proper system operation and component safety. The objective of this work is to perform a review of control methods for DC-DC power converters.

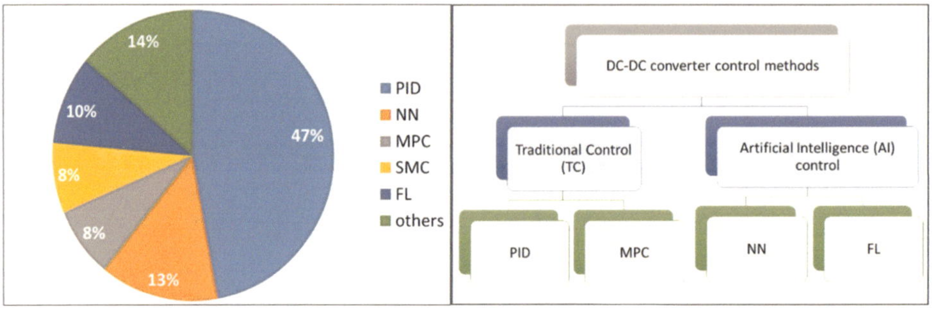

Figure 1.
(Left) Control Methods pie chart, (Right) Proposed classification of control methods.

In order to provide a meaningful classification, a sample of papers from various journals was analyzed. For each of them, the type of control and circuit topology were registered. The control type changes according to the circuit topology. Converters with elementary topologies require simpler control methods. More complex topologies involving a larger number of components need additional control loops to regulate power flow. Also, the control technique varies according to the number of inputs/outputs. Single input-output might work with a single feedback loop, while multi-input/output topologies incorporate additional loops to regulate simultaneous power flow from multiple sources.

The left side of **Figure 1** shows the type of control in a pie chart for the papers reviewed. The most popular methods found were proportional-integral-derivative (PID), neural network control (NN), and fuzzy logic control (FL). Other methods, such as model predictive control (MPC) and sliding mode control (SMC), were also applied. Based on this, a classification is proposed, dividing the methods into traditional control (TC) and artificial intelligence (AI) control. The right side of **Figure 1** shows the proposed flow diagram for the review. PID and MPC are the main methods grouped into the TC section, while NN and FL are the main methods for the AI section. Therefore, the contribution of this work is to provide insight into the control techniques currently applied for DC-DC converters. This will permit the identification of methods, scope, and limits of current works. Detecting the trends in the state-of-the-art will allow for defining directions for future research.

2. Traditional control methods

2.1 Proportional integral derivative

The most popular control method for industrial applications is PID. It is easy to implement and works even if a rigorous model of the plant is unavailable. Besides being the most in-demand method, it is also the benchmark for comparison of new approaches. Then, basic details about this technique have been included to serve as the benchmark comparison for the next techniques. The block diagram for this method is presented in **Figure 2(a)**. The input of the PID block e is defined as the difference between the output voltage of the converter V_{out} and the desired reference voltage

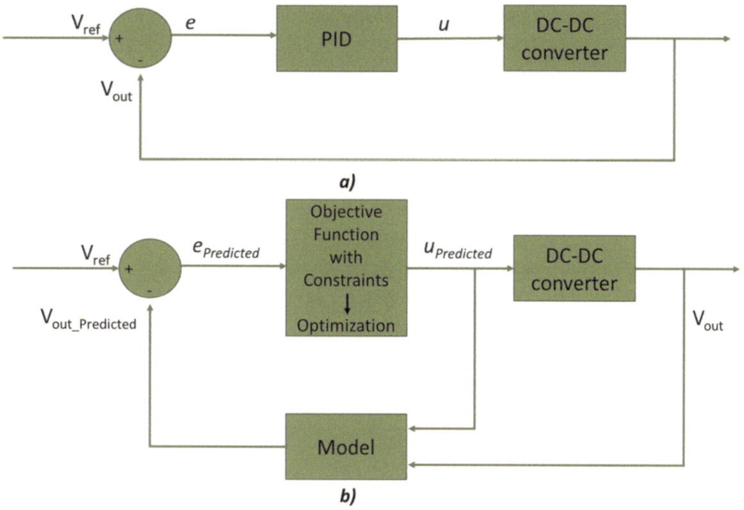

Figure 2.
Block diagram of (a) PID control and (b) MPC control.

V_{ref}. The control action u is defined in Eq. (1) with K_p the proportional gain, T_i the integral time and T_d the derivative time [1]:

$$u(t) = K_p e(t) + \frac{K_p}{T_i}\int_0^t e(t)dt + K_p T_d \frac{de(t)}{dt} \tag{1}$$

2.2 Model predictive control

For this approach, a model of the converter predicts a series of output voltages based on the current output and duty cycle. These predicted output values are compared against the desired reference to estimate the predicted difference. With these values, the control action that minimizes the objective function, within certain constraints, must be found. **Figure 2(b)** shows the block diagram of the MPC control. An example of the objective function has the form [2]:

$$J = \sum_{j=0}^{N-1}\left(\left\| V_{ref} - \left(V_{outk+j|k} - \hat{d}_k\right)\right\|_Q^2 + \left\|\Delta u_{k+j|k}\right\|_S^2\right. \tag{2}$$

where N is the prediction horizon, the subscript $k + j|k$ indicates prediction up to time j including time k, and the matrices Q and S are weighting matrices. The term $\hat{d}_k$ is defined as the difference between the output of the converter and the model prediction at time k. Additionally, a model of the converter is required to predict future values. Accurate models will deliver better predictions but at the cost of increased complexity. A quick comparison against the simple PID model in Eq. (1) shows the increased complexity of this approach, which will result in additional computational burden. However, when comparing the performance of both methods, MPC reaches the steady state faster without voltage overshoots [3, 4]. Also, the constraints on the objective function help control the current without additional feedback loops [5].

3. Artificial intelligence control methods

3.1 Neural networks control

System modeling for industrial processes is a complicated process with many variables and unknown parameters. Identifying a plant model is costly compared to the benefits. A general approach for system modeling based on existing data, easy to apply, and applicable in different scenarios is found in neural networks [2]. Even though more complicated NNs have been developed, the basic setup form of feed-forward neural networks is the most used for DC-DC converter control. Also, it has been proved that an NN with a single hidden layer and enough neurons can model any nonlinear function with random accuracy [6]. The approach in [7] is based on the principle of MPC using a NN with four inputs and one hidden layer with eight neurons to model the nonlinear behavior of the converter. Their control action is defined in Eq. (3),

$$
\begin{aligned}
H &= ReLU[W_{IH}X(k) + B_H] \\
u_{RL}(k+1) &= ReLU[W_{IH}X(k) + B_H]
\end{aligned} \tag{3}
$$

where H is the output from the hidden layer, $X(k) = [V_{in}(k), i_L(k), v_O(k), i_O(k)]$, W_{IH}, B_H, W_{HO}, B are the weight matrices, and ReLU is the activation function. A reward function is optimized to update the weights of the network and account for parameter deviations. Practical validation with an FPGA is provided; however, a comparison with other approaches is missing. **Figure 3(a)** shows a block diagram of this approach. The input of the NN is the sampled variables $X(k)$ to derive the duty cycle control action.

3.2 Fuzzy logic control

Fuzzy logic is a method in which a computer delivers results accounting for the natural ambiguity in real-world scenarios. For example, instead of providing an

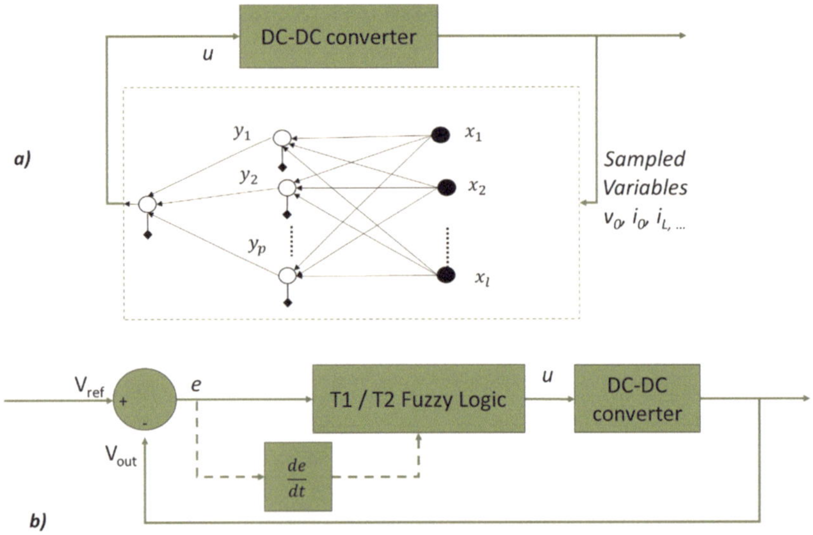

Figure 3.
Block diagram of (a) neural network control and (b) fuzzy logic control.

output of 0 or 1, it will provide an output in the interval [0, 1]. **Figure 3(b)** shows the block diagram for type 1/2 fuzzy logic control (T1/2-FLC); type 1 FLC is considered a special case of T2-FLC where the input membership functions are one-dimensional functions without considering uncertainties [8, 9]. The FLC process consists of a fuzzifier, a set of rules, an inference engine, and output processing. For T2-FL, the latter consists of a type reducer and defuzzifier. The advantage of this method is that the complex process of deriving a model of the system is avoided. Instead, knowledge of the system is expressed as a set of rules that relate the input to the desired control action.

4. Simulations

To better assess the advantages and disadvantages of each control method, simulations were performed for the elementary buck and boost topologies shown in **Figure 4**. The values of the components are $L = 29\ \mu H, C = 1.1\ \mu F, R = 5\Omega$ and $V_{in} = 48\ VDC$ which agrees with the ones reported in [6] for a switching frequency of 1 MHz operating under continuous conduction mode (CCM). The output voltage range for the buck converter is $V_{out} = 18 - 24\ VDC,$ and for the boost converter $V_{out} = 80 - 96\ VDC$. A high switching frequency allows for increasing the power density of the converter by reducing the size of the passive components. Still, it poses additional challenges for the control methods, which have a shorter amount of time to compute the required control action.

Of-the-shelve elementary algorithms from [10] were used to implement PID, MPC, NN, and FL for the elementary buck topology. For the PID, the small-signal transfer functions in [11] were used for tuning. Once the transfer function is obtained, traditional control analysis can be performed such as root-locus and frequency analysis. For this work, frequency analysis was implemented to vary the cut-off frequency and thus modify the transient response of the converter.

The averaged state-space model was used for the design of the MPC control. This model has been used extensively for converter control designs [3, 6, 11–14]. For the buck converter, it is defined as

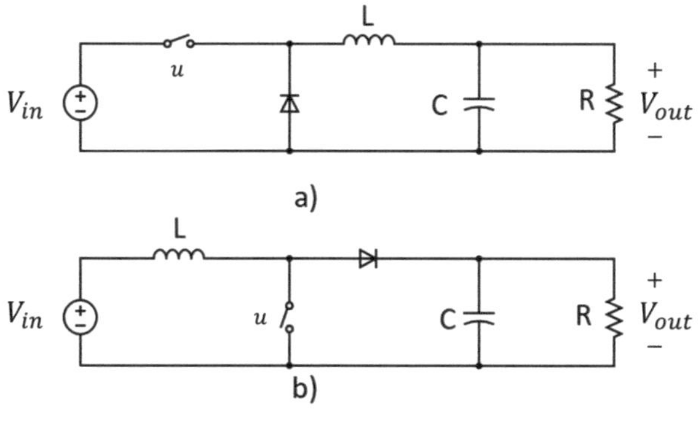

Figure 4.
(a) Buck and (b) boost DC-DC converter topologies. $L = 29\ \mu H, C = 1.1\ \mu F, R = 5\Omega$ *and* $V_{in} = 48$ VDC, $V_{outBuck} = 24$ VDC, $V_{outBoost} = 96$ VDC.

$$\begin{bmatrix} \dot{x}_1 \\ \dot{x}_2 \end{bmatrix} = \begin{bmatrix} 0 & -\dfrac{1}{L} \\ \dfrac{1}{C} & -\dfrac{1}{RC} \end{bmatrix} \begin{bmatrix} x_1 \\ x_2 \end{bmatrix} + \begin{bmatrix} \dfrac{V_{in}}{L} \\ 0 \end{bmatrix} u \qquad (4)$$

$$y = \begin{bmatrix} 0 & 1 \end{bmatrix} \begin{bmatrix} x_1 \\ x_2 \end{bmatrix}$$

And for the boost converter is

$$\begin{bmatrix} \dot{x}_1 \\ \dot{x}_2 \end{bmatrix} = \begin{bmatrix} 0 & -\dfrac{1-d}{L} \\ \dfrac{1-d}{C} & -\dfrac{1}{RC} \end{bmatrix} \begin{bmatrix} x_1 \\ x_2 \end{bmatrix} + \begin{bmatrix} \dfrac{V_{in}}{L} \\ 0 \end{bmatrix} u \qquad (5)$$

$$y = \begin{bmatrix} 0 & 1 \end{bmatrix} \begin{bmatrix} x_1 \\ x_2 \end{bmatrix}$$

where x_1 and x_2 are the inductor current i_L and the voltage of the capacitor v_C which, are the states of the model. R, L, and C correspond to the passive elements in the topologies of **Figure 4**. Finally, u is the control action. **Figure 5** shows the state space implementation in MATLAB/Simulink of the buck and boost converters, and **Figure 6** shows the open loop simulation results. Startup conditions and a step change in the duty cycle are included for both converters at 0.1 ms.

From **Figure 6**, overshoots can be appreciated during the startup condition and step change of the buck converter. Also, undershoot is observed for the boost converter at the duty cycle step change. Then, the objective of the control algorithms is to improve the transient response of the converter. That is, to reduce the time required to reach 90% of the output target (raise time) and the time to stay within 2% of the output voltage (settling time) while minimizing the overshoots. Formally, the raise time is measured as time it takes to increase from 10 to 90% of the output voltage, and the settling time can be selected to be within 2 or 5% of the output voltage. PID, MPD,

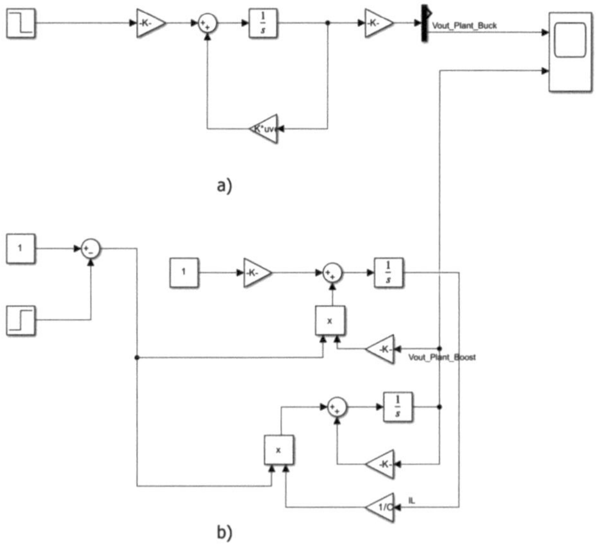

Figure 5.
State space model for (a) buck and (b) boost converters in Simulink.

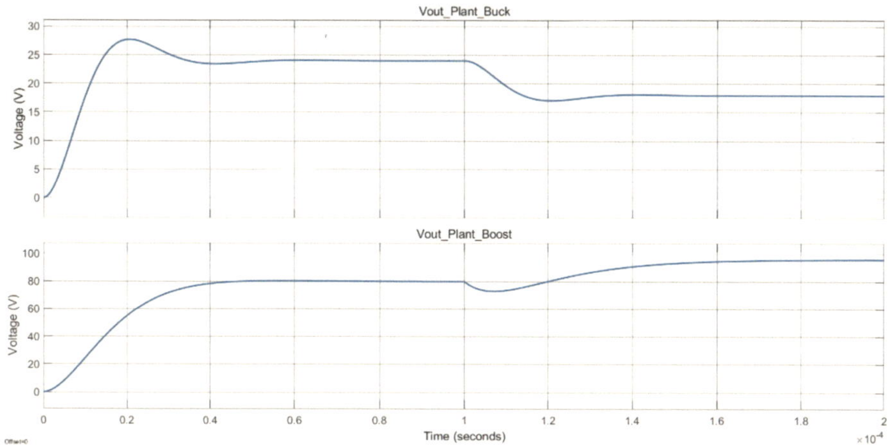

Figure 6.
Open loop simulation of (top) buck converter and (bottom) boost converter during startup and step change in duty cycle at 0.1 ms.

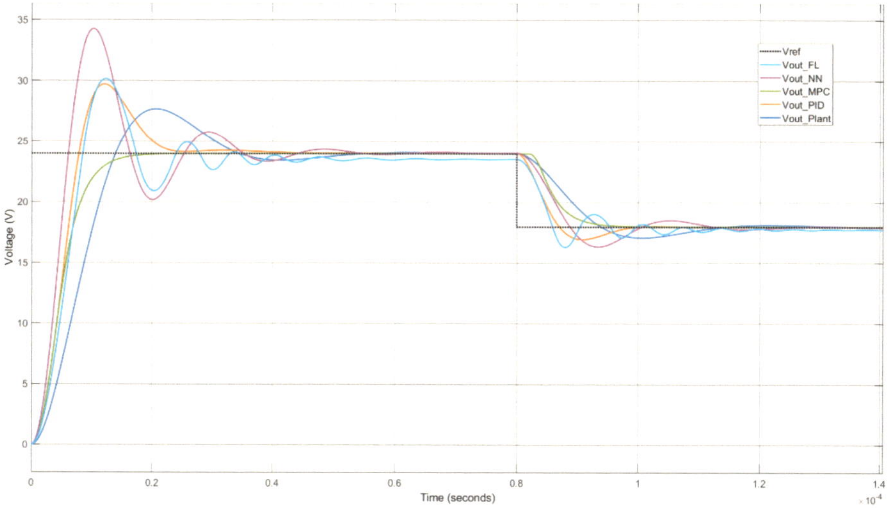

Figure 7.
Buck converter simulation results – startup conditions and step change in Ref. voltage at t = 0.08 ms.

NN, and FL control algorithms were implemented in a closed loop with the state space models of the converters in Simulink. Results of the simulations are presented in **Figure 7** for the buck converter and **Figure 8** for the boost case. All the mentioned figures include the PID, MPC, NN, and FL while also showing the open loop response and input reference voltage. The input voltage in all scenarios is 48 VDC, and the switching frequency is 1 MHz.

Figure 7 shows the startup conditions of the buck converter with an output voltage reference of 24 VDC, which corresponds to a duty cycle equal to 0.5. It can be observed that the open loop response (blue line) has the largest settling time with an overshoot of 15.13%, a settling time of 44.57 μs and a raise time of 9.1 μs. The traditional control methods show faster transient response: PID has a raise time of 5.165 μs and settling time of 21–7 μs while MPC achieves a settling time of 14.3 μs and

	Start up			Step change		
	Rise time	**Settling time**	**Overshoot**	**Rise time**	**Settling time**	**Overshoot**
	(μs)	(μs)	(%)	(μs)	(μs)	(%)
Open loop	9.42	31.15	15.3	9.4	29.1	5.1
PID	5.25	21.83	32.1	4.74	16.26	5.83
MPC	7.7	14.21	0	8	11.82	0
NN	3.9	41.13	42.7	6	28.5	9.22
FL	5.637	65.8	25.4	4.8	36.7	9.41

Table 1.
Performance metrics for buck converter control.

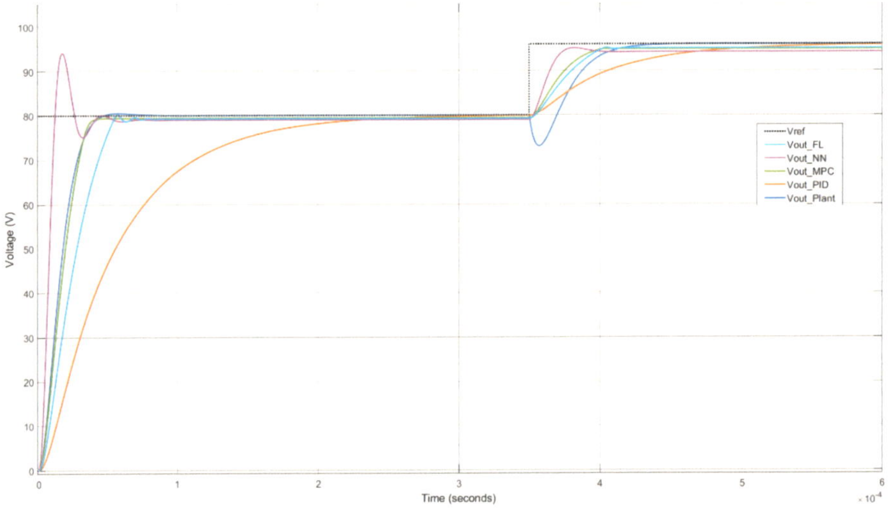

Figure 8.
Boost converter simulation results – startup conditions and step change in Ref. voltage at t = 0.35 ms.

raise time of 7.64 μs. Still, the fastest raise time of PID comes at the expense of an increased overshoot of 23.64%.

Also, in **Figure 7**, at time t = 0.8 ms, a step change in the voltage reference from 24 to 18 VDC shows the transient response of the converter. It can be appreciated that PID and MPC still provide faster response with MPC also showing zero overshoot. Then, MPC data were used to train the neural network. The raise time, settling time, and overshoot for all the control methods in buck mode are summarized in **Table 1**.

PID, MPC, NN, and FL were also simulated for the boost case. **Figure 8** shows the results of the simulation, and **Table 2** summarizes the performance metrics. It can be observed a very slow transient response for PID control as a result of a zero in the right-hand plane (RHP) for the boost control-to-output transfer function. Attempts to reduce the transient response with PID resulted in an unstable system. The fastest response during startup is achieved by the neural network control with 8.4 μs and settling time of 39.8 μs. However, this method also has the largest overshoot at 17.6%. The response of MPC and FL is similar to the open loop behavior and then of little improvement.

	Startup			Step change		
	Rise time	Settling time	Overshoot	Rise time	Settling time	Overshoot
	(μs)	(μs)	(%)	(μs)	(μs)	(%)
Open loop	25.4	40	0	36.4	55.4	−7.25
PID	113	209	0	113.13	115.34	0
MPC	25.8	37.8	0	38.7	43.4	0
NN	8.4	39.8	17.6	19.01	62.7	0
FL	39	53.45	0	39.7	46.7	0

Table 2.
Performance metrics for boost converter control.

Also, in **Figure 8**, a step change in the voltage reference is shown at t = 0.35 ms. The open loop response shows an undershoot of 7.25% and a settling time of 55 μs. PID, MPC, FL, and NN are all able to eliminate any under and overshoots. PID also presents the slowest transient response in this scenario with a settling time of 115 μs.

MPC presents a slightly higher rise time than the open loop case but with a reduction of 21% in the settling time and zero over or undershoot. Then, the data obtained with MPC control was used to train the neural network. The reference voltage and its difference from the output voltage of the converter were selected as inputs to the neural network. The output was the control action u which regulates the switching action in the converter. Feed-forward neural networks with at least a single hidden layer can represent nonlinear relationships found in data. Then, this type of network was selected for the neural network control. The topology of the network is a two-layer feed-forward with 13 neurons in the hidden layer, a sigmoid activation function, and a single output with a linear activation function. Levenberg-Marquard algorithm was used as the training algorithm, and the mean squared error (MSE) was used as the performance metric. The data were divided into 70% for training and 30% for testing and validation. Neural network control obtained the fastest rise time response in this scenario without any under or overshot. However, a settling time of 62.7 μs decreases its performance against the other methods. This type of control is strongly dependent on the quality of the training data and then the best-performing methods are selected for training.

The data from MPC were also selected for the neural network control in the buck scenario. MPC is also a popular method for DC-DC converter control as it is able to improve the transient response while minimizing the voltage overshoots. Constraints can be added to the model to maintain additional parameters under the desired limits, that is, maintain the output or inductor current within a desired operating zone to avoid damage to the components. However, this increases the complexity of the model and, hence, its computational burden. Simulations on the computer have no problem with this issue, but when transferring this model to real-time devices, such as microcontrollers or FPGAs, the computational capabilities are limited by the hardware resources. Also, increasing the switching frequency reduces the time required to perform the MPC procedure in real time. Explicit MPC has been proposed to solve this issue. The approach consists of performing comprehensive offline simulations covering a broad range of operating scenarios. Control actions are generated and stored in a database. Real-time operating is then reduced to identifying the appropriate control action for the current converter operating point [6].

A Mandami fuzzy logic control was designed to differentiate the reference and output voltage signals as the single input to the system. Three triangular membership functions defined the degree of membership of this input into three possible values: low, medium, and high. The control action u is the output of the fuzzy system. Three triangular membership functions (slow, medium, and fast) are defined for the output. The relationships between the input and output membership functions are defined by the inference rules. These rules are of the type *if-then*, that is, *if* the input is low *then* the output is slow. The centroid method is used to convert the output of the membership functions into a crisp value, which is sent to the PWM modulator of the DC converter.

One of the advantages of fuzzy logic control is that it can be designed without the need for a complex mathematical model. The inference rules express the knowledge required to control the system. These rules can be deduced from intuitive knowledge of the system under control or from practical experience. Nonlinear systems can be controlled with fuzzy logic without having to perform complex linearization procedures that are only valid around the linearization point. Deviations from this point will result in degradation of the performance of the control system. Then, multiple operating points must be taken into account, each with a respective linearization. For the basic topologies used in this work, mathematical analysis and linearization procedures are well defined. However, when DC-DC converter topologies increase in complexity, so does its linearization procedure. States in the model also increase with the addition of components. Then, fuzzy logic simplifies the procedure for control design of nonlinear systems.

A comparison of the transient response with MPC and FL for the boost converter shows that similar results can be obtained with both methods. Still, MPC requires a linear model of the system and the optimization of a cost function. Neural networks are able to control the nonlinear behavior of the converter based on training data. If data are available from measurements of a physical system, a mathematical model is also not required. However, if the data is obtained from simulations of a high-performing control method, such as MPC, formal analysis will be required. Even though PID had the slowest transient response, its implementation requires simple arithmetic operations, which makes it suitable for high switching frequencies in real-time devices such as microcontrollers or FPGAs. Then, even though high-performing control methods exist, their implementation in real-time will be limited to the hardware capabilities. An increase in computational burden will require more advanced resources, which will also result in higher prices.

5. Conclusions

In this work, a review of control methods for DC-DC power converters has been performed. The classification proposed divided the control algorithms into Traditional (PID, MPC, SMC, and others) and Artificial Intelligence methods (NN, FL, and others). To better compare each control method, simulations were performed for the elementary buck and boost topologies with four basic off-the-shelve control algorithms. PID is the most popular control method due to its relative ease of implementation. Control design and stability analysis by frequency or root locus can be performed once the converter has been linearized around an operating point. Deviations from the operating point will result in the degradation of the control algorithm. MPC showed superior performance than PID in the settling times with zero voltage overshoot. However, its design requires a linear model of the system and the optimization of a cost function. Nonlinear MPC can also be implemented at the cost of increasing computational burden.

Neural networks are able to extract the nonlinear relationships inherent in large amounts of data. Then, NN performance is strongly related to the input data used for training. More advanced networks can adapt online to changes in the parameters of the system, thus eliminating the need to retrain the network every time there are changes in the operating parameters. FL was able to control both topologies with a simple configuration of one input/output and three inference rules. However, its processing speed was considerably slower than the other methods, which is an issue for real-time implementation at high switching frequencies. The advantage of AI control over traditional methods is that they can perform without the need for a precise linear model of the system. However, traditional stability analysis requires rigorous mathematical study.

Acknowledgements

The authors would like to acknowledge the support from Universidad de las Américas-Puebla (UDLAP) and Consejo Nacional de Humanidades, Ciencias y Tecnologías (CONAHCyT) for their support throughout the course of this work.

Author details

Erik Martínez-Vera* and Pedro Bañuelos-Sánchez
Department of Computing, Electronics, and Mechatronics, Universidad de las Américas Puebla – UDLAP, Ex hacienda Sta. Catarina Mártir S/N, San Andrés Cholula, Puebla, México

*Address all correspondence to: erik.martinezva@udlap.mx

References

[1] Ogata K. Modern control engineering. In: Instrumentation and Controls Series. Upper Saddle River, New Jersey, USA: Prentice Hall; 2010

[2] Henson MA, Seborg DE. Nonlinear Process Control. Upper Saddle River, New Jersey, USA: Prentice Hall PTR; 1997

[3] Liu Z, Xie L, Bemporad A, Lu S. Fast linear parameter varying model predictive control of Buck DC-DC converters based on FPGA. IEEE Access. 2018;**6**:52434-52446. DOI: 10.1109/ACCESS.2018.2869043

[4] Zhu Y, Wang Z, Yang T, Dragicevic T, Bozhko S, Wheeler P. Model predictive control for DC offset suppression of dual active bridge converter for more-electric aircraft applications. In: IEEE International Symposium on Industrial Electronics. Vol. 2021. USA: Institute of Electrical and Electronic Engineers (IEEE); 2021. pp. 0-5. DOI: 10.1109/ISIE45552.2021.9576199

[5] Shi K, Bui T, Marco J. Optimal control of bidirectional active clamp forward converter with synchronous rectifier based cell-to-external-storage active balancing system. Journal of Energy Storage. 2021; **41**:102851. DOI: 10.1016/j.est.2021.102851

[6] Chen J, Chen Y, Tong L, Peng L, Kang Y. A backpropagation neural network-based explicit model predictive control for DC-DC converters with high switching frequency. IEEE Journal of Emerging and Selected Topics in Power Electronics. 2020;**8**(3):2124-2142. DOI: 10.1109/JESTPE.2020.2968475

[7] Chen J, Chen Y, Kang Y. A real-time self-learning control for megahertz GaN-based DC-DC converter. In: IEEE Workshop on Wide Bandgap Power Devices and Applications in Asia, WiPDA Asia. Vol. 2021. USA: Institute of Electrical and Electronic Engineers (IEEE); 2021. pp. 157-161. DOI: 10.1109/WiPDAAsia51810.2021.9656043

[8] Castillo O. Type-2 fuzzy logic in intelligent control applications. In: Studies in Fuzziness and Soft Computing. Vol. 272. Berlin, Germany: Springer; 2012. DOI: 10.1007/978-3-642-24663-0

[9] Chelladurai B, Sundarabalan CK, Santhanam SN, Guerrero JM. Interval Type-2 fuzzy logic controlled shunt converter coupled novel high-quality charging scheme for electric vehicles. IEEE Transactions on Industrial Informatics. 2021;**17**(9):6084-6093. DOI: 10.1109/TII.2020.3024071

[10] "Matlab Help Center," Documentation; 2025. Available from: https://www.mathworks.com/help/ [Accessed: February 24, 2025]

[11] Choi B. Pulsewidth Modulated DC-to-DC Power Conversion. Hoboken, NJ, USA: John Wiley & Sons, Inc; 2013. DOI: 10.1002/9781118772188

[12] Wang J, Xu D, Zhou H, Zhou T. Adaptive fractional order sliding mode control for boost converter in the battery/supercapacitor HESS. PLoS One. 2018; **13**(4):1-10. DOI: 10.1371/journal.pone.0196501

[13] Bennaoui A, Saadi S, Ameur A. Invasive weed optimization algorithm for tuning transitioning from type-1 to interval type-2 fuzzy logic controller for boost DC-DC converters. Journal Europeen des Systemes Automatises. 2020;**53**(2):195-202. DOI: 10.18280/jesa.530205

[14] Bennaoui A, Saadi S. Type-2 fuzzy logic PID controller and different uncertainties design for boost DC–DC converters. Electrical Engineering. 2017; **99**(1):203-211. DOI: 10.1007/s00202-016-0412-3

Chapter

Frame Reconstruction and Object-Tracking Assessment for Video Surveillance Anomaly Detection

Ernesto Cruz-Esquivel and Zobeida Jezabel Guzman-Zavaleta

Abstract

This chapter evaluates methodologies for frame reconstruction and object tracking in the context of video surveillance anomaly detection tasks. Contemporary models for anomaly detection in video surveillance are characterized by extensive resource consumption and come across challenges in identifying temporal anomalies. This research tests the combination of frame reconstruction and object tracking to enhance the detection of temporal anomalies without escalating resource usage. The investigation employs a top-heavy autoencoder for frame reconstruction, which is noted for resource efficiency. Additionally, a Gaussian Mixture Model, enhanced by data derived from the DeepSORT algorithm, is utilized to augment temporal anomaly detection. These methodologies are evaluated using well-known datasets, such as UCSD Ped2, CUHK Avenue, and ShanghaiTech, ensuring a robust comparative analysis. The findings suggest the feasibility of object tracking for temporal anomaly detection, but it has several limitations.

Keywords: anomaly detection, spatiotemporal features, reconstruction, tracking, video surveillance

1. Introduction

Video surveillance anomaly detection is a critical component in contemporary security systems, addressing a vast spectrum of applications from theft deterrence to predictive policing measures. Despite its critical role, the deployment of intelligent video surveillance anomaly detection systems is limited by considerable computational demands, which create barriers to real-time processing and broad implementation. Initiatives to decrease computation expenses while preserving system accuracy have widely fluctuated depending on the foundational model employed. Techniques such as convolutional autoencoders, generative adversarial networks (GANs) [1], variational autoencoders [2], and long short-term memory networks [3] have been explored. This work narrows its focus on the convolutional autoencoder paradigm.

Convolutional autoencoders adeptly condense video data into a more compact latent representation. The strength of this architecture lies in the semi-supervised nature of its training regimen, facilitating anomaly detection. Nonetheless, while useful at identifying spatial discrepancies, autoencoders have found temporal anomaly detection more challenging, representing a critical area needing further exploration.

Therefore, our research is at the intersection of these concerns: reducing computational costs and improving the temporal features analysis and extraction. Our dual contributions are as follows: we dissect the prevailing deficiencies in video anomaly detection frame reconstruction models using public video surveillance anomaly detection datasets, and we assess an innovative combination of a custom-designed autoencoder, termed the top-heavy autoencoder [4], in tandem with an object-tracking framework, to rectify these deficiencies.

We chose an object-tracking schema for its versatility and minimal computational footprint to remedy certain identified constraints. This tracking methodology is inspired by the velocity estimation technique delineated by [5]. The top-heavy autoencoder and the object-tracking algorithm are assessed individually and combined across several prominent public video surveillance anomaly detection datasets. The following sections will present our methods, experimental framework, and outcomes, accentuating how our approach propels the domain forward in computational efficiency and anomaly detection proficiency.

The remainder of this document is organized as follows: Section 2 surveys related works in video analysis techniques. Section 3 delineates the models under evaluation. Section 4 shows the dataset and metrics used in the assessment. Section 5 interprets the experimental findings, and Section 6 summarizes the conclusions drawn.

2. Related work

Classical methodologies in video anomaly detection (VAD) predominantly utilized handcrafted features, such as optical flow histograms and scale-invariant feature transformation (SIFT) for feature extraction [6, 7]. Anomalies were generally identified using reconstructive techniques [6] and one-class classification [8]. Yet, these traditional methods struggled to capture the intricate dynamics within video data, limiting their effectiveness in practical applications.

The proliferation of deep learning has precipitated a paradigm shift within the VAD domain. Initial methods that delved into deep learning for VAD employed two-dimensional convolutional layers to extract spatial features from static frames. However, these models often faltered in recognizing temporal correlations [9]. An approach involving the application of two-dimensional convolutions across successive frames was explored with models like R2D [9], but the results were not entirely satisfactory.

Subsequently, three principal approaches have crystallized in the deep learning-centric VAD landscape: reconstruction-based, prediction-based, and auxiliary classification-based methods. Autoencoders, pivotal in reconstruction-based frameworks, delineate the boundaries of normal behavior during training and endeavor to reconstruct frames, identifying anomalies by deviation from this baseline [9]. Prediction-based models strive to anticipate subsequent frames or flow maps, with certain architectures combining both strategies to enhance detection capability [10]. The introduction of self-supervised learning tasks, such as image colorization [11], puzzle assembly [12], and video frame prediction [10], has augmented these methods, especially in data-deficient contexts.

Temporal feature extraction remains a critical facet of anomaly detection, with several approaches incorporating it, particularly in predictive models. Despite the frequent use of optical flow as an auxiliary data type [1, 2, 4, 5], the pre-processing requirement of optical flow imagery is recognized as a resource-intensive drawback.

Recent advances have steered the focus toward object-level anomaly detection, analyzing individual objects within frames [5, 13, 14]. This refined focus has demonstrated substantial improvements over the conventional frame-level detection techniques. However, deep learning models, particularly those incorporating volumetric three-dimensional convolutions or multiple network streams, continue to impose significant computational demands. In recognition of these developments, we combine the computational moderation of autoencoders with the descriptive power of semantic attributes, such as object tracking, to assess the benefits the combination could provide for VAD models.

3. Description of designed models

3.1 The top-heavy design

In pursuit of an optimal balance between performance and computational efficiency, we adopt the autoencoder architecture conceived by Cruz-Esquivel and Guzman [4]. Termed the top-heavy autoencoder, this architecture is characterized by a hybrid convolutional scheme, integrating two-dimensional and three-dimensional convolutions in the encoder and reciprocating with three-dimensional and two-dimensional transpose convolutions in the decoder. This innovative multidimensional approach is strategically engineered to enhance the model's performance, particularly when confronted with the limited data scenarios prevalent in video surveillance anomaly detection environments.

The two-dimensional convolutional layers are strategically implemented to augment computational efficiency while concurrently reducing the tendency for overfitting, positioning the top-heavy autoencoder as a lighter and more resource-conscious alternative to the standard three-dimensional autoencoder. The encoder comprises a sequence of 2 two-dimensional followed by 2 three-dimensional convolutional layers. At the same time, the decoder architecture inversely mirrors this with a three-dimensional transpose convolutional layer followed by 2 two-dimensional transpose convolutional layers.

Activation functions are pivotal for non-linear transformations, and in this case, Leaky ReLU functions are employed across both the convolutional and transpose convolutional layers. The output normalization, essential for the model's functioning, is achieved via a final three-dimensional convolutional layer with a sigmoid activation function. Batch normalization is systematically applied after the decoding phase to enhance the model's learning efficiency.

This autoencoder is tailored to process sets of 16 frames, a choice reflecting the intention to capture temporal features robustly. It should be noted that while our structure shares its roots with the architecture delineated by Cruz-Esquivel and Guzman [4], it is distinctively adapted to PyTorch.

3.2 Object tracking for temporal anomaly detection

This work evaluates an object-tracking framework based on Ref. [5], which utilizes a velocity estimator to identify temporal anomalies. Our framework commences with the YOLOv3 object detector [15] as a foundation for detecting objects and delineating their bounding boxes. The object detector implements a selective threshold to reduce false positives, excluding certain useless detections. After this initial detection, the model employs Deep-Sort [16], an enhancement of the SORT (Simple Online and Real-time Tracking) algorithm that integrates deep learning for improved tracking fidelity. Deep SORT harnesses YOLOv3's precise detections to extract high-dimensional feature vectors through a convolutional neural network, facilitating accurate differentiation of objects even under occlusion conditions or similar movement patterns.

With the completion of feature extraction, Deep SORT utilizes a Kalman filter to forecast each object's track state in the following frame, a critical process for the continuity of identity, especially when occlusions or detector errors occur. The association process of Deep SORT is governed by a metric that synthesizes the Kalman filter's predictions and the deep-learned appearance features, globally optimized through the Hungarian algorithm to maintain existing tracks or create new ones. Deep SORT also integrates an intricate track management system, adeptly initiating and deleting tracks to balance persistent tracking and minimizing false positives.

Each tracked object is characterized by a bounding box, classification, and a unique ID, which facilitate the computation of variances for objects sharing the same ID. These variances across the bounding box points of the same ID are instrumental in establishing the object's trajectory:

$$\text{Difference} = \left[\left(x_{t+1} - x_t\right)\left(y_{t+1} - y_t\right)\left(w_{t+1} - w_t\right)\left(h_{t+1} - h_t\right) \right] \tag{1}$$

In tracking scenarios, the state of an object at any given time is represented by its bounding box coordinates, delineating the top-left and bottom-right corners. Specifically, at time t, the object's state is denoted as $\left(x_t + 1y_t + w_t + 1h_t\right)$, where $\left(x_t, y_t\right)$ are the coordinates of the top-left corner and $\left(w_t, h_t\right)$ are the coordinates of the bottom-right corner, encapsulating the last known position. At the subsequent time step, $t + 1$, the object's updated state is represented by $\left(x_{t+1},,, y_{t+1},,, w_{t+1},,, h_{t+1}\right)$, corresponding to the current position. A differential calculation is performed for each coordinate to quantify the spatial displacement of the object between these two time steps. The resulting displacement vector is computed as observed in Eq. 1, where each element of difference signifies the positional shift along the respective coordinate axis. This vector conveys not only the magnitude of movement but also directional information, which is critical for understanding the trajectory and velocity of the tracked object.

The differences derived from these comparisons are aggregated into arrays, with the potential for multiple arrays per frame corresponding to the number of objects and their appearances across frames. A Gaussian Mixture Model is then trained on these arrays, akin to the approach in the velocity estimator [5], to define and learn the typical trajectory variances. The model's ability to discern anomalous trajectories hinges on evaluating log-likelihood, with lower values indicating deviations from the established norm.

3.3 Combination of both approaches

The analytical outcomes from the top-heavy autoencoder and the object-tracking Gaussian mixture model are synthesized through an additive equation. This integration is performed using Eq. 2, where S represents the final anomaly score, RL the reconstruction loss from the top-heavy autoencoders, and $GMM(t)$ the log-likelihood obtained from the object-tracking Gaussian mixture model:

$$S(t) = RL(t) + GMM(t) \tag{2}$$

4. Dataset and metrics in the assessment

In this section, we delineate the datasets employed for the training and evaluation of the proposed architecture. Furthermore, the evaluation criteria employed to ascertain the efficacy of the models are explicated.

The research incorporated a compendium of three publicly accessible datasets tailored for anomaly detection, specifically the CUHK Avenue dataset [17], the ShanghaiTech Campus dataset [18], and the StreetScene dataset [19]. The characteristics of the aforementioned datasets are encapsulated in **Table 1** and detailed herein:

- The CUHK Avenue dataset is focused on pedestrian dynamics and primarily documents irregular pedestrian activities and the occurrence of uncommon objects such as bicycles. This dataset comprises 15,328 training frames and 15,324 frames for testing, each with a 640 × 360 pixels resolution.

- The ShanghaiTech Campus dataset spans 13 scenes, each distinguished by varying camera perspectives and illumination conditions. It encompasses pedestrian zones with anomalies generally involving non-pedestrian elements, such as cyclists or individuals maneuvering strollers. This dataset boasts 274,515 frames dedicated to training and 42,883 for testing, each measuring 856 × 480 pixels in resolution.

- The StreetScene dataset features a singular scenario that includes a variety of daylight conditions, capturing a bi-directional lane flanked by pedestrian pathways. Automobiles, bicycles, and pedestrians are designated as normal entities in this setting. The unconventional conduct of these entities defines anomalies. The dataset includes 56,847 frames for training and 146,445 for testing, with a resolution of 1280 × 720 pixels.

4.1 Evaluation metrics

The model's results are evaluated by deploying two metrics: The area under the curve-receiver operating characteristic (AUC-ROC) and the number of the model's parameters. These metrics are paramount in evaluating models' performance and size within video anomaly detection.

Dataset	Training frames	Testing frames	Total frames	Frames' size
CUHK Avenue	15,328	15,324	30,652	640 x 360
ShanghaiTech	274,515	42,883	317,398	856 x 480
StreetScene	56,847	146,445	203,292	1280 x 720

Table 1.
Number of training, testing, and total frames for the selected datasets.

The AUC-ROC metric is indispensable for binary classification tasks and measures the model's discriminative power. It is derived by plotting the True Positive Rate against the False Positive Rate, with the resultant curve encapsulating the model's classification effectiveness. An AUC value approximating 1 indicates exemplary classification accuracy, whereas a value near 0.5 suggests random guessing.

The number of trainable parameters corresponds to the variables adjusted during the training phase via backpropagation. This metric is emblematic of the model's complexity and its memory footprint. While the quantity of trainable parameters inherently influences the memory requisites, it does not directly correlate with the model's inference time or ultimate detection capabilities.

4.2 Experimental setup

The implementation of the architectures and the processing of the datasets delineated in the preceding sections were conducted utilizing PyTorch version 1.12.1 and OpenCV version 4.8.0. In the training regimen of all models, a consistent learning rate of $1 \times 10^{-}5$ was employed and operationalized through an Adam Optimizer, renowned for its efficiency in gradient descent optimization. The computational training exercises were performed on an Nvidia Tesla V100 graphics card with CUDA version 10.2.

5. Testing of proposed architecture

5.1 Top-heavy autoencoder results

The Top-Heavy autoencoder is evaluated separately on three different public datasets. Results can be observed in **Table 2**. For the Avenue and ShanghaiTech datasets, the autoencoder yields satisfactory results. The nature of anomalies in these datasets typically involves spatial inconsistencies, such as the presence of unexpected objects or unusual movements in pedestrian areas, which can be reasonably captured through reconstruction loss.

However, for StreetScene, the results suggest that reconstruction loss is not as effective. This could be due to the dataset's definition of normalcy, which includes cars and bicycles—elements not typically present in pedestrian datasets like Avenue and ShanghaiTech. In StreetScene, the challenge lies in detecting temporal anomalies, which are more subtle and harder to capture with just reconstruction loss of grayscale frames. In StreetScene, the presence of cars and bicycles is not inherently anomalous, but their unusual movement patterns are.

This outcome indicates that different datasets, due to their inherent characteristics and definitions of normal behavior, may require different approaches to anomaly

Dataset	Macro AUC-ROC
Avenue	79.7
ShanghaiTech	80.9
StreetScene	57.8

Table 2.
Macro AUC-ROC results of the frames reconstruction using top-heavy autoencoder.

detection. While frame reconstruction loss may work well for spatial anomalies, it may not be sufficient for datasets with a strong temporal component.

5.2 Anomalies results

The results from the anomaly detection using the top-heavy autoencoder in the four datasets were further examined to find the most challenging situations with the dataset. Three of them were found:

- Normal appearance with anomalous movement: The most common difficulty for the StreetScene dataset, highlighting the limitation of reconstruction loss in capturing anomalies solely based on movement when the appearance remains normal. For StreetScene, the issue arises with vehicles that typically belong in the scene but move anomalously.

- A movement without significant spatial displacement, such as unusual arm movements or stationary dancing, is observed in **Figure 1**. These anomalies affect fewer pixels, making them harder to detect with reconstruction-based methods.

- Anomalous movements with high displacement: Certain anomalies are characterized by movements that are not typical for the setting, such as walking in circles. While walking is a normal activity, the pattern or trajectory of the movement (i.e., circling) can be the anomaly.

- Various distances to the camera: The effectiveness of anomaly detection can be significantly impacted by the distance of the strange object from the camera. Objects closer to the camera tend to affect more pixels and, thus, are more likely to be detected by frame reconstruction loss methods.

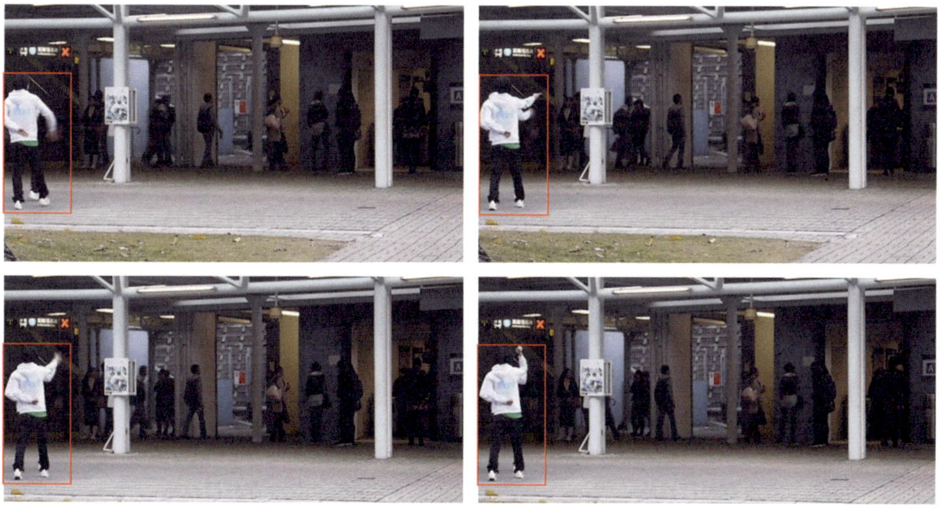

Figure 1.
Person in the red square is dancing throughout several frames without displacing to other parts of the frame.

Dataset	Macro AUC-ROC
Avenue	78.2
ShanghaiTech	72.4
StreetScene	54.6

Table 3.
Macro AUC-ROC results of the velocity estimator using GMM.

Dataset	Macro AUC-ROC
Avenue	80.3
ShanghaiTech	76.2
StreetScene	53.3

Table 4.
Macro AUC-ROC results of the combination of the top-heavy autoencoder and GMM.

5.3 Object-tracking model

The object tracking model was tested in the three selected public datasets. Frames were resized to 512 x 512 for object detection and the Deep-SORT application. The results can be observed in **Table 3**.

In evaluating anomaly detection models, the datasets from Avenue and ShanghaiTech demonstrated modest performance, with accuracies exceeding 70%. However, these outcomes did not signify an enhancement over prior benchmarks. Similarly, the binary classification results obtained using the StreetScene dataset remained almost random, consistent with earlier findings. This suggests that integrating a Gaussian mixture model (GMM) coupled with object tracking offers a cost-effective approach but falls short of being highly effective. The ineffectiveness of this solution in improving dataset outcomes can be attributed primarily to two factors. Firstly, the process of tracking objects in a dataset such as StreetScene is impeded by the substantial number of objects present within a single frame, leading to noisy tracking outputs. Secondly, this method does not adequately address the challenges posed by objects exhibiting minimal movement or those located at varying distances from the camera.

Combining the top-heavy autoencoder with the Gaussian mixture model (GMM) fails to enhance overall performance outcomes, as observed in **Table 4**. This lack of improvement is mainly attributed to the GMM's ineffectiveness when trained with object-tracking techniques. While there are instances, such as with the Avenue dataset, where this combination marginally betters results due to a slight enhancement in capturing temporal aspects, this trend does not consistently extend across all three datasets under consideration.

6. Conclusions

In conclusion, this study examines frame reconstruction and object-tracking techniques in video anomaly detection. It emphasizes the capabilities and constraints of the top-heavy autoencoder and the Gaussian mixture model, which is trained using object-tracking data. The analysis conducted with the selected datasets highlights the necessity of accounting for unique spatial and temporal anomalies in model development. The frame reconstruction approach is somewhat restricted by four specific types of anomalies: Normal appearance with anomalous movement, anomalous movements with low displacement, anomalous movements with high displacement, and various distances to the camera. These anomalies are sufficiently recurrent to affect the accuracy of the top-heavy autoencoder. While the Gaussian mixture model enhances the extraction of temporal features from the DeepSORT data, it does not entirely succeed in detecting all temporal anomalies within the datasets. Therefore, an alternative strategy is required to address all four shortcomings within a singular model, or a combination of models should be developed for a comprehensive solution to these challenges.

Author details

Ernesto Cruz-Esquivel*† and Zobeida Jezabel Guzman-Zavaleta†
Universidad de las Américas, Puebla, San Andrés Cholula, Mexico

*Address all correspondence to: ernesto.cruzel@udlap.mx

† These authors contributed equally.

References

[1] Ravanbakhsh M, Sangineto E, Nabi M, Sebe N. Training adversarial discriminators for cross-channel abnormal event detection in crowds. In: Proceedings—2019 IEEE Winter Conference on Applications of Computer Vision (WACV). Piscataway, NJ, USA: IEEE; 2019. pp. 1896-1904

[2] Fan Y, Wen G, Li D, Qiu S, Levine M, Xiao F. Video anomaly detection and localization via Gaussian mixture fully convolutional variational autoencoder. Computer Vision and Image Understanding. Amsterdam, Netherlands: Elsevier. 2018;**2020**(195):102920

[3] Duman E, Erdem O. Anomaly detection in videos using optical flow and convolutional autoencoder. Piscataway, NJ, USA: IEEE Access. 2019;**7**:183914-183923

[4] Cruz-Esquivel E, Guzman-Zavaleta Z. An examination on autoencoder designs for anomaly detection in video surveillance. Piscataway, NJ, USA: IEEE Access. 2022;**10**:6208-6217

[5] Reiss T, Hoshen Y. Attribute-based representations for accurate and interpretable video anomaly detection. Transactions on Machine Learning Research. 2025. Available from: https://openreview.net/forum?id=XL1N6iLr0G

[6] Hasan M, Choi J, Neumann J, Roy-Chowdhury A, Davis L. Learning temporal regularity in video sequences. In: Proceedings of the 2016 IEEE Conference on Computer Vision and Pattern Recognition (CVPR). Piscataway, NJ, USA: IEEE; 2016. pp. 733-742

[7] Lowe D. Distinctive image features from scale-invariant keypoints. International Journal of Computer Vision. Cham, Switzerland: Springer. 2004;**60**:91-110

[8] Schölkopf B, Williamson R, Smola A, Shawe-Taylor J, Platt J. Support vector method for novelty detection. In: Proceedings of the 13th International Conference on Neural Information Processing Systems. Cambridge, MA, USA: MIT Press; 1999. pp. 582-588

[9] Tran D, Wang H, Torresani L, Ray J, Lecun Y, Paluri M. A closer look at spatiotemporal convolutions for action recognition. Proceedings of the 2018 IEEE/CVF Conference on Computer Vision and Pattern Recognition (CVPR). Piscataway, NJ, USA: IEEE; 2018. pp. 6450-6459

[10] Zhao Y, Deng B, Shen C, Liu Y, Lu H, Hua X. Spatio-temporal AutoEncoder for video anomaly detection. In: Proceedings of the 2017 ACM International Conference on Multimedia. New York, NY, USA: ACM; 2017. pp. 1933-1941

[11] Larsson G, Maire M. Shakhnarovich G. Learning representations for automatic colorization. In: Computer Vision–ECCV 2016. Lecture Notes in Computer Science. Vol. 9908. Cham, Switzerland: Springer; 2016. pp. 577-593

[12] Noroozi M, Favaro P. Unsupervised learning of visual representations by solving jigsaw puzzles. In: Computer Vision – ECCV 2016. Lecture Notes in Computer Science, vol. 9910. Springer, Cham, Switzerland; 2016:69-84

[13] Hirschorn O, Avidan S. Normalizing flows for human pose anomaly detection. In: Proceedings of the 2023 IEEE/CVF International Conference on Computer

Vision (ICCV). Piscataway, NJ, USA: IEEE; 2023. pp. 13499-13508

[14] Wang G, Wang Y, Qin J, Zhang D, Bao X, Huang D. Video anomaly detection by solving decoupled spatio-temporal jigsaw puzzles. In: Computer Vision – ECCV 2022. Lecture Notes in Computer Science. Vol. 13670. Cham, Switzerland: Springer; 2022. pp. 497-514

[15] Redmon J, Farhadi A. Yolov3: An incremental improvement. arXiv preprint arXiv:1804.02767. 2018

[16] Wojke N, Bewley A, Paulus D. Simple online and realtime tracking with a deep association metric. In: Proceedings of the 2017 IEEE International Conference on Image Processing (ICIP). Piscataway, NJ, USA: IEEE; 2017. pp. 3645-3649

[17] Lu C, Shi J, Jia J. Abnormal event detection at 150 FPS in MATLAB. In: Proceedings of the 2013 IEEE International Conference on Computer Vision (ICCV). Piscataway, NJ, USA: IEEE; 2013. pp. 2720-2727

[18] Liu W, Luo W, Lian D, Gao S. Future frame prediction for anomaly detection – a new baseline. In: Proceedings of the 2018 IEEE/CVF Conference on Computer Vision and Pattern Recognition (CVPR). Piscataway, NJ, USA: IEEE; 2018. pp. 6536-6545

[19] Ramachandra B, Jones M. Street scene: A new dataset and evaluation protocol for video anomaly detection. In: Proceedings of the 2020 IEEE Winter Conference on Applications of Computer Vision (WACV). Piscataway, NJ, USA: IEEE; 2020. pp. 2558-2567

Chapter

Natural Coagulants from Fruit Waste for Surface Water Treatment for Human Consumption

Guillermo Díaz-Martínez, Ricardo Navarro-Amador, José-Luis Sánchez-Salas and D. Xanat Flores-Cervantes

Abstract

The provision of adequate clean water to communities is a problem throughout the world, particularly in areas where populations rely on surface water from rivers, dams, and streams for domestic use, which may contain pathogen microorganisms or other organic and inorganic contaminants. As a result, water from most surface water sources must be treated and purified to meet drinking water standards. Coagulation is one of the oldest processes that is still widely used in many water and wastewater treatment plants, using large quantities of inorganic/synthetic coagulants and flocculants that cause environmental problems and adverse effects on human health. Natural coagulants and flocculants obtained from fruit waste are presented here as an alternative option that is safe for human health, non-polluting, and biodegradable and generates smaller sludge volumes in comparison to synthetic options, among other advantages. The proposed solution provides a viable alternative for not only water treatment but also to fruit waste use, a significant global issue with economic, environmental, and social implications.

Keywords: coagulation and flocculation, surface water treatment, chemical coagulants, fruit waste, natural coagulant

1. Introduction

Water is vital for life, an integral element of natural ecosystems, and essential for sustaining and reproducing life on the planet. Water is essential not only to health but also to achieving poverty reduction, food security, peace, human rights, ecosystems, and education.

Fresh water for human consumption is mainly available in the form of surface sources (lakes, rivers, springs, streams, ponds, wetlands, swamps, dams, etc.), soil moisture, and shallow groundwater.

However, water scarcity has grown at an alarming rate and has now become a global concern for sustainable development. In addition to water being scarce, water can be contaminated and, if not treated properly, can cause water-borne illnesses such as cholera, dysentery, typhoid fever, or poliomyelitis, among other diseases [1]. As a result, every day, more than 700 children under the age of 5 years die from diarrhea related to unsafe water, sanitation, and poor hygiene. These infant deaths occur mostly in developing countries, which represents a significant burden for the scarce resources available for public health [1].

With few exceptions, surface water needs to be treated to remove particles, impurities, and color before it is distributed to consumers to meet regulatory quality standards. This treatment is done in a drinking water treatment plant, which is a complex system responsible for subjecting surface water to various processes to ensure that it is suitable for human consumption and use in the daily activities of the population. A physicochemical treatment, called coagulation-flocculation, is an essential part of these treatments to remove suspended and colloidal impurities in preparation for sedimentation, which is probably the second most used method after chlorination in water purification processes [2].

According to the latest 2021 inventory prepared by the Mexican National Water Commission (CONAGUA), there are 984 municipal drinking water treatment plants in Mexico, and a total of 377 plants (38.3%) are conventional clarification plants that include the coagulation-flocculation process. This indicates the indispensable role that the coagulation-flocculation process plays in the supply of clean water for human consumption.

Coagulation can be achieved by adding a coagulating agent, capable of neutralizing the electrostatic charges of colloids suspended in water, allowing their agglomeration to form easily settleable macro particles (flocs). To complement the coagulation process, flocculation is often used to facilitate the agglomeration of the destabilized particles, and it is carried out by a high molecular weight compound. During the flocculation process, the collisions between the submicroscopic small particles (micro flocs) expand the particle size, thereby becoming visible suspended particles. The size of pin flocs continues to increase, additional bonds are formed, and collisions occur, increasing the particle size [3]. The agents required for coagulating/flocculating processes are known as coagulants and flocculants.

The mechanisms by which the flocculation process occurs are mainly divided into two: (1) charge neutralization and (2) polymer bridging (as shown in **Figure 1**).

The charge neutralization mechanism can take place if the coagulant or polymer has an opposite charge to that on the surface of the colloidal particles (**Figure 1a**). In this case, the particle surface charge density is reduced by the adsorption of the coagulant, which results in the destabilization of this particle (repulsive electrostatic interactions are replaced by attractive forces). A certain variation of this process can be explained by the so-called electrostatic patch model, which involves partial neutralization of the charge in the presence of a polyelectrolyte of not very high molecular weight. This process involves incomplete neutralization, thus the formation of positively and negatively charged fragments on the surface of the same molecule. Such patches or "islands" with different charges cause attraction and the precipitation of neighboring clots. Flocs created in this way are more strongly bonded than in the case of ordinary charge neutralization. Another mechanism described in the literature is sweep flocculation, which is a different kind of charge neutralization process combined with the transfer of colloidal particles to the sludge (which resembles sweeping). Initially, the negative charge of colloid particles is neutralized, and then positively charged large aggregates (sweep flocs) are formed. The mutual attraction between aggregates and still-present colloidal particles leads to their attachment and settling.

During polymer bridging (**Figure 1b**), some polymer segments are adsorbed on the surface of colloidal particles, resulting in loops and tails suspended in the

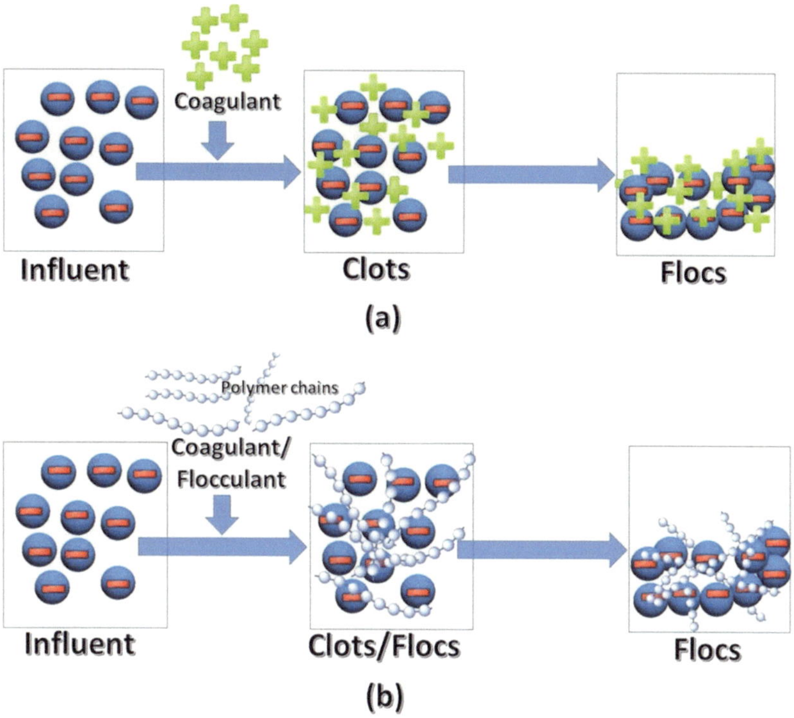

Figure 1.
Scheme of coagulation-flocculation mechanisms: (a) charge neutralization and (b) polymer bridging.

solution, which can attach to adjacent particles to form larger aggregates. The polymer can be adsorbed because of Van der Waals forces, hydrogen bond formation, or chemical reaction between the functional groups of the macromolecules and the colloidal particles. This mechanism is particularly effective for high molecular weight polymers, which form large aggregates that settle down at much larger velocities than the original suspended particles.

In addition to these two basic processes, an intermediate mechanism must be considered, which is adsorption, where, in the case of polymers, they provide a free surface to adsorb colloid particles and form larger particles that are easier to settle down [4]. In the case of the bridging flocculation process, a very large floc can be produced due to the adsorption of high molecular weight linear-chain compounds. For this reason, the adsorption process could be used to remove and recover heavy metals in wastewater, even at a low concentration [5].

2. Environmental and health risk impacts of conventional chemical coagulants/flocculants

Concerns regarding the toxicity of conventional coagulants/flocculants used in our everyday lives are emerging [4]. These concerns are mainly due to the health risks and environmental pollution potential of chemical compounds used in our daily lives as coagulants/flocculants. The most widely used coagulants and flocculants in the world are inorganic/synthetic chemicals, and the most common are aluminum-based and iron-based. Aluminum-based coagulants include aluminum chloride, aluminum sulfate, sodium aluminate, and aluminum polychloride (PAC). Iron-based coagulants include ferric chloride, ferric chloride sulfate, ferric sulfate, and ferrous sulfate. The efficacy of the mentioned chemicals such as aluminum and iron salts is well known, but there are disadvantages associated with their use, such as high operating costs, detrimental effects on human health, and the fact that they significantly affect the pH of treated water [6].

Aluminum-based coagulants are the most widely used around the world and present the greatest disadvantage for human health as there is strong evidence linking aluminum-based coagulants and residual aluminum present in drinking water to the development of Alzheimer's or other neurodegenerative diseases in human beings [7, 8]. Aluminum can also accumulate in bone tissue, causing excessive brittleness and softening, resulting in osteomalacia with osteodystrophy fracture. Atrophic lateral sclerosis and parkinsonism associated with dementia are the following diseases whose cause is largely associated with aluminum [9]. The presence of aluminum in treated water may be one of the main routes of ingestion of this element.

In addition, the utilization of synthetic chemical coagulants can lead to the production of large quantities of non-biodegradable sludge, generating important environmental problems that have to do with the final disposal of this sludge or its use as fertilizer for crops. Hydrolysis processes produce ionic species or amorphous metal hydroxides that can be toxic in sludge.

An example of the presence of aluminum in treated water is the water treatment plant in the city of Taxco, Guerrero, Mexico. The water processed at the plant comes from the Chontalcuatlan River, the Chacuhalco and Teneria springs, and the San Marcos and El Sombrerito dams. This plant uses, on average, between 10 and 20 tons of solid aluminum sulfate per month. Sánchez-Montoya and collaborators [10] reported an average concentration value of 1.82 mg/L, and a maximum of 2.68 mg/L of aluminum in the water leaving the Taxco treatment plant for distribution, which is more than nine times the limit of 0.2 mg/L set by the Official Mexican Standard NOM-127- SSA1–2021, "Water for human use and consumption. Permissible limits of water quality."

Another example of the extensive use of aluminum-based coagulant is the largest water treatment plant "Los Berros" in the State of Mexico (Cutzamala system) with a capacity of 20 m^3/s, and an average consumption of 2500 tons per month of aluminum sulfate. Although the output concentrations have not been previously reported, it is important to consider that part of that aluminum will be left in the residual form in the treated water, and the rest will be in the form of non-biodegradable sludge.

As an alternative to chemical coagulants, we have synthetic polymers, mostly polyacrylamides (PAMs), used for water coagulation-flocculation, to improve the performance of conventional clarification and for sludge dewatering processes. However, synthetic polymers also have disadvantages, they may increase head losses in granular media filters, biodegrade slowly, and are potentially toxic [11].

3. Natural coagulants from fruit waste

Fruit waste refers to the portion of fruits that is discarded or not used for consumption. This waste can occur at various stages of the food supply chain, from production and harvesting to distribution, retail, and household consumption (**Figure 2**). Fruit waste is a significant global issue with economic, environmental, and social implications.

Figure 2.
Generation of fruit waste.

Tropical fruits encompass a substantial amount of rind or skin, and seed which are disposed of as wastes. These fruit wastes are high in moisture content and rich in biodegradable organic ingredients, which result in the waft of an unbearable stench during decomposition. The long-term disposal of these wastes to the environment not only results in greenhouse gas emission during decomposition but also facilitates a breeding ground for bacteria, pests, and mice [12].

Wasted fruits represent a financial loss for farmers, retailers, and consumers, and therefore, reducing fruit waste can help save money and resources. Important alternative uses can be classified into food and nonfood applications. Food applications are their uses for obesity remedies, food additives, and edible coatings and films. The major uses of fruit wastes in nonfood applications are biosorbents to remove pollutants, heavy metals, and dyes from wastewater and biogas production [12].

As seen in **Figure 2**, in agriculture, during harvest and production, not all fruits that are grown and harvested meet the quality and appearance standards set by markets and consumers. As a result, a significant portion of fruits is left in the field or orchard, contributing to waste. This is often due to cosmetic imperfections or size discrepancies. It also occurs due to mechanical damage, climatic events, pests, and diseases. Fruits may also be wasted during transportation, storage, and at retail locations. Factors such as inadequate storage conditions, temperature fluctuations, and overstocking can lead to spoilage and waste. Additionally, fruits that do not sell due to esthetic imperfections may be discarded by retailers. Later, during the processing of fruits into various products, such as fruit juices, canned fruits, and dried fruits, waste is generated in the form of peels, cores, and other by-products. Subsequently, at the consumer level, fruits are frequently wasted due to over-purchasing, improper storage, or neglect. Fruits may rot or become inedible if not consumed promptly or if stored incorrectly. Finally, what is not consumed in homes, restaurants, and dining rooms, or what is not consumable, such as shells, cores, and leaves, becomes waste.

A novel, alternative use for fruit waste is to consider it a source for natural coagulants as an alternative to chemical coagulants used in water treatment. Because of their nature, many of the environmental and health problems associated with synthetic coagulants could be eliminated with natural coagulants, especially those related to aluminum-based coagulants.

Natural coagulants are considered safe for human health and less polluting than chemicals due to their biodegradability [13]. In addition, these coagulants are more accessible to emerging economies in developing countries than inorganic/synthetic ones and the treated water without harmful chemicals facilitates its use in systems such as irrigation of agricultural crops [14].

Natural coagulants are sustainable, respectful of the environment, can be produced locally, are less toxic than chemical coagulants, and are economical during their use as well, as the volumes generated are smaller and the disposal costs can be reduced [4]. In addition, because of their biodegradability, sludge utilization is an option. The further treatment of sludge through anaerobic digestion is applicable and produces valuable gas as a by-product. Natural coagulants have low toxic effects on the microorganisms in anaerobic digestion and will thus not interrupt their performance. The current trend also indicates the potential of the utilization of generated sludge from wastewater treatment as a soil conditioner/fertilizer. The organic content in sludge can be further composted and used as a soil conditioner, while the available nutrients will act as a soil fertilizer. The utilization of the produced sludge will not only reduce the potential harm to the environment but could also benefit the agricultural sector. **Figure 3** presents a summary of the advantages of using natural coagulants obtained from fruit waste:

To produce natural coagulants, it is suggested to utilize local resources to prevent future scarcity of raw materials. This option will benefit the local community economy by optimizing the potential of local resources. Biological waste from local

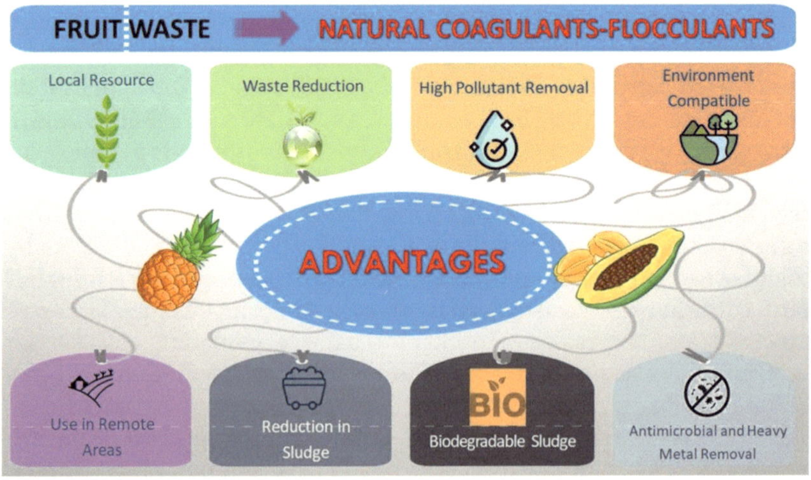

Figure 3.
Advantages of natural coagulants.

resources is normally abundant in amount and often has no alternative potential uses. For example, fruit waste can be considered as an alternative without competing with human consumption and product availability in food markets while reducing waste disposal processes and costs.

The use of local resources to produce natural coagulants is also an advantage for remote areas because the distribution of synthetic homologous hardly reaches these areas or is too expensive, while local resources are normally high and inexpensive. The application of natural coagulants in rural communities can, therefore, provide a viable economic alternative for drinking water treatment, improving their health and, subsequently, their living conditions.

The use of natural coagulants is not only environmentally friendly but also has reliable performance in terms of pollutant removal, which is comparable to the performance of synthetic chemical coagulants in terms of turbidity removal. In addition, natural coagulants have also been reported to have antimicrobial and heavy metal removal properties, which are effective in high-turbidity water [4]. For example, Hoa and Hue reported that the use of a moringa (*Moringa oleifera*) based natural coagulant removed 100% of the *Escherichia coli* in lake water while sustaining 99% reduction of turbidity and Chemical Oxygen Demand (COD) [15]. Conversely, poly-aluminum chloride coagulant only achieved 40–60% removal of these impurities. The high removal of bacteria could be associated with the high removal of suspended solids and the cationic protein of moringa that can potentially destroy *E. coli* cells.

4. Removal mechanisms in water treatment using natural coagulants obtained from fruit waste

The mechanisms of coagulation and flocculation are highly related to the various types of functional groups present in the structure of natural coagulants-flocculants [3]. The polysaccharides, protein, carbonyl, carboxyl, and hydroxyl groups are related to and influence the bridging mechanism in coagulation processes. The amines, carboxyl, and hydroxyl groups are correlated to the charge neutralization mechanism. The amides, amines, carbonyl, and methylene groups participate in charge neutralization and patch flocculation mechanism during coagulation-flocculation processes [3]. The existence of carboxyl, hydroxyl, and amine groups in the natural coagulant facilitates the further attachment of suspension to the aggregated particles, forming a bigger floc.

In addition, positively charged natural coagulants may assist the process by performing charge neutralization or bridging mechanisms, while negatively charged natural coagulant assists in the process by bridging, sweep coagulation or patch flocculation mechanisms.

Finally, an increase in positively charged functional groups in natural coagulants allows for more interactions with the negatively charged suspended particles, improving the binding capabilities.

Table 1 presents the examples of natural coagulants derived from fruits, detailing the activity of the coagulant, the functional groups detected, and the coagulation-flocculation mechanisms that work to purify water.

The hydroxyl group is the most mentioned functional group/compound that contributes to the coagulation process in the natural coagulants, followed by the amine and carboxyl groups and protein. Bridging is the most mentioned mechanism to occur during flocculation using natural coagulant followed by charge neutralization, patch flocculation and adsorption [3].

Natural coagulant	Compound/functional groups	Removal efficiency (%)	Mechanism	Reference
Banana	Alkanes group	Turbidity 98.14%	Adsorption	[3]
Musa paradisica (peel)	Amines group		Bridging	
	Carboxyl group			
	Cyanide group			
	Hydroxyl group			
	Carbohydrate			
	Protein			
Nopal	carbonyl group	Turbidity 70%		[16]
Opuntia ficus indica (mucilage)	carboxyl group	90% for iron and manganese	Adsorption	
	hydroxyl groups		Bridging	
Moringa	Amides group	Oil 81.89%	Charge neutralization Patch flocculation	[3]
Moringa oleifera (seed)	Amines group			
	Carbonyl group			
	Methylene group			
Papaya *Carica papaya* (seeds)	−OH of polysaccharides groups OH groups methyl group (C−OH) C=O carbonyl group of carboxylic acid −C−O group ether, ester, or phenol groups	88% turbidity	Charge neutralization	[17]
Nopal	N/A	77.8% turbidity	Adsorption	
Opuntia ficus indica (cladiode)		51.8% color	Bridging	[18]
		26.8% Chemical Oxygen Demand (COD)		
		31.0% Suspended solids (SS)		
		26.4% total solids (TS)		
		99.96% fecal coliforms (FC)		
Nopal		Turbidity 98%	Adsorption	[3]
Opuntia ficus indica (cladiode)	Amines group		Bridging	
	Carbonyl group			

Natural coagulant	Compound/functional groups	Removal efficiency (%)	Mechanism	Reference
	Carboxyl group			
	Ether group			
	Hydroxyl group			
	Methylene group			
Pineapple	N/A	88.40% turbidity		
Ananas comosus (leafs)		94.60% Chemical Oxygen Demand (COD)	Charge neutralization	[19]
Banana	hydroxyl group (−OH)	85.20% turbidity	Bridging	[20]
Musa paradisica (peel)	amino group (−NH_2)	76.50 5 color		
	saturated C−H in cellulose and hemicellulose			
	CO_2			
	carbonyl and benzene ring skeleton		Adsorption	
	methyl and methylene		Bridging	
	C−H, O−H and C−O in glycosides			
	R−OH in alcohol			
	C−H on the aromatic ring of lignin			
	C=H or aromatic rings.			
	C−O−C in pyranose ring skeleton			

Note: N/A information not available.

Table 1.
Examples of compounds/functional groups in plant-based natural coagulants and their mechanism.

Kurniawan et al. reported several major compounds that could perform as bio coagulants or bio flocculants inside the natural coagulants; for example, in the lentil vegetable (*Lens culinaris*), they reported hydroxyl and carboxyl groups, in the yucca or cassava amino acids, carboxyl group, and hydroxyl group, and in the moringa seed (*Moringa oleifera*) cationic protein, starch, glucose, fatty acids, and phenolic compounds [4].

5. Examples of the most common fruits used to obtain natural coagulants

The best-known and most studied primary coagulant of plant origin today is the moringa seed (*Moringa oleifera*). Moringa extracts consist of water-soluble cationic coagulant proteins. The seeds contain around 34% proteins, 15% carbohydrates and 15.5% lipids. This high protein amount ushered that these agents can be one of the most active coagulating compounds. The coagulation activity is carried out through the mechanism of charge neutralization and adsorption [21–23].

The nopal (*Opuntia ficus indica*) is a plant belonging to the Cactaceae family and commonly known as prickly pear in North America and nopal in México. It is native to the arid and semi-arid regions of México. Nopal cladodes are well recognized for their viscous liquid, mucilage, which is composed of polysaccharides. In addition, it has been reported that polygalacturonic acid is also present in the nopal and might be the main active compound which contributes alone to a high coagulating activity with more than 50% for turbidity in comparison with the cladode polysaccharide content (arabinose, galactose, rhamnose, and xylose), which enabled only 30% of turbidity reduction. It is known that the combination of these two active compounds accelerates the sedimentation of the sludge [23, 24].

Another tropical fruit belonging to the Anacardiaceae family is the mango (*Mangifera indica*). While the mango pulp is consumed, the pits are usually discarded as fruit waste. Several studies using mango pits as coagulants have reported that they exhibited significant coagulant activity attributed to their content. The kernel, commonly known as the pit or embryo, accounts for about 70% of the seed weight, and the coagulating active components are believed to be the proteins that account for 58% of the total seed content. In addition to the pit, mango peels have also been tested to act as natural coagulants and have great potential to replace chemical coagulants for water treatment [6, 23].

One of the most cultivated and well-known tropical fruits is the banana (*Musa paradisiaca*) due to its high nutritional value and its important addition to the diet. Worldwide, the production and consumption of this plant are very high, and therefore, previous studies have also tested the capacity of the waste produced as a natural coagulant and its potential for the removal of Chemical Oxygen Demand (COD), Total Suspended Solids (TSS) and turbidity from contaminated water. The results confirmed that the banana peel can reduce water turbidity with different efficiencies [25, 26].

Finally, the two last popular tropical fruits that have also been tested as natural coagulants include the pineapple (*Ananas comosus*) and papaya (*Carica papaya*). The pineapple leaves contain p-coumaric acid, flavonoids, tannins, bromelain, glycosides, proteins, and ascorbic acid, relevant for the coagulation process, showing effective COD, turbidity, and heavy metal removal from contaminated water [19]. Similarly, papaya seed extracts proved to be effective in the removal of total coliforms (up to 96.32%), as well as turbidity. Furthermore, the papaya seeds removal efficiency was good within the whole pH range but especially at neutral and slightly basic pH conditions. The presence of many functional groups indicates its potential to adsorb a wide range of contaminants [17, 27].

6. Potential for using fruit waste to purify surface water

To determine the quantity required for coagulant production, various tests were carried out in the laboratory to determine the average quantity of water treated with 1 kg of different types of fruits considering an application of 40 ppm of the natural coagulant (a concentration commonly used in surface water treatment) to moderately contaminated water. The results are presented in **Figure 4**.

Considering that in the year 2021, the global production of fruits was around 909.64 million metric tons, that it has steadily increased for the past 60 years [28],

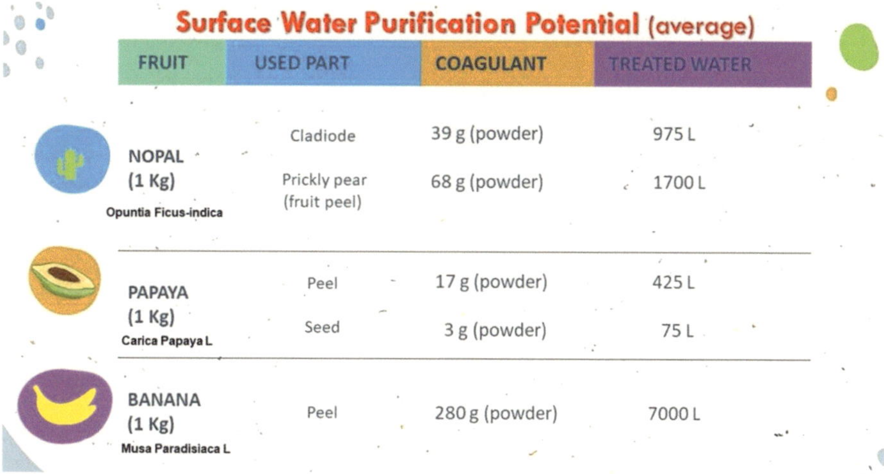

Figure 4.
Water purification potential with fruit waste.

and that it will continue to increase in the future, the utilization of plant-based waste from various fruits can be potentially used as a suitable source of natural coagulant with continuous supply in the future [29].

Taking the banana as an example, we will show that the availability of fruit waste is sufficient, taking into consideration that the banana is the most cultivated tropical fruit, with 114 million tons of banana produced in 2017 and approximately 5.6 million hectares of land are dedicated to banana production globally [30]. With data from the National Crusade against Hunger for Mexico, the [31] reported that 53.76% of banana production is wasted. Considering the above data related to the production and the amount wasted, as well as our estimates from lab experiments presented in **Figures 4**, 336,200 tons of natural coagulant can be obtained. Furthermore, with only 10% of this amount, water could be purified in one of the largest water treatment plants in Mexico, like the one in "Los Berros," in the State of Mexico, México.

7. Conclusions

The use of synthetic chemical coagulants in water purification represents serious environmental and health problems. Natural coagulants derived from fruits are a viable alternative with multiple advantages, with fruit waste being an important source. The large amounts of fruit waste make it an inexhaustible source for the requirements of water treatment plants in many parts of the world, mainly developing countries. The raw material for natural coagulants also has the advantage of being widely available, being plant-based waste from the food and agricultural sectors. The future commercialization of bio-based waste for natural coagulants might also boost the social and economic aspects of communities, especially rural communities, providing them with additional sources of income, as well as an additional solution to wastewater and biomass waste pollution. Furthermore, natural coagulants from renewable sources guarantee a continuous supply of raw materials. Natural coagulants and flocculants are becoming an interesting option that is effective, safe for human health, non-polluting, locally available, economical, and biodegradable.

Acknowledgements

Support provided by the Consejo Nacional de Ciencia y Tecnología (CONACYT) with a scholarship. Additional support was received from the Universidad de las Américas Puebla (UDLAP) through a graduate fellowship.

Conflict of interest

The authors declare no conflict of interest.

Author details

Guillermo Díaz-Martínez[1*], Ricardo Navarro-Amador[2], José-Luis Sánchez-Salas[2] and D. Xanat Flores-Cervantes[3]

1 Department of Civil Engineering, Engineering School, Universidad de las Américas Puebla, Puebla, México

2 Department of Chemical Biological Sciences, School of Sciences, Universidad de las Américas Puebla, Puebla, México

3 Department of Chemical Food and Engineering, Engineering School, Universidad de las Américas Puebla, Puebla, México

*Address all correspondence to: guillermo.diazmz@udlap.mx

References

[1] ONU. Organización de las Naciones Unidas (ONU). Objetivos de Desarrollo Sostenible. ONU; 2022. Available from: https://www.un.org/sustainabledevelopment/es/water-and-sanitation/

[2] Saritha V, Karnena MK, Dwarapureddi BK. "Exploring natural coagulants as impending alternatives towards sustainable water clarification" – A comparative studies of natural coagulants with alum. Journal of Water Process Engineering. 2019;**32**:100982. DOI: 10.1016/j.jwpe.2019.100982

[3] Kurniawan SB, Imron MF, Chik CENCE, Owodunni AA, Ahmad A, Alnawajha MM, et al. What compound inside biocoagulants/bioflocculants is contributing the most to the coagulation and flocculation processes? Science of the Total Environment. 2022;**806**:150902. DOI: 10.1016/j.scitotenv.2021.150902

[4] Kurniawan S, Abdullah S, Imron M, Said N, Ismail N, Hasan H, et al. Challenges and opportunities of biocoagulant/bioflocculant application for drinking water and wastewater treatment and its potential for sludge recovery. International Journal of Environmental Research and Public Health. 2020;**17**(24):9312. DOI: 10.3390/ijerph17249312

[5] Sukmana H, Bellahsen N, Pantoja F, Hodur C. Adsorption and coagulation in wastewater treatment – Review. Progress in Agricultural Engineering Sciences. 2021;**17**(1):49-68. DOI: 10.1556/446.2021.00029

[6] Zaidi NS, Muda K, Loan LW, Sgawi MS, Abdul Rahman MA. Potential of fruit peels in becoming natural coagulant for water treatment. International Journal of Integrated Engineering. 2019;**11**(1). DOI: 10.30880/ijie.2019.11.01.017

[7] Adelodun B, Ajibade FO, Ogunshina MS, Choi K-S. Dosage and settling time course optimization of *Moringa oleifera* in municipal wastewater treatment using response surface methodology. Desalination and Water Treatment. 2019;**167**:45-56. DOI: 10.5004/dwt.2019.24616

[8] Trindade S, Rouxinol MI, Nabais J. Evaluation of the potential of *Opuntia ficus-indica* cladodes as a natural flocculant for wastewater treatment through simple procedures. Journal of Ecological Engineering. 2021;**22**(5):249-257. DOI: 10.12911/22998993/135959

[9] Krupińska I. Aluminium drinking water treatment residuals and their toxic impact on human health. Molecules. 2020;**25**(3):641. DOI: 10.3390/molecules25030641

[10] Sánchez-Montoya G, Talavera-Mendoza O, Hernández-Flores G, Díaz-Villaseñor E, Ramírez-Guzmán AH, Galarza-Brito Z. Potentially toxic elements determination and chemical-microbiological analysis of potable water in Taxco de Alarcón, Guerrero. Revista Mexicana de Ciencias Geológicas. 2019;**36**(2):147-158. DOI: 10.22201/cgeo.20072902e.2019.2.927

[11] Lapointe M, Barbeau B. Substituting polyacrylamide with an activated starch polymer during ballasted flocculation. Journal of Water Process Engineering. 2019;**28**:129-134. DOI: 10.1016/j.jwpe.2019.01.011

[12] Cheok CY, Mohd Adzahan N, Abdul Rahman R, Zainal Abedin NH,

Hussain N, Sulaiman R, et al. Current trends of tropical fruit waste utilization. Critical Reviews in Food Science and Nutrition. 2016;**58**:335-361. DOI: 10.1080/10408398.2016.1176009

[13] Lugo-Arias J, Lugo-Arias E, Ovallos-Gazabon D, Arango J, de la Puente M, Silva J. Effectiveness of the mixture of nopal and cassava starch as clarifying substances in water purification: A case study in Colombia. Heliyon. 2020;**6**(6):e04296. DOI: 10.1016/j.heliyon.2020.e04296

[14] Czerwionka K, Wilinska A, Tuszynska A. The use of organic coagulants in the primary precipitation process at wastewater treatment plants. Water. 2020;**12**(6):1650. DOI: 10.3390/w12061650

[15] Hoa NT, Hue CT. Enhanced water treatment by *Moringa oleifera* seeds extract as the bio-coagulant: Role of the extraction method. Journal of Water Supply: Research and Technology-Aqua. 2018;**67**(7):634-647. DOI: 10.2166/aqua.2018.070

[16] Vargas-Solano SV, Rodríguez-González F, Martínez-Velarde R, Morales-García SS, Jonathan MP. Removal of heavy metals present in water from the Yautepec River Morelos México, using *Opuntia ficus-indica* mucilage. Environmental Advances. 2022;**7**:100160. DOI: 10.1016/j.envadv.2021.100160

[17] Amran AH, Zaidi NS, Syafiuddin A, Zhan LZ, Bahrodin MB, Mehmood MA, et al. Potential of *Carica papaya* seed-derived bio-coagulant to remove turbidity from polluted water assessed through experimental and modeling-based study. Applied Sciences. 2021;**11**(12):5715. DOI: 10.3390/app11125715

[18] Chuiza-Rojas M, Calderón SH, Vargas-Escobar J, Borja-Mayorga D, Renola L. Clarification of a dairy wastewater by coagulation using tuna (*Opuntia ficus indica*). Revista Ciencia e Ingeniería, Universidad de los Andes. 2019;**40**(3):245-252

[19] Husen R, Idris J, Wakimin ND, Mijim J, Luncha LESM, Diman JL, et al. High cod and turbidity removal in the treatment of polluted pond water using low dosage of pineapple leaf coagulant: A preliminary study. Journal of Asian Scientific Research. 2021;**11**(4):42-49. DOI: 10.18488/journal.2.2021.114.42.49

[20] Fu Y, Meng XJ, Lu NN, Jian HL, Di Y. Characteristics changes in banana peel coagulant during storage process. International journal of Environmental Science and Technology. 2019;**16**(12): 7747-7756. DOI: 10.1007/s13762-018-02188-0

[21] Nimesha S, Hewawasam C, Jayasanka D, Murakami Y, Araki N, Maharjan N. Effectiveness of natural coagulants in water and wastewater treatment. Global Journal of Environmental Science and Management, Online First. 2021;**8**(1):101-116. DOI: 10.22034/GJESM.2022.01.08

[22] Rianos K, Meza Leones MC, Mercado Martinez ID. Clarification of the water of wetlands using a mixture of natural coagulants. DYNA. 2019;**86**(209):73-78. DOI: 10.15446/dyna.v86n209.73687

[23] Othmani B, Rasteiro MG, Khadhraoui M. Toward green technology: A review on some efficient model plant-based coagulants/flocculants for freshwater and wastewater remediation. Clean Technologies and Environmental Policy. 2020;**22**(5):1025-1040. DOI: 10.1007/s10098-020-01858-3

[24] Ang WL, Mohammad AW. State of the art and sustainability of natural coagulants in water and

wastewater treatment. Journal of Cleaner Production. 2020;**262**:121267. DOI: 10.1016/j.jclepro.2020.121267

[25] Alwi H, Idris J, Musa M, Ku Hamid KH. A preliminary study of banana stem juice as a plant-based coagulant for treatment of spent coolant wastewater. Journal of Chemistry. 2013;**2013**:1-7. DOI: 10.1155/2013/165057

[26] Kiew PL, Chong KH. Development of fruit-based waste material as bioflocculant for water clarification. Journal of Mechanical Engineering, SI. 2017;**4**(5):1-10

[27] Yimer A, Dame B. Papaya seed extract as coagulant for potable water treatment in the case of Tulte river for the community of Yekuset district, Ethiopia. Environmental Challenges. 2021;**4**:100198. DOI: 10.1016/j.envc.2021.100198

[28] Statista. Statista. 2023. Available from: https://www.statista.com/statistics/262266/global-production-of-fresh-fruit/#statisticContainer

[29] Bahrodin MB, Zaidi NS, Hussein N, Sillanpää M, Prasetyo DD, Syafiuddin A. Recent advances on coagulation-based treatment of wastewater: Transition from chemical to natural coagulant. Current Pollution Reports. 2021;**7**(3):379-391. DOI: 10.1007/s40726-021-00191-7

[30] Food and Agriculture Organization of the United Nations. 2024. Available from: https://www.fao.org/economic/est/est-commodities/oilcrops/bananas/bananafacts/en/

[31] Food and Agriculture Organization of the United Nations. Food Losses and Waste in Latin America and the Caribbean, Bulletin No 2. Food and Agriculture Organization of the United Nations; 2015

Chapter

k-Mers Counting for Protein-Nucleic Acid Interactions Classification by Machine Learning Techniques

Jesús Cabello-Lima, Patricio A. Zapata-Morin and J. Horacio Espinoza-Rodríguez

Abstract

Protein interactions have significant importance in cell functions in all species. Proteins that interact with nucleic acids develop transcription and translation processes which are critical for expressing genetic information. Interactions of this kind are possible by the shapes that proteins conform in folding of the amino acids sequence, exposing interactive regions. With Machine Learning (ML), it is possible to identify interaction classes by processing amino acids in a suitable numeric representation. In this work, a classification method based on *k*-mers is conducted. Proteins are segmented by different number of amino acids (from 2 to 5 mers) to resemble protein regions or motifs. Obtained regions are transformed into a matrix representation used to train supervised machine learning methods. Five classifiers were considered: Support Vector Machine (SVM), K-Nearest Neighbors (KNN), Decision Tree (DT), Random Forest (RF), and Multilayer Perceptron (MLP). The best results were obtained using two consecutive segments of residues (2,2) in the MLP training, resulting in 84.3% accuracy, 84.4% precision, 84.3% recall, and 84.3% F1. The feature extraction of protein sequence regions is suitable for interaction classification. The results suggest that *k*-mers counting can be used for binary classification of protein-nucleic acid interactions, where the size of the protein region has a significant impact on the performance of the classifier.

Keywords: proteins, nucleic acids, *k*-mers, binary classification, machine learning, protein regions, encoding

1. Introduction

Biological sequence analysis is a major component of the proteomics and genomics research fields. In terms of protein interactions, biological sequences play a crucial role in many cellular processes, including antimicrobial resistance [1], protein–protein interactions [2] and gene expression [3, 4]. In this way, protein functions and regions of interest must be identified to facilitate experimental activities related to protein characterization. Hence, a successful biological test depends on selecting the correct candidate proteins, which reduces the time and costs incurred. Since proteins regulate living organisms by inhibiting, activating genes, or transporting information [5], it is vital to identify protein interactions with greater precision to improve disease treatment and better understand the functioning of organisms [6]. Although protein interactions can be identified, there are some limitations; for example, simply processing protein sequences takes a great deal of time and effort [7]. In addition, Machine Learning (ML) can be used to classify, identify, and process protein data. ML algorithms require substantial amounts of information to be effective, so protein groups with fewer incidences may be discarded. Protein sequences must be formatted in a specific manner in order for ML algorithms to be applied to them [8]. A numerical representation of sequences is required to encode and process them [9]. The FASTA format is the most commonly used method of representing protein sequences [10]. In this format, each amino acid is represented by a letter. The result is a text string that contains the positions of each amino acid. Protein and nucleotide sequences have been characterized successfully using this representation [11]. Even so, there is still a need for improvement in the processing of these text strings, taking into account the most relevant protein regions and restrictions. This is due to the fact that proteins do not have homogeneous sizes, despite the fact that they may perform similar functions. Protein sequence sizes can vary greatly. In the absence of regions of protein interest (domains or motifs), learning algorithms are limited in their ability to obtain information from proteins. Therefore, it is necessary to consider a numerical representation that can characterize all sequence sizes [12, 13]. An important challenge in classification is the incorporation of a biological context. Protein information can be translated into algorithms while maintaining a reference to the origin of protein information [9, 12]. This results in encoding that makes sense for both biological and computational elements. Thus, the encoding process is crucial for identifying sequences and ensuring replicability across all protein cases.

k-mers based methods are well known in the genomic field, mostly in applications to analyze large sequences of nucleic acids (DNA and RNA) [14, 15]. In protein analysis, these methods have demonstrated satisfactory results in processing specific regions of proteins, which can serve as inputs to machine learning classifiers (MLCs) with excellent results [9, 13–15]. The *k*-mers counting method establishes a matrix where word incidences are counted. In this case, protein sequences are divided into segments with variable lengths, which can be considered as words of the same length. Since protein regions have a strong biological significance, *k*-mers counting is an attractive method for classifying proteins using MLC [16]. In addition, *k*-mers encoding ensures the same number of features per protein, which is necessary for MLC training.

Accordingly, the main objective of this work is to evaluate different ranges of lengths into $k-$ mers counting method in order to train ML classification, as well as determine which range of *k*-mers performs better for identifying protein interactions, while varying the lengths of protein segments. The approach focuses on factors that are necessary for classifying protein information, such as the study of regions of interest in proteins and how these regions influence classification. Five classifiers were evaluated, namely SVM, KNN, DT, RF, and MLP. Thus, the classification results are presented and compared in order to assess the viability of the amino acids encoding method.

2. Methodology

2.1 Data extraction

Human proteins were extracted from UNIPROT database, using SPARQL tool [17]. For this study, we identified labels corresponding to specific protein interactions based on molecular function annotations, derived from reported proteins described by gene ontology (GO) information [18]. A preprocessing step is considered to filter and use labels from molecular functions. All proteins that bind DNA (deoxyribonucleic acid) or RNA (ribonucleic acid) were retrieved in FASTA format [10].

In the datasets obtained, proteins have lengths ranging from 16 to 5540 residues. We selected proteins whose sequences ranged from 55 to 800 residues, following the example provided in protein classification proposals [11]. This selection was based on the fact that UNIPROT considers chains of amino acids that contain fewer than 20 or 30 amino acids [19]. Additionally, the database incorporates protein complexes, which are multiple proteins in the same residue chain. In total, 7058 proteins were considered for interactions. The DNA and RNA classes contain 4323 and 2735 amino acids, respectively. At this point, we have unbalanced classes among the categories considered. A balance between the two classes must be maintained. However, by excluding proteins from one class, sequence sizes may be biased toward larger or smaller chains.

To balance both classes (DNA and RNA interactions), considering the limitations of the sequences and not intervening too much in the specific selection of sequence sizes, a uniform random sample of 2735 elements was generated from the DNA-interacting group, made with the random function of Python [20]. This allowed for an equal number of occurrences in both classes while maintaining a more homogeneous distribution among sequence sizes. **Figure 1** shows a comparison of the two classes (balanced and unbalanced), where Y represents the number of proteins and X represents the number of amino acids. **Figure 1a** illustrates the unbalanced class of proteins where it is clear that protein-DNA interactions are more frequent. Similarly, **Figure 1b** illustrates the balanced class of proteins, in which there is an equal relationship between proteins that interact with DNA or RNA.

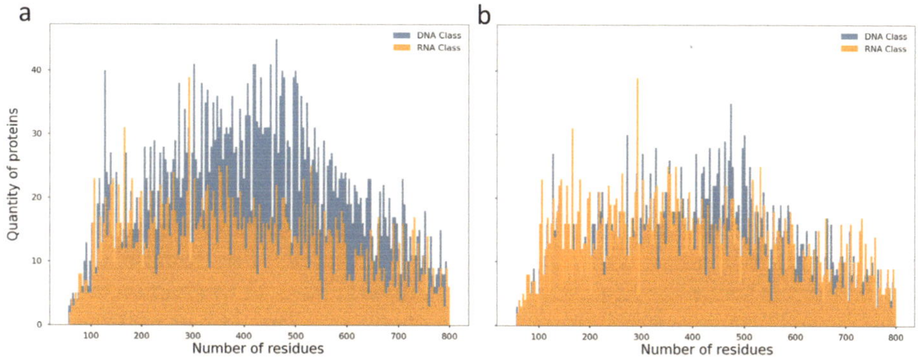

Figure 1.
Class distribution histogram of proteins that interact with DNA or RNA: (a) unbalanced, and (b) balanced.

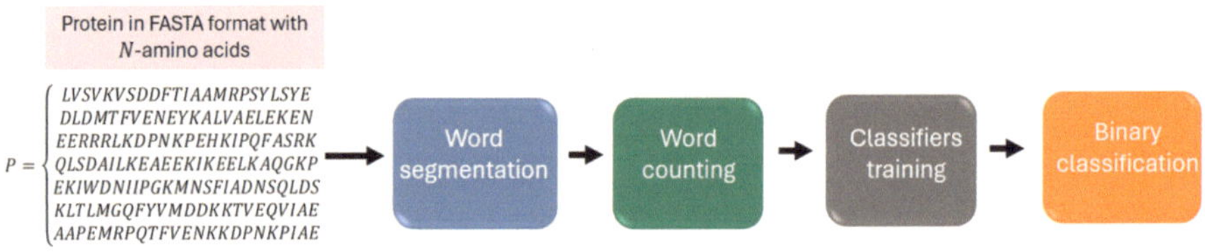

Figure 2.
k-mers counting implementation for protein encoding and classification.

2.2 *k*-mers-based protein encoding

A diagram of the protein sequence processing flow is shown in **Figure 2**. This method begins with a text string in FASTA format, where each sequence is processed using word segmentation. In order to convert the sequence into words, the entire sequence was looped through, and all possible occurrences were sampled based on the word size of six characters since that was the optimal value. Now, it is important to mention that the protein regions (e.g., domains) contain approximately 50 to 100 amino acids [21]. However, due to the size of the proteins in the dataset, smaller sizes of regions of interest were considered. The protein sequences were then divided into segments of six residues according to their optimal character value.

The CountVectorizer method [22] from the Scikit-learn library in Python was used to count the occurrences of words. There were a number of combinations of words considered, with the sequence ranging from 2 to 5 consecutive words that must occur within the sequence in order to be counted. The purpose of combining word occurrences within a sequence and counting them is to uncover intrinsic relationships between regions and their potential impact on protein characterization. Protein sequences are transformed into a segmented representation, where a window of n residues is scrolled m residues to separate sequence segments, in this case $n = 6$ and $m = 1$. Those segments are counted and placed into a matrix. It represents the incidence or occurrence of segmented regions of a protein. The matrix contains the same number of predictors for any protein in both classes (interactions with DNA and RNA). After counting all word occurrences in a matrix, the classification process is initiated. By using this matrix, the same number of features can be provided to the classification algorithms. There were five different classifiers considered (SVM, KNN, DT, RF, and MLP). In this study, we present a comparison of the results in terms of performance based on both the encoding used and the parameters selected for word counts.

2.3 Classification methods

In machine learning, classifiers refer to models or approaches for training and testing predictions, in which the label of a class is estimated, for example, whether an encoded protein sequence (e.g., counting k-mers) will have an interaction with a type of nucleic acid (DNA or RNA). Now, for this, the comparison of various learning classifiers is important in order to evaluate the viability of the study encoding method, which will allow us to determine the behavior in the training of the data (for example, the protein sequence encoded). This can be done by evaluating the reproducibility in class classification. To achieve this, we used five well-known supervised classification methods: Support Vector Machine (SVM), K-Nearest Neighbors (KNN), Decision Tree (DT), Random Forest (RF), and Multilayer Perceptron (MLP). The following paragraphs provide a brief description of each.

2.3.1 Decision tree

A decision tree (DT) is a classification algorithm based on if-then rules, which defines classes of inputs based on a range of values [23]. The method automatically stratifies data based on the features provided. DT is analogous to expert discrimination and allows for a deeper analysis of the training process if necessary. The method can be applied to binary and multi-class problems, making it a general-purpose classifier. Information can be quickly classified using decision trees. One of the most significant limitations of DT is the order in which information is provided, which can result in different results every time. In spite of this, the results are relatively stable.

2.3.2 Support vector machine

A support vector machine (SVM) is also a machine learning technique for classification and regression. This algorithm defines a hyperplane or group of hyperplanes in a high-dimensional space. Therefore, the higher the margin between the hyperplane and the nearest training points in each class, the more accurate the classification. Proteomic classification problems have been solved using SVM [24], making it suitable to be applied to this problem as well. In general, SVM is used to solve a minimization problem defined by the following equations:

Minimize

$$w, b, \xi \ \frac{1}{2} w^T w + C \sum_{i=1}^{N} \xi_i \tag{1}$$

subject to

$$y_i \left(w^T x_i + b \right) \leq 1 - \xi_i \tag{2}$$

where

$$\xi_i \geq 0 \tag{3}$$

Here, the training vectors are represented by $X_i \in \mathbb{R}^p$, being the encoded proteins with *k*-mers counting. y_i represents the class of proteins to be categorized, where 1 represents proteins that interact with DNA, and –1 represents proteins that interact with RNA. Variable *i* represents the number of proteins in the dataset. Vector *w* is perpendicular to the separation hyperplane, and ξ_i is the slake variable to consider cases where training data can be allowed to be in the separation margin. Finally, a cost parameter C is introduced to set a trade-off in training error [25].

2.3.3 K-nearest neighbors

KNN is a supervised learning algorithm that uses proximity measures to make classification and prediction, assuming that related inputs are close to one another [26]. The Euclidean distance between points is commonly used to determine the proximity of neighbors (see Eq. (4)). KNN stores training data in memory, and often, small datasets are more appropriate to use; however, it is also capable of handling large amounts of data:

$$d(X_i, X_j) = \sqrt{\sum_{r=1}^{p} (X_{ri} - X_{rj})^2} \tag{4}$$

Here, KNN calculates the Euclidean distance d between inputs (numerical values converted from proteins). The distance estimate is based on the features of proteins X_{ri} and X_{rj}. p refers to the number of features.

2.3.4 Random forest

Random Forest (RF) is a learning classification method that uses decision trees [27]. There are a defined number of decision trees in this forest, and each decision tree classifies a subset of the data. After subsets training is complete, classes from each DT are recorded. As a result of classification, a majority decision determines the output class. **Figure 3** illustrates the basic configuration of Rando Forest, where several DTs receive the same information to achieve the best results. The green nodes represent the different ways in which data is classified for each tree. An output is produced by considering each best result.

In this study, 100 DTs were considered based on the default parameters of the Scikit-Learn function.

2.3.5 Multilayer perceptron

Multilayer perceptrons (MLPs) are the architecture of deep learning, also known as feed-forward artificial neural networks. A perceptron resembles a biological neuron and is capable of separating information in two-dimensional feature spaces [28]. The problem of separability can be solved by a new architecture of multiple perceptrons when more dimensions are taken into account. A representation of the architecture of a neural network is shown in **Figure 4**. A basic architecture consists of an input layer X, a hidden layer H, and an output layer Y. The layers are interconnected, passing information forward, and adjusting weights w_j in response to features. For binary classification tasks, a perceptron can be considered as a linear classifier. Therefore, perceptrons are represented in the output layer by a discrimination function a, which depends on the input x_j, the weight w_j, and a bias term b.

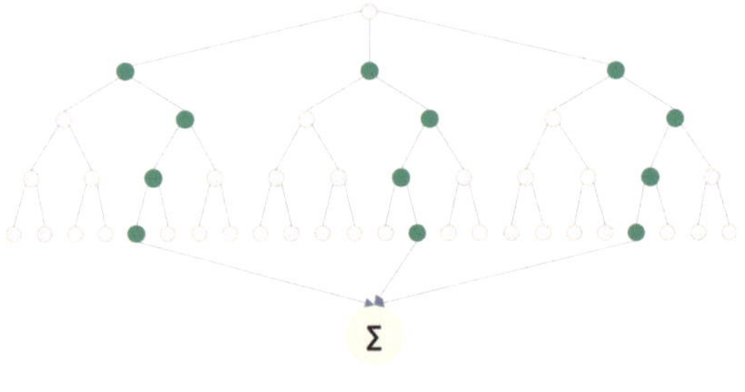

Figure 3.
A basic representation of a Random Forest.

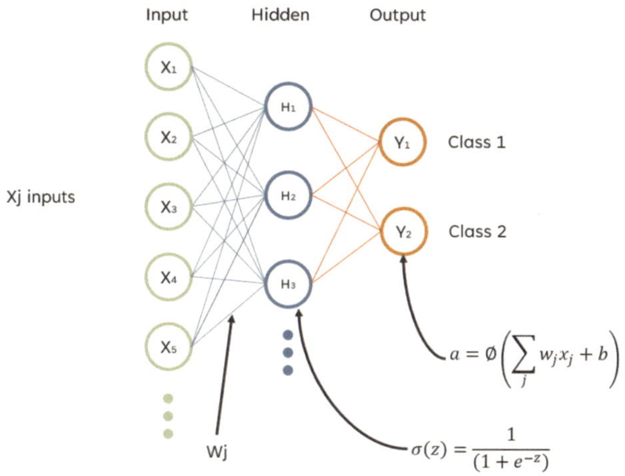

Figure 4.
An architecture based on multilayer perceptrons.

For the purpose of the present study, we implemented only three layers: Input, Hidden, and Output (similar to that shown in **Figure 4**). The input layer is composed of a different number of neurons depending on the number of features available. A hidden layer is composed of 100 perceptrons. The output layer is based on two neurons since it is a binary classification problem. All neurons follow the sigmoid function as the activator, and the neurons are fully interconnected, resulting in a fully connected neural network.

2.4 Performance metrics

Statistical metrics are used to evaluate the performance of classifiers [29] by calculating the number of correct and incorrect classifications based on four possible outcomes:

- True positive rate (TP): Indicates the proportion of positive samples correctly classified.
- True negative rate (TN): Indicates the proportion of negative samples correctly classified.
- False positive rate (FP): Indicates the proportion of positive samples incorrectly classified. Also known as Type-I error.
- False negative rate (FN): Indicates the proportion of negative samples incorrectly classified. Also known as Type-II error.

Figure 5 displays how the four parameters (TP, TN, FN, and FP) can be represented in a tabular format, known as a confusion matrix. In this study, -1 indicates RNA-protein interaction, while 1 indicates DNA-protein interaction.

*A confusion matrix is presented in **Figure 6** for each of the five classifiers (SVM, KNN, DT, RF, and MLP) for which two consecutive words (2-Mers) were counted.* In the case of the MLP classifier, we see that 564 proteins were predicted to interact with RNA and

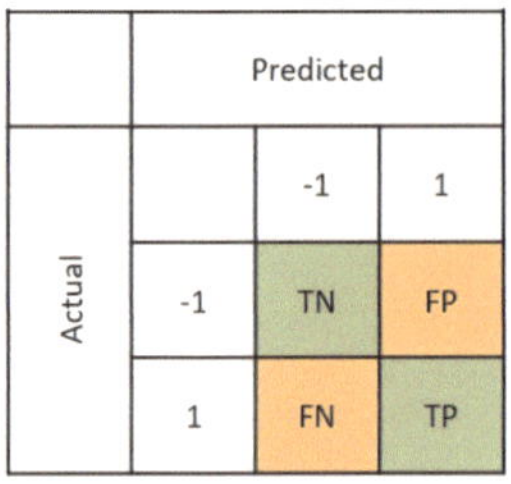

Figure 5.
An illustration of a confusion matrix for binary classification.

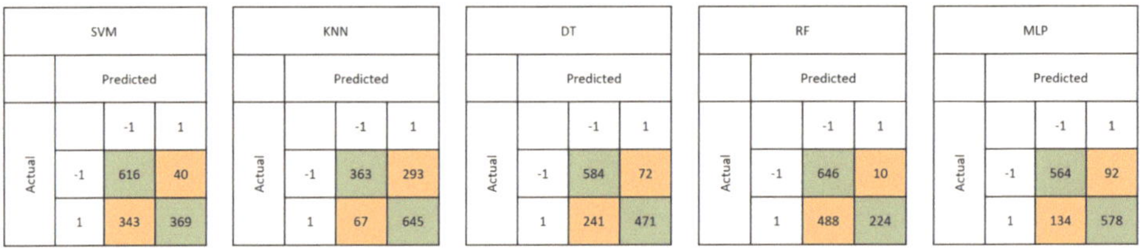

Figure 6.
Comparison of confusion matrices for the five classifiers based on 2,2 vector counts.

actually did so, while 134 proteins were predicted to interact with RNA but did not actually interact, and at the same time, 92 proteins were predicted to interact with DNA, but did not, and 578 proteins were predicted to interact with DNA, but actually did.

There are four common measures reported as a result of confusion matrix analysis: Precision, Accuracy, Recall, and F1 score. Often, these evaluation metrics are used to assess the performance of learning classifiers [23], as shown in this study:

Accuracy is the number of correct classifications divided by the total number of classifications, which is defined as:

$$Accuracy = \frac{TP + TN}{TP + TN + FP + FN} \tag{5}$$

Precision is defined as the ratio between the true positives and the total positives, and it is calculated as follows:

$$Precision = \frac{TP}{TP + FP} \tag{6}$$

Recall, also known as sensitivity or hit rate, is the ratio of true positives to all the positives. Here is how it is calculated:

$$Recall = \frac{TP}{TP + FN} \tag{7}$$

F1 score can be used to compare precision and recall metrics. In other words, it represents the harmonic mean of precision and recall. The F1 score is a good indicator of imbalanced classification problems. This can be expressed as follows:

$$F1\,score = 2 * \frac{Precision * Recall}{Precision + Recall} \tag{8}$$

2.5 Computational resources

Python language was used in this work, with Scikit learn package. All classifiers were implemented using Scikit learn. *k*-mers counting encoding was applied using the "Count Vectorize" function of Scikit learn. All experiments were conducted on a computer running the Ubuntu 22.04.3 LTS (Jammy Jellyfish) distribution with an Intel® Core™ i7-9700K CPU (3.60 GHz × 8) processor, with 64 GB of DDR4 RAM in dual channel configuration.

3. Results

3.1 Feature extraction from *k*-mer segments

The number of features in each protein sequence is determined automatically by segmenting the proteins into words. For this purpose, each protein was divided into six letter words, then the occurrences of each word were counted with respect to selected ranges (2,2, which considers two consecutive words, 3,3, which considers three consecutive words, and so on up to 6,6 consecutive characters). The number of occurrences of each feature is calculated in a matrix, and its dimensions determine the indicator referred to as "Features" shown in **Table 1** (see second column).

3.2 Evaluation of five classifiers using *k*-mer counting

This study compares the performance of five machine learning classifiers based on the *k*-mer counting technique to study the interactions of proteins with nucleic acids, by segmenting protein sequences based on six different words. In order to train the classification algorithms, datasets were divided into training and testing sets: 75% were used for training and 25% for testing [30]. **Table 1** shows the performance of five machine learning classifiers using features selected based on the value of *k* (between 2 and 5).

A decrease in the number of top-selected features is shown in **Table 1** as *k*-length increases. Consequently, the best results are obtained when MLP is used, specifically in the ranges 2,2 and 3,3 (see bolded values). The results indicate that neural network algorithms perform better than classical algorithms. It should be noted that more than a million parameters (features) are gathered for classification, which consumes a large amount of memory. In order to resolve this issue, 25 gigabytes of swap memory were used in addition to 64 gigabytes of RAM. Additionally, it can be observed that, as the ranges of information selected have increased, the features found in segmentation and counting have also increased. Despite the fact that this amount of information may seem excessive, it is usually necessary for the training of learning algorithms.

4. Discussion

k-mers counting is a technique used in Natural Language Processing (NLP), which assumes that the information used is related to texts. As a result, in order to develop

Range	Features	Classifier	Accuracy	Precision	Recall	F1 score
2,2	1,082,510	SVM	0.731	0.787	0.731	0.721
		KNN	0.689	0.731	0.689	0.678
		DT	0.784	0.801	0.784	0.782
		RF	0.67	0.801	0.67	0.637
		MLP	**0.843**	**0.844**	**0.843**	**0.843**
3,3	1,103,238	SVM	0.719	0.781	0.719	0.707
		KNN	0.691	0.735	0.619	0.68
		DT	0.761	0.788	0.761	0.757
		RF	0.627	0.781	0.627	0.577
		MLP	**0.838**	**0.846**	**0.838**	**0.837**
4,4	1,124,465	SVM	0.702	0.766	0.702	0.687
		KNN	0.674	0.724	0.674	0.66
		DT	0.763	0.789	0.763	0.76
		RF	0.626	0.787	0.626	0.574
		MLP	0.784	0.797	0.784	0.78
5,5	1,129,579	SVM	0.687	0.763	0.687	0.668
		KNN	0.661	0.721	0.661	0.645
		DT	0.75	0.778	0.75	0.746
		RF	0.617	0.784	0.617	0.561
		MLP	0.825	0.839	0.825	0.824

Table 1.
Classifiers performance.

NLP algorithms, a large corpus of data is necessary [15]. Therefore, the information extracted from protein chains is considered to be relatively limited when compared to entirely text-based applications. In order to increase the number of protein samples, it is possible to incorporate information from more protein families, and even from different species. Due to the fact that this study only examined human proteins, the interaction groups were DNA and RNA interactions (see **Figure 1**), which also influenced the classification result by limiting it to those two types of interactions. In view of these considerations, *k*-mer counting has shown promising results, especially in terms of the ability to characterize proteins solely based on amino acid positions [1, 12, 13]. Therefore, amino acid positions are the primary source of information. To apply the method (see **Figure 2**), standard protein sizes must be applied, as well as balancing the binary classes (DNA and RNA), since these are the factors relevant to improving machine learning metrics. After that, it is necessary to complete the protein segmentation processes (*k*-mers) in order to generate the appropriate words. In this way, it is possible to identify regions within the protein with a higher number of occurrences (word count). Consequently, a large number of features, depending on the *k*-mer count, are extracted from the protein chains, which are then fed into machine learning algorithms to determine whether the binary class can be separated. As proteins possess available sources of information, increasing the separability between classes can be considered as an opportunity for improvement. Despite the promising classification results, further improvements can be made in protein interaction analysis.

The current problem with word occurrence counting is that there are many zero-valued features due to the absence of these words in certain proteins. In this regard, the method has a generalization problem and is limited in its ability to separate classes. Moreover, the increase in feature space leads to performance issues, resulting in a longer computation time. The problem can be partially addressed by applying feature reduction techniques such as PCA. Nevertheless, it is necessary to explore the possibility of improving the features to allow them to be further separated without directly increasing the amount of information to be analyzed.

5. Conclusions

In this study, k-mers are used as an encoding method to extract features from protein sequences. Protein regions (motifs) are segmented by amino acid number (from 2 to 5 mers). The obtained regions are transformed into a matrix representation to train supervised machine learning algorithms. A total of five classifiers were considered in our analysis, including Support Vector Machines (SVMs), K-Nearest Neighbors (KNNs), Decision Trees (DTs), Random Forests (RFs), and Multilayer Perceptrons (MLPs). The proposed approach demonstrated better performance for MLPs, achieving 84.3% precision, 84.4% precision, 84.3% recall, and 84.3% F1. These results suggest that *k*-mers counting can be applied to binary classifications of protein-nucleic acid interactions, in which protein region size impacts classifier performance significantly, suggesting that certain regions are more relevant to protein-DNA/RNA interactions, leading to better classification results.

Our study suggests that exploring regions within proteins may improve machine learning classification algorithms. It may be helpful to consider the biophysical aspects of amino acids that are relevant to specific interactions (such as DNA or RNA) as enhancers of regions associated with biological processes. For future work, we propose a novel method for protein encoding that can be compared with this technique (*k*-mer counting) for extracting automatic features prior to binary protein classification (with DNA or RNA).

Acknowledgements

The authors acknowledge the support and resources provided by the Universidad de las Américas Puebla to conduct this research. Jesus Cabello-Lima wishes to thank CONAHCYT for funding his doctoral scholarship (CVU: 1002979).

Conflict of interest

The authors declare no conflict of interest.

Author details

Jesús Cabello-Lima[1*], Patricio A. Zapata-Morin[2] and J. Horacio Espinoza-Rodríguez[1*]

1 Department of Computing, Electronics and Mechatronics, Universidad de las Americas Puebla, San Andres Cholula, Puebla, Mexico

2 School of Biological Sciences, Universidad Autonoma de Nuevo Leon, San Nicolas de los Garza, Nuevo Leon, Mexico

*Address all correspondence to: jesus.cabellola@udlap.mx and juan.espinoza@udlap.mx

References

[1] ValizadehAslani T, Zhao Z, Sokhansanj BA, Rosen GL. Amino acid k-mer feature extraction for quantitative antimicrobial resistance (AMR) prediction by machine learning and model interpretation for biological insights. Biology. 2020;**9**(11):365

[2] Ren X, Wang YC, Wang Y, Zhang XS, Deng NY. Improving accuracy of protein-protein interaction prediction by considering the converse problem for sequence representation. BMC Bioinformatics. 2011;**12**(1):409

[3] Kähärä J, Lähdesmäki H. Evaluating a linear k-mer model for protein-DNA interactions using high-throughput SELEX data. BMC Bioinformatics. 2013; **14**(10):S2

[4] Yi HC, You ZH, Cheng L, Zhou X, Jiang TH, Li X, et al. Learning distributed representations of RNA and protein sequences and its application for predicting lncRNA-protein interactions. Computational and Structural Biotechnology Journal. 2019;**18**: 20-26

[5] Berg JM, Tymoczko JL, Stryer L. Protein Structure and Function. Biochem. 5th ed. [Internet]. 2002. Available from: https://www.ncbi.nlm.nih.gov/books/NBK21177/ [Accessed: April 5, 2021]

[6] Gupta R, Srivastava D, Sahu M, Tiwari S, Ambasta RK, Kumar P. Artificial intelligence to deep learning: Machine intelligence approach for drug discovery. Molecular Diversity. 2021; **25**(3):1315-1360

[7] Raimondi D, Orlando G, Vranken WF, Moreau Y. Exploring the limitations of biophysical propensity scales coupled with machine learning for protein sequence analysis. Scientific Reports. 2019;**9**(1):16932

[8] ElAbd H, Bromberg Y, Hoarfrost A, Lenz T, Franke A, Wendorff M. Amino acid encoding for deep learning applications. BMC Bioinformatics. 2020; **21**(1):235

[9] Ho B, Baryshnikova A, Brown GW. Unification of protein abundance datasets yields a quantitative Saccharomyces cerevisiae proteome. Cell Systems. 2018;**6**(2):192-205.e3

[10] BankIt Submission Help: Protein FASTA [Internet]. Available from: https://www.ncbi.nlm.nih.gov/WebSub/html/help/protein.html [Accessed: October 19, 2023]

[11] Qu YH, Yu H, Gong XJ, Xu JH, Lee HS. On the prediction of DNA-binding proteins only from primary sequences: A deep learning approach. PLoS One. 2017;**12**(12):e0188129

[12] Yu N, Li Z, Yu Z. Survey on encoding schemes for genomic data representation and feature learning—From signal processing to machine learning. Big Data Mining and Analytics. 2018;**1**(3):191-210

[13] Anastassiou D. Genomic signal processing. IEEE Signal Processing Magazine. 2001;**18**(4):8-20

[14] Jin W, Brannan KW, Kapeli K, Park SS, Tan HQ, Gosztyla ML, et al. HydRA: Deep-learning models for predicting RNA-binding capacity from protein interaction association context and protein sequence. Molecular Cell. 2023;**83**(14):2595-2611.e11

[15] Ofer D, Brandes N, Linial M. The language of proteins: NLP, machine

learning and protein sequences. Computational and Structural Biotechnology Journal. 2021;**19**: 1750-1758

[16] Manekar SC, Sathe SR. A benchmark study of k-mer counting methods for high-throughput sequencing. GigaScience. 2018;**7**(12):giy125

[17] SPARQL-W3C Wiki [Internet]. Available from: https://www.w3.org/wiki/SPARQL [Accessed: April 3, 2023]

[18] Gene Ontology Resource [Internet]. Gene Ontology Resource. Available from: http://geneontology.org/ [Accessed: April 3, 2023]

[19] Pan XM, Niu WD, Wang ZX. What is the minimum number of residues to determine the secondary structural state? Journal of Protein Chemistry. 1999;**18**(5):579-584

[20] Python Documentation [Internet]. Random — Generate Pseudo-Random Numbers. Available from: https://docs.python.org/3/library/random.html [Accessed: April 5, 2023]

[21] Xu D, Nussinov R. Favorable domain size in proteins. Folding & Design. 1998; **3**(1):11-17

[22] Scikit-Learn [Internet]. sklearn.feature_extraction.text.CountVectorizer. Available from: https://scikit-learn/stable/modules/generated/sklearn.feature_extraction.text.CountVectorizer.html [Accessed: February 1, 2024]

[23] Rokach L, Maimon O. Decision trees. In: Maimon O, Rokach L, editors. Data Mining and Knowledge Discovery Handbook [Internet]. Boston, MA: Springer US; 2005. pp. 165-192. DOI: 10.1007/0-387-25465-X_9

[24] Chakraborty A, Mitra S, De D, Pal AJ, Ghaemi F, Ahmadian A, et al. Determining protein–protein interaction using support vector machine: A review. IEEE Access. 2021;**9**:12473-12490

[25] Pisner DA, Schnyer DM. Chapter 6—Support Vector Machine. In: Mechelli A, Vieira S, editors. Machine Learning: Methods and Applications to Brain Disorders. 1st ed. London: Academic Press; 2020. pp. 101-121. DOI: 10.1016/B978-0-12-815739-8.00006-7

[26] Taunk K, De S, Verma S, Swetapadma A. A brief review of nearest neighbor algorithm for learning and classification. In: 2019 International Conference on Intelligent Computing and Control Systems (ICCS); 15-17 May 2019; Madurai, India: IEEE; 2020. pp. 1255-1260

[27] What is Random Forest?|IBM [Internet]. Available from: https://www.ibm.com/topics/random-forest [Accessed: October, 23 2023]

[28] Haykin SS. Neural Networks: A Comprehensive Foundation/Simon Haykin. 2nd ed. 1999

[29] Vakili M, Ghamsari M, Rezaei M. Performance Analysis and Comparison of Machine and Deep Learning Algorithms for IoT Data Classification

[30] Dobbin KK, Simon RM. Optimally splitting cases for training and testing high dimensional classifiers. BMC Medical Genomics. 2011;**4**(1):31

Chapter

Wastewater Treatment Technologies for Paper Industry Discharges: A Comprehensive Review (2018–2023)

Laura Leal-Vázquez and Jose Luis Sánchez-Salas

Abstract

Wastewater from the paper industry has a great impact into the environment, emphasizing the discharge of pollutants during the papermaking process, which uses substantial amounts of water and results in effluents containing heavy metals, cellulose, dyes, and various organic and inorganic compounds, leading to high levels of chemical oxygen demand (COD). This review focuses on the necessity to explore effective alternatives to manage discharge and meet regulatory standards. Researchers have been focused on treating industrial waste from the paper industry, using several methods, each with its pros and cons. The chapter discusses common treatment methods and combination techniques, exploring their applications in addressing the challenges of the paper industry. The review covers diverse treatment approaches, including physical, chemical, and biological methods, with an emphasis on their effectiveness in removing specific pollutants from paper industry wastewater. With stricter regulations, significant technological advancements, and the emergence of novel approaches, a review focused on the past 5 years (2018–2023) is crucial to capture the latest developments and their potential impact.

Keywords: paper industry wastewater, wastewater treatment, industrial wastewater management, environmental impact, treatment alternatives, pollutants, chemical oxygen demand

1. Introduction

Along with the rapid population growth, the problem of wastewater pollution has become increasingly prominent. Over the last few decades, high quantities of industrial wastewater, containing diverse organic and inorganic pollutants, have been released into the environment, sometimes without adequate treatment.

Water is an essential part in the papermaking process. The water is used for the disintegration of the raw material, transporting fibers, and finally the formation of the paper.

Paper is essentially made of fibers to which various chemicals are added to modify its properties and quality. In addition to fibers and chemicals, papermaking requires large amounts of water and energy. The pulp and paper industry uses approximately 30 tons of water for every produced ton of pulp and paper [1]. As can be seen, the papermaking process comes with significant environmental challenges. This raises concerns about the increasing impact that waste can have in the future. This is why the demand for innovative wastewater treatment technologies has emerged due to the implementation and enforcement of discharge guidelines into the environment. If we leave the effluent untreated, it can harm the environment surrounding the paper industry.

Due to the rapid technological advances that have occurred in life, it is important to carry out a comprehensive review focused on the period 2018–2023. This allows the evaluation of the current wastewater treatment technologies of this industry as well as the regulatory policies, providing an understanding of the technological advances that are currently being managed. The study focused on answering the following research query: Which wastewater treatment methods for the paper industry have been explored in the past 5 years?

Also, environmental laws in several countries have become more restrictive in recent years, highlighting the importance of water reuse and the development of "zero-discharge" paper mills [2]. Considering the increasingly strict standards of environmental protection and discharge limitations, other treatment methods are needed to meet these rigorous standards [3].

Nowadays, in the papermaking industry, the main treatment process involves clarification as a primary process and biological treatment as a secondary process. However, these treatments could not meet the water quality requirements since high COD, toxicity, and color remain in the effluent [3].

In the upcoming review, we will delve into various technologies primarily focused on chemical oxygen demand (COD) removal from paper industry wastewater since COD is an important indicative measure of the number of oxidizable pollutants in wastewater [4]. While our main emphasis will be on COD removal, it is essential to acknowledge the broader landscape, including technologies related to other purposes.

As part of the physicochemical characteristics of the paper mill effluents, it can be mentioned the following [5–7]:

- COD in a range of 600–4900 mg/L
- Biochemical oxygen demand (BOD) 560–2565 mg/L
- pH between 6.1 and 7.8
- Total suspended solids (TSS) 1900–3900 mg/L
- Alkalinity of 300–1230 mg/L
- Nitrogen 28 mg/L

Nowadays, the Chinese government has initiated measures to reduce freshwater consumption and act in wastewater discharge from the paper industry. A new technology was introduced in the paper industry to reuse tissue paper mill effluent water as intake water from corrugated paper mill [8] with the purpose of reuse and trying not to discharge more contaminated water.

Another aspect to highlight is the characterization of pulp and paper mill waste to produce waste-based cement [9]. This broader perspective ensures a comprehensive overview of the diverse strategies employed in the treatment of paper industry wastewater, encompassing both environmental and energy-related considerations.

Researchers are also addressing energy concerns, exemplified by a recent study in the pulp and paper industry in Brazil, focused on the country's transition to renewable energy, and the researchers propose a systemic biorefinery framework using up-flow anaerobic sludge blanket (UASB) reactors [10]. These reactors treat wastewater, emphasizing energy recovery through biogas production converted into heat and power. Results indicate that UASB adoption addresses environmental concerns, aligns with the national renewable energy agenda, and offers a rapid return on investment. The UASB-based biorefinery has the potential to help the industry, contributing significantly to Brazil's decarbonization goals.

Another study in Brazil investigated turning pulp and paper mill waste into environmentally friendly cement. They want to create ecological cement by using lime mud, biomass ash, and wastewater treatment plant (WWTP) sludge, which shows good strength [9]. Tests confirmed its non-hazardous nature, offering a sustainable solution for waste management and eco-friendly cement production.

Also, studies investigate the potential of buffalo grass and couch grass artificial floating islands (AFIs) for removing total nitrogen (TN) and total phosphorus (TP) from simulated paper mill wastewater. This research aims to evaluate AFIs' capability to address nutrient issues, particularly nitrogen and phosphate, which can contribute to eutrophication in receiving waters [11]. A study made in Indonesia focuses on treating leachate wastewater from the pulp and paper industry, known for containing heavy metals, particularly lead (Pb) [12]. The contamination of heavy metals in industrial wastewater poses a significant environmental threat, necessitating effective treatment to meet legal quality standards. The research employs an adsorption method, and the most favorable result is a 92.45% reduction in Pb concentration after 180 minutes of contact time. This highlights the potential of the adsorption method for efficient Pb removal in the treatment of pulp and paper wastewater [12].

Effective COD removal safeguards public health by mitigating the risks associated with the discharge of pollutants into water bodies. Beyond regulatory and health considerations, this review plays a fundamental role in preserving water quality and preventing oxygen depletion in aquatic ecosystems, aligning with sustainability objectives, and minimizing the overall environmental impact. The economic implications are noteworthy as well, with optimized COD removal technologies translating into reduced operational costs and improved financial viability for industries. Lastly, the continuous review of these technologies encourages innovation and environmentally friendly approaches for a more sustainable future in the wastewater treatment practices.

2. Methodology

The Preferred Reporting Items for Systematic Reviews and Meta-Analyses (PRISMA) framework was followed to ensure a systematic and transparent review process. First, the database Elsevier's Scopus was the only database selected for the literature search. Then, the search string was defined including the keywords "paper industry" AND "wastewater treatment" as to include all the possible wastewater treatments only for the paper industry. The search string was applied to article titles, abstracts, and keywords. After the search string was applied, 1377 documents were obtained. After excluding documents that were not written in English and Spanish, 1361 documents were obtained. The search was limited to papers published only between 2018 and 2023, and 319 documents were left. Subject areas excluded from the search included (mathematics, computer science, social sciences, physics and astronomy, economics, and business). This decision was made to focus on specific aspects of wastewater treatment within the paper industry; after these steps, 275 documents were left. Initial screening involved evaluating titles and abstracts to determine relevance based on the inclusion and exclusion criteria. Seventy documents were left behind due to the content in the abstract. When each document was sought to be recovered, 38 were not obtained. A full-text assessment was performed on potentially relevant papers identified during the initial screening. Inclusion or exclusion decisions were made based on a more detailed review of the content. Data extracted from each included paper included author details, publication year, study design, key findings, and relevant information related to wastewater treatment in the

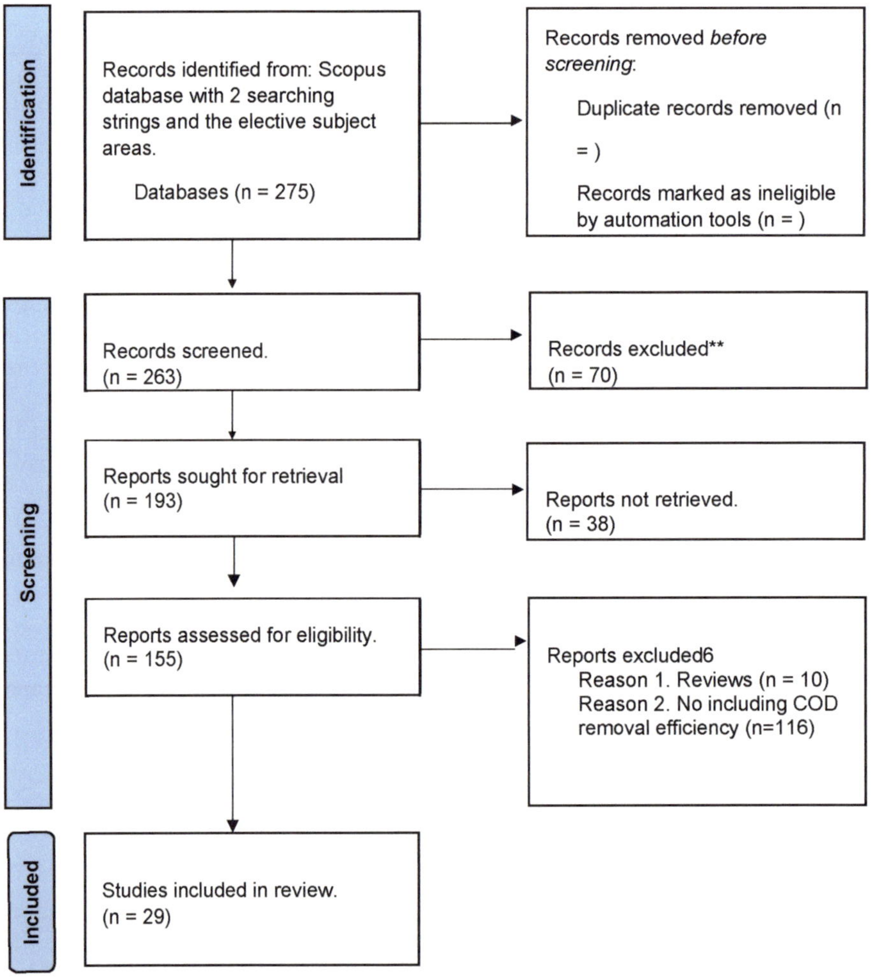

Figure 1.
PRISMA flowchart for studies selection.

paper industry. The quality of selected papers was assessed based on criteria such as study design, methodology, and the robustness of findings. Papers were not scored or ranked; instead, the focus was on ensuring a comprehensive understanding of the methodological quality of each paper. Lastly, documents that were reviewed and did not include information about chemical oxygen demand (COD) removal efficiency were left behind. A qualitative synthesis approach was employed to analyze and integrate findings from the selected papers. Potential biases were considered in the synthesis, and heterogeneity among the selected papers was addressed by carefully comparing methodologies and study designs. At the end, 29 studies are included in this review as shown in **Figure 1**.

3. Results and discussion

Tables 1–3 present the summary of all the included studies categorized by the type of treatment, biological, physicochemical, and advanced oxidation processes (AOPs). In the subsequent Results section, a more detailed discussion will be provided on the specific findings related to these three types of treatments, as identified during the systematic review (PRISMA). Certainly, biological treatments were the most frequently studied in this review.

3.1 Advanced oxidation processes (AOPs)

As a common transition metal, manganese has received attention in the Fenton system because of its benign environmental effect [3]. Becuase several studies [3] show that Fenton processes have produced better overall results treating pulp and paper mill effluents than other AOP-based treatment technologies, ranging from 67% to 81% with Mn(II) with an optimal value of pH of 4.

Another treatment method included in this review from the paper and pulp industry involves ozonation alone and a combined ozone/UV process. The study investigates the removal of color and chemical oxygen demand (COD) from wastewater. The results reveal that increasing the solution pH enhances COD and color removal, with the highest efficiencies, 56% COD reduction, and 71% decolorization, achieved at pH 9.8 using the O_3/UV process with an ozone dose of 0.2 mg/mL [13]. The UV-ozone process is found to be more effective than ozonation alone in this case. Findings suggest that the electro-activated persulfate process is a suitable treatment alternative for COD removal from paper industry wastewater [14]. **Table 1** displays the studies related to advanced oxidation processes.

3.2 Physicochemical treatment

Researchers address the urgent need for integrated treatment in recycling paper industrial wastewater, particularly emphasizing zero liquid discharge (ZLD) coupled with solar energy [21]. Various coagulants, including alum, ferric chloride, lime, and cationic polymer, are employed. The removal efficiencies for chemical oxygen demand (COD) range from 75% to 84%, and for total suspended solids (TSS), they range from 97.5% to 99.6%. The treated industrial wastewater becomes suitable for recycling, aligning with the principles of ZLD. Additionally, one technique involving a solar dryer effectively diminishes the negative impact of chemical sludge, producing dry sludge and purified water for recycling in the industrial process. Solar energy

Study no.	Type of effluent	Process	Parameter	Removal percentage (%)	Retention time/additional comments	Country	Reference number
1	Paper and pulp industry	Ozonation and UV process	COD removal	56	180 minutes At pH 9.8 with the O_3/UV process with ozone dose of 0.2 mg/mL	Pakistan	[13]
			Color removal	71			
2	Paper industry	Electrocoagulation (EC) process with the addition of the strong oxidant persulfate	COD removal	72.8	25 minutes Persulfate was activated by the dissolution of iron (Fe) and aluminum (Al)	Turkey	[14]
3	Pulp and paper wastewater	Catalytic ozonation process	COD removal	71	At natural pH 6.8	Pakistan	[15]
			Color removal	88			
4	Paper mill	Electro-Fenton process	COD removal	74.82	300 minutes Central composite design	Spain	[16]
5	Papermaking wastewater	Fenton and manganese	COD removal	81	30 minutes The COD removal efficiency was enhanced in the presence of Mn(II), which increased by 19% compared with that of the Fenton system alone.	China	[3]
6	Recycled paper production	Advanced oxidation process (AOP)	COD removal	94	93 minutes Response surface methodology (RSM) was applied to optimize the operational parameters	India	[17]
7	Pulp and paper mill wastewater	UV/H_2O_2 and photo-Fenton	Phenols	88.2		Brazil	[18]
			COD removal	90.4			
			Color removal	93.4			
8	Paper industry	Electrochemical peroxidation process	Color degradation	100	40 minutes pH = 3	Indonesia	[19]
			COD removal	83			

Table 1.
The studies included advanced oxidation processes (AOPs).

Study no.	Type of effluent	Process	Parameter	Removal percentage (%)	Retention time/additional comments	Country	Reference number
9	Recycled paper production	Electrocoagulation using aluminum and iron electrodes	Turbidity	92.7	32.8–39.5 minutes	Mexico	[20]
			Total suspended solids	91.3			
			Chlorides	70.4			
			Sulfate	66.6			
			COD removal	64			
10	Paper recycling mill industry	Chemical coagulation—flocculation	COD removal	84	30 minutes Sludge from the coagulation can be dried to zero water content *via* solar drier and it can recycle in the same industry	Egypt	[21]
			Total suspended solids	99.6			
11	Paper industry wastewater		Total dissolved solids	51	60 minutes Optimum pH 8 and 20 V (voltage)	India	[22]
			Turbidity	65			
			TOC	68			
			COD removal	70			
12	Esparto paper industry wastewater	Powdered activated carbon	COD removal	94	60 minutes pH = 6, an adsorbent dose of 75 g/L, initial COD concentration of 9.3 g/L	Tunisia	[23]
			Color removal	90			
13	Recycled paper industry	Electrocoagulation	Color degradation	100	80 minutes The energy used is 13.56 kWh/L	Indonesia	[24]
			COD removal	50			
14	Raw wastewater from pulp and paper industry	Fenton and photo-Fenton enhanced by membrane oxidation reactor	TOC	69.7	60 minutes By response surface analyses of Taguchi experimental design	Turkey	[25]
			COD removal	82.1			

Table 2.
Studies included in the physicochemical treatment.

Study no.	Type of effluent	Process	Removal percentage	(%)	Retention time/additional comments	Country	Reference number
15	Paper and cardboard recycling industry	*Candida glabrata*	COD removal	71	96 hours Economic and eco-friendly option	Iran	[26]
		Candida albicans	COD removal	81.8			
16	Pulping wastewater	Sequential batch reactor (SBR) and biomass-based fly ash	COD removal	96	4 hours	China	[27]
			Color removal	95			
17	Paper recycling plant	Bench-scale horizontal-flow anaerobic immobilized biomass	COD removal	90.8	12 hours	Brazil	[2]
			Sulfate	94.8			
18	Paper and pulp industry	Oleaginous *Rhodococcus opacus* and microfiltration	Toxicity removal	94.5	96 hours; Polynomial model was generated to estimate the optimum conditions	India	[28]
			COD removal	87.6	144 hours		
19	Paper and pulp wastewater	Sequential batch reactor	COD removal	73.2	15 hours	India	[5]
			AOX removal	57.6			
20	Recycled pulp and paper wastewater	Anaerobic treatment	COD removal	78	40 days pH 7.24 and initial COD of 5540 mg/L	Morocco	[29]
			Volatile solids	52			
			Total solids	48			
21	Raw paper mill wastewater	*Penicillium citrinum*	Decolorization efficiency	84	72 hours	India	[6]
			COD removal	75			
			Phenolic reduction	70			
22	Paper industry wastewater	Two microalgae and two bacteria	Total nitrogen	99.95	240 hours Photobioreactor under batch mode	India	[30]
			COD removal	95.16			
23	Synthetic wastewater from pulp and paper industry	Biofilm SBR	COD removal	99.21	4 days Biofilm sequencing batch reactors (SBR)	Egypt	[31]

DOI: http://dx.doi.org/10.5772/intechopen.115472

Study no.	Type of effluent	Process	Removal percentage	(%)	Retention time/additional comments	Country	Reference number
24	Recycled paper mill effluent	Multimedia sequencing batch biofilm reactor (MM-SBBR) system	Pentachlorophenol (PCP) removal	95–99%	24 hours The stability and performance of the removal was influenced differently by the increasing PCP load	Malaysia	[32]
			COD removal	93–99%			
			Ammoniacal nitrogen (NH_3-N) removal	92–100%			
25	Paper and cardboard recycling industry wastewater	*Candida glabrata*	COD removal	71	96 hours The optimal pH for *C. albicans* was 6 and for *C. glabrata* was 5. They can be considered as an economic and eco-friendly option for the treatment	Iran	[26]
		Candida albicans	COD removal	81.8			
26	Pulp and paper industry wastewater and domestic	Activated sludge bioreactor system	COD removal	88.4	2 days	Malaysia	[33]
27	Pulp and paper industry	*Aspergillus uvarum*	COD, color, and phenolic compounds tested	61%, 81% and 67%	96 hours The effluent was treated with and without solids and with and without pH control	Spain	[34]
		Phanerochaete chrysosporium	COD, color, and phenolic compounds	65%, 37% and 56%			
28	Paper recycling industry	Up flow anaerobic sludge blanket (UASB) reactor	Chemical oxygen demand (COD)	87	24 hours At 37°C and pH of 7.21	Bahrain	[35]
			Biological oxygen demand (BOD)	93			
			Total dissolved solids (TDS), total suspended solids (TSS), total solids (TS)	79, 88, and 82			
			Heavy metals (Hg, Pb, Cd, Cr, Cu)	91, 93, 98, and 98			

Study no.	Type of effluent	Process	Removal percentage	(%)	Retention time/additional comments	Country	Reference number
29	Domestic and pulp and paper industry wastewater	Vertical flow-constructed wetland	Suspended solids (TSS) and total dissolved solids (TDS)	100 and 83	3 days 40–45°C and flow rate of 10 mL/minute	United States	[36]
			COD removal	80			
			BOD removal	81			
			Sulfates, phosphates, and total nitrogen	81, 63, and 61			

Table 3.
Studies included in the biological treatment.

demonstrates significant potential for sludge drying with competitive costs and environmental cleanliness [21].

Also, activated carbon can be used as an adsorbent for treating pulp and paper mill wastewater. Through batch adsorption experiments, optimal conditions were determined for variables such as adsorbent dose, pH, time, agitation speed, and temperature [37]. The study resulted in a substantial reduction of 96.48% in color and 94.25% in lignin content in the pulp and paper mill wastewater under the optimized conditions. The activated carbon proved effective in significantly reducing color and lignin content in the wastewater under optimized conditions, showcasing its potential for wastewater treatment [37]. **Table 2** displays the studies related to physicochemical treatment.

3.3 Biological treatment

The biological treatment is viable option that involves the utilization of microorganisms including fungi, bacteria, algae, and enzymes as individual or in combination with other treatment technologies such as primary and tertiary treatments. The objective of researching new technologies is to meet the permissible effluent discharge limits laid by regulatory authorities. Due to excessive use of chlorine in the manufacturing paper process, various chlorophenolic compounds are formed and need appropriate treatment before discharge. These compounds discharged in the effluent stream are not characterized specifically as phenolic compounds and are generally measured as adsorbable organic halides (AOX). These compounds along with several other contaminants have been reported for their ill health effects, being carcinogenic to mutagenic [5].

A sustainable process was demonstrated by using microalgae together with bacteria [28], and the symbiotic growth of microalgae and bacteria represents a significant advancement in enhancing both total biomass concentration and nutrient removal efficiency. In a study, the consortium of two microalgae and two bacteria exhibited superior outcomes compared to the growth of two bacteria or two microalgae alone. This synergistic collaboration highlights the importance of the collective potential of microorganisms for more effective and sustainable wastewater treatment processes and can lead us to new emerging fields of study.

Also, the sequencing batch reactor (SBR) process has proven to be flexible and provides high-performance treatment technology for wastewater treatment, especially for pulp and paper wastewater. However, providing some modifications to SBR such as addition of bio-media (BSBR) may improve its performance [31].

Anaerobic digestion is an efficient waste treatment technology that uses natural anaerobic decomposition to reduce the volume of waste while producing biogas [29]. The results demonstrate promising outcomes with significant COD removal efficiency (78%) and favorable stability throughout the operational cycle. However, the study identifies the need for further research to optimize microbial metabolism and reactor functionality for enhanced biogas production.

Another biological treatment was the use of batch biodegradation; first, it is carried out using wastewater as the potential substrate for oleaginous *Rhodococcus opacus* bacteria; subsequently, a microfiltration system was applied to recover the bacterial biomass grown as well as residual COD removal from the effluent. The combined process showed results in terms of COD removal by degradation from the industrial wastewaters, with the values 56.8%, 46.1%, and 68.9% for dairy, paper, and pulp and biomass gasification wastewaters, respectively, by biodegradation. These values were further improved to 92.7%, 87.6%, and 88.2%, respectively, following the microfiltration step performed by employing a low-cost ceramic membrane [28].

Pentachlorophenol (PCP) is a type of chlorophenol that has been well-documented as a wood preservative, pesticide, herbicide, disinfectant, and biocide [32]. Surprisingly, paper mill effluent still surpasses the allowable limit of PCP even after industrial wastewater treatment [32]. Degradation of PCP is, however, problematic because of its strong aromatic ring system and high chloride content, continuing with the discussion on biological treatments [32], worked with a multimedia sequencing batch biofilm reactor (MM-SBBR) system to treat recycled paper mill effluent containing a high level of pentachlorophenol. Lab-scale MM-SBBRs were used, demonstrating highly effective removals of PCP, chemical oxygen demand (COD), and ammoniacal nitrogen (NH_3-N) ranging from 95% to 99%, 93% to 99%, and 92% to 100%, respectively.

In another process to treat pulp and paper industry wastewater, a study investigated the use of native herbs for the phytoremediation of heavy metals. The native plants reduced heavy metal contents in the wastewater by nearly 60%, with iron accumulating the most in plant tissues [38]. **Table 3** displays the studies related to biological treatment.

4. Conclusion

Furthermore, various researchers have reported that effluents from paper and pulp mills and other industries can sometimes contain carcinogenic substances, leading to toxicity. Despite technological improvements that reduce these harmful effects, trace amounts may persist after the completion of treatment processes. That is why the choice of adopting a relevant treatment process and its optimization for successful application depends upon the influent type and its concentration as well. The treatments should be designed exclusively for each type of effluent, considering the site-specific characteristics. It is crucial to be very specific and study each case thoroughly. There is an ongoing need for sustainable and cost-effective treatment methods every day, and the importance of continuous technological improvements will lead us to a sustainable future. It is crucial to research and implement new technologies for wastewater treatment to address these challenges effectively.

There is still much room for study within the treatment of wastewater, both from the paper industry and recycled paper. Several research studies on coagulation-flocculation have been observed, although this can serve as the foundation to understand which direction and type of treatment, we can explore whether it is a physicochemical, biological, advanced oxidation processes, and membrane processes, among others.

Acknowledgements

Support provided by the Consejo Nacional de Humanidades Ciencia y Tecnología (CONAHCyT) is appreciated. Additional support was received from the Universidad de las Américas Puebla (UDLAP) through a fellowship.

Conflict of interest

The authors declare no conflict of interest.

Author details

Laura Leal-Vázquez[1] and Jose Luis Sánchez-Salas[2]*

1 Departamento de Ingeniería Civil y Ambiental, Escuela de Ingeniería, Universidad de las Américas Puebla, Puebla, Mexico

2 Departamento de Ciencias Químico-Biológicas, Escuela de Ciencias, Universidad de las Américas Puebla, Puebla, Mexico

*Address all correspondence to: jluis.sanchez@udlap.mx

References

[1] From-Aldaron M, Sandberg M, Granström KM. Low dosage chemical treatment for improved oxygenation of pulp mill effluents. Journal of Environmental Engineering. 2018;**144**(3):06017012, 1-7. DOI: 10.1061/(asce)ee.1943-7870.0001320

[2] Damianovic MHRZ, de Godoi LAG, Saia FT, Foresti E. Horizontal-flow anaerobic immobilized biomass (HAIB) reactor for organic matter and sulfate removal from paper recycling plant wastewater with simultaneous conversion of sulfide into elemental sulfur. Journal of Environmental Chemical Engineering. 2018;**6**(1):964-969. DOI: 10.1016/j.jece.2018.01.024

[3] Wang Y, Wang C, Shi S, Fang S. Improved removal performance and mechanism investigation of papermaking wastewater treatment using manganese enhanced Fenton reaction. Water Science and Technology. 2018;**77**(10):2509-2516. DOI: 10.2166/wst.2018.208

[4] Shen W, Huang F, Zhang X, Zhu Y, Chen X, Akbarjon N. On-line chemical oxygen demand estimation models for. Water Science and Technology. 2018;**78**(2):310-319. DOI: 10.2166/wst.2018.299

[5] Khan NA et al. Performance evaluation of column-SBR in paper and pulp wastewater treatment: Optimization and bio-kinetics. Desalination and Water Treatment. 2019;**156**:204-219. DOI: 10.5004/dwt.2019.23775

[6] Namboodiri MMT, Pakshirajan K. Sustainable and green approach of chitosan production from *Penicillium citrinum* biomass using industrial wastewater as a cheap substrate. Journal of Environmental Management. 2019;**240**:431-440. DOI: 10.1016/j.jenvman.2019.03.085

[7] Mojab H, Raj E, Pacheco-Ruiz S. Granular sludge bed anaerobic treatment systems for resource recovery. In: Waste Biorefinery: Integrating Biorefineries for Waste Valorisation. Elsevier; 2020. pp. 107-124. DOI: 10.1016/B978-0-12-818228-4.00004-6

[8] Tao J, Long X, Li Z, Ying G. Reusing tissue paper mill effluent water as corrugated paper mill intake water: Case study of a new clean production measure. Environmental Progress & Sustainable Energy. 2018;**37**(3):934-941. DOI: 10.1002/ep.12747

[9] Simão L, Hotza D, Raupp-Pereira F, Labrincha JA, Montedo ORK. Characterization of pulp and paper mill waste for the production of waste-based cement. Revista Internacional de Contaminacion Ambiental. 2019;**35**(1):237-246. DOI: 10.20937/RICA.2019.35.01.17

[10] Ferreira SF, Buller LS, Berni MD, Bajay SV, Forster-Carneiro T. An integrated approach to explore UASB reactors for energy recycling in pulp and paper industry: A case study in Brazil. Biofuel Research Journal. 2019;**6**(3):1039-1045. DOI: 10.18331/BRJ2019.6.3.4

[11] Ayres JR, Awad J, Burger H, Marzouk J, Van Leeuwen J. Investigation of the potential of buffalo and couch grasses to grow on AFIs and for removal of nutrients from paper mill wastewater. Water Science and Technology. 2019;**79**(4):779-788. DOI: 10.2166/wst.2019.098

[12] Aprianti T, Miskah S, Moeksin R, Sisnayati S, Nasir S. Pb removal in pulp

and paper industry leachate wastewater using activated carbon-ceramic composite adsorbent. In: IOP Conference Series: Earth and Environmental Science. Institute of Physics Publishing. Bristol, United Kindom: IOP Publishing; 2019. DOI: 10.1088/1755-1315/298/1/012011

[13] Munir HMS, Feroze N, Ikhlaq A, Kazmi M, Javed F, Mukhtar H. Removal of colour and COD from paper and pulp industry wastewater by ozone and combined ozone/UV process. Desalination and Water Treatment. 2019;**137**:154-161. DOI: 10.5004/dwt.2019.23186

[14] Varank G, Guvenc SY, Demir A, Kavan N, Donmez N, Onen ZT. Modeling and optimizing electro-persulfate processes using Fe and Al electrodes for paper industry wastewater treatment. Water Science and Technology. 2020;**81**(2):345-357. DOI: 10.2166/wst.2020.115

[15] Munir HMS et al. Fe-zeolite catalyst for ozonation of pulp and paper wastewater for sustainable water resources. Chemosphere. 2022;**297**:134031, 1-11. DOI: 10.1016/j.chemosphere.2022.134031

[16] Buftia G, Rosales E, Pazos M, Lazar G, Sanromán MA. Electro-Fenton process for implementation of acid black liquor waste treatment. Science of the Total Environment. 2018;**635**:397-404. DOI: 10.1016/j.scitotenv.2018.04.139

[17] Kumar V, Punathil S, Ghosh P. Optimization of COD removal by advanced oxidation process through response surface methodology from pulp & paper industry wastewater. Journal of Scientific and Industrial Research. 2019:386-390

[18] Carvalho Neves L et al. Phytotoxicity indexes and removal of color, COD, phenols and ISA from pulp and paper mill wastewater post-treated by UV/H_2O_2 and photo-Fenton. Ecotoxicology and Environmental Safety. 2020;**202**:110939, 1-8. DOI: 10.1016/j.ecoenv.2020.110939

[19] Marlina E, Purwanto P, Sudarno S. Decolorization of industrial wastewater using electrochemical peroxidation process. Journal of Electrochemical Science and Engineering. 2022;**12**(2):373-382. DOI: 10.5599/jese.1017

[20] Rodríguez-Rosales MDJ, Betancourt-Frías AE, Valle-Cervantes S, Bailón-Salas AM, Quiroga MG, Ordaz-Díaz LA. Design and evaluation of electrocoagulation system for the treatment of effluent from recycled paper production. BioResources. 2019;**14**(1):2113-2132. DOI: 10.15376/biores.14.1.2113-2132

[21] El-Awady MH, El-Ghetany HH, Aboelghait KM, Dahaba AA. Zero liquid discharge and recycling of paper mill industrial wastewater via chemical treatment and solar energy in Egypt. Egyptian Journal of Chemistry. 2019;**62**:37-45. DOI: 10.21608/EJCHEM.2019.13949.1866

[22] Hugar GM, Marol CK. Feasibility of electro coagulation using iron electrodes in treating paper industry wastewater. Sustainable Water Resources Management. 2020;**6**(4):70, 1-10. DOI: 10.1007/s40899-020-00433-4

[23] Boubaker F, Ridha BC. Esparto paper industry wastewater treatment by adsorption onto powdered activated carbon. International Journal of Environmental Analytical Chemistry. 2021;**101**(15):2817-2834. DOI: 10.1080/03067319.2020.1711897

[24] Marlina E, Purwanto P, Sudarno S. COD removal, decolorization, and energy consumption of

electrocoagulation as a wastewater treatment process. In: IOP Conference Series: Earth and Environmental Science. Bristol, UK: IOP Publishing Ltd; 2021. DOI: 10.1088/1755-1315/896/1/012043. Available from: https://iopscience.iop.org/article/10.1088/1755-1315/896/1/012043

[25] Dogan EC et al. Fenton and photo-Fenton processes integrated with submerged ultrafiltration for the treatment of pulp and paper industry wastewater. Journal of Environmental Chemical Engineering. 2021;**9**(5):105878, 1-17. DOI: 10.1016/j.jece.2021.105878

[26] Ebrahimi AA, Ghaneian MT, Ghafourzadeh M, Gholami M. Bioremediation of paper and cardboard recycling industry wastewater by native yeasts. Desalination and Water Treatment. 2021;**233**:217-225. DOI: 10.5004/dwt.2021.27583

[27] Chen X, Si C, Fatehi P. Enhancement in biological treatment of pulping wastewater by fly ash. Chemosphere. 2018;**210**:1-9. DOI: 10.1016/j.chemosphere.2018.06.160

[28] Goswami L, Kumar RV, Pakshirajan K, Pugazhenthi G. A novel integrated biodegradation—Microfiltration system for sustainable wastewater treatment and energy recovery. Journal of Hazardous Materials. 2019;**365**:707-715. DOI: 10.1016/j.jhazmat.2018.11.029

[29] Bakraoui M, Hazzi M, Karouach F, Ouhammou B, El Bari H. Experimental biogas production from recycled pulp and paper wastewater by biofilm technology. Biotechnology Letters. 2019;**41**(11):1299-1307. DOI: 10.1007/s10529-019-02735-w

[30] Goswami G, Makut BB, Das D. Sustainable production of bio-crude oil via hydrothermal liquefaction of symbiotically grown biomass of microalgae-bacteria coupled with effective wastewater treatment. Scientific Reports. 2019;**9**(1):15016, 1-12. DOI: 10.1038/s41598-019-51315-5

[31] Adel S, Ibrahim WA, Fayed M, Aly SA. Evaluation the performance of sequencing batch reactor and bio-film sequencing batch reactor for pulp and paper wastewater treatment. Desalination and Water Treatment. 2020;**208**:136-147. DOI: 10.5004/dwt.2020.26479

[32] Muhamad MH, Abdullah SRS, Abu Hasan H, Abu Bakar SNH. Multimedia-sequencing batch biofilm reactor in treating recycled paper mill effluent containing high level of pentachlorophenol: Long-term performance, mechanism and kinetic studies. Journal of Water Process Engineering. 2020;**37**:101522, 1-17. DOI: 10.1016/j.jwpe.2020.101522

[33] Jagaba AH et al. Combined treatment of domestic and pulp and paper industry wastewater in a rice straw embedded activated sludge bioreactor to achieve sustainable development goals. Case Studies in Chemical and Environmental Engineering. 2022;**6**:100261, 1-10. DOI: 10.1016/j.cscee.2022.100261

[34] Díaz AI, Laca A, Lima N, Díaz M. Treatment of Kraft black liquor using basidiomycete and ascomycete fungi. Process Safety and Environmental Protection. 2022;**168**:67-76. DOI: 10.1016/j.psep.2022.09.065

[35] Iftikhar A et al. Anaerobic biological treatment of wastewater from paper recycling industry by UASB reactor. Sains Malaysiana. 2023;**52**(4):1087-1099. DOI: 10.17576/jsm-2023-5204-05

[36] Rasool T, Ishtiaq Ali M, Jamal A, Huang Z, Urynowicz M. Performance

evaluation and microbial profiling of integrated vertical flow constructed wetland (IVFCW) for simultaneous treatment of domestic and pulp and paper industry waste water. Saudi Journal of Biological Sciences. 2023;**30**(12):103850, 1-12. DOI: 10.1016/j.sjbs.2023.103850

[37] Gupta V, Bhardwaj NK, Rawal RK. Removal of colour and lignin from paper mill wastewater using activated carbon from plastic mix waste. International Journal of Environmental Science and Technology. 2022;**19**(4):2641-2658. DOI: 10.1007/s13762-021-03263-9

[38] Sharma P, Tripathi S, Purchase D, Chandra R. Integrating phytoremediation into treatment of pulp and paper industry wastewater: Field observations of native plants for the detoxification of metals and their potential as part of a multidisciplinary strategy. Journal of Environmental Chemical Engineering. 2021;**9**(4):105547, 1-16. DOI: 10.1016/j.jece.2021.105547

Chapter

Carbon Nanomaterials in Drug Delivery Systems

Luis Alberto De Sales Palestina, Sergio Alberto Bernal Chávez and Jessica Rosaura Campos Delgado

Abstract

Nanotechnology is a multidisciplinary science responsible for the development of systems with the use of materials of at least one dimension on the nanometric scale. In medicine, the application of these nanomaterials has an impact on the implementation of new diagnostic devices, tissue engineering, and even the innovation of therapeutic alternatives to the usual treatments as a drug delivery system. On the other hand, carbon nanomaterials are a very broad group of nanomaterials that have very varied physicochemical characteristics depending on the type of structure. These have been introduced as potential raw materials in biomedical use, where main exponents are carbon nanotubes, derivatives of graphene, and fullerenes, which can be modified to form the basis of drug delivery systems. These materials are reviewed to give an overview of the advantages and challenges faced in this biomedical application.

Keywords: carbon material, carbon nanotube, graphene, biomedical application, targeted drug delivery

1. Introduction

A strict definition of nanotechnology refers to the use, engineering, and design of structures approximately in the size regime of 1–100 nanometers (nm) in at least one of their dimensions [1]. Despite this size restriction, nanomedicine commonly refers to structures that exceed the size range and that are developed by top-down engineering, starting from a large bulk material and after treatment, the nanoscale material is reached, or bottom-up techniques that consist of the coupling or assembly of atoms or molecules for the construction of the nanoscale material [2, 3].

Nanomedicine leads to the development of better devices, drugs, and other applications for early diagnosis or treatment of a wide range of diseases with high specificity, efficacy, and personalization, with the aim of improving the quality of life of patients [1]. Nanomedicine differs from other types of medicine in that it involves the development and application of materials and technologies with nanometer-length scales. In this sense, the use of various nanomaterials has been reflected in the detection of biomarkers of different diseases, as part of the composition of implants and an important part in the use of drug release and delivery systems, as well as in the design of scaffolds and surfaces for biosensors [4–6].

In essence, a drug delivery system is one that serves as a carrier for a drug and protects it from certain physiological conditions so that it can exert the desired therapeutic effect [7]. The main objectives of the design and improvement of formulations from the use of nanostructures as drug delivery systems are improving the aqueous solubility and chemical stability of the active compounds, improving the pharmacological characteristics and reducing side effects, and taking advantage of the characteristics of physicochemical properties of the same materials to promote the adsorption of different molecules and administration of active ingredients, which generate a complex system useful as drug delivery system [8].

On the other hand, there is a wide range of materials used to create drug delivery systems, such as lipid-based, polymeric, and inorganic systems [4]. Lipid-based systems such as liposomes and lipid nanoparticles stand out based on the use of phospholipids and physiological lipids that are in a solid state at body temperature. They have the characteristic of forming vesicles with an internal matrix that helps encapsulate drugs: hydrophilic and hydrophobic. These nanoparticles stand out for their low toxicity and high biocompatibility [9]. An advantage of these materials is the ease of being superficially modified; for example, the use of polymers generates better biocompatibility and prevents the formation of protein biocorona, as well as the addition of target molecules such as antibodies or peptides makes the system direct. To the target cell, and thanks to stimuli such as pH or temperature, they release the drug [10].

Polymeric nanomaterials are three-dimensional structures that function as nanocapsules or nanospheres in which drugs of different chemical nature are loaded. They are formed from natural raw materials (such as alginate or chitosan) or synthetic (polyethylene glycol or polyethyleneimine), and the final structure is defined by the type of polymer to be used. The therapeutic agent can be encapsulated at the core, trapped in the polymer matrix, chemically conjugated to the polymer, or attached to the surface of the nanoparticle. This allows the delivery of various payloads regardless of the chemical characteristics of the drug such as solubility and molecular weight. Within these structures, micelles and dendrimers can be found [11, 12].

There is another group made up of inorganic nanoparticles, the most important of which are gold nanoparticles, iron oxides, and silica, which can be designed to have different geometric shapes. Gold nanoparticles can be designed to have different geometric arrangements and have the ability to serve as drug delivery systems where the ability to anchor covalently or non-covalently is observed. Iron oxide derivatives such as magnetite (Fe_3O_4) contain important magnetic properties and behave as a contrast medium for *in vivo* imaging studies. Furthermore, mesoporous silica nanoparticles are highly used for drug delivery systems due to their high surface area. These types of particles with specific characteristics can be used for photothermal therapy, although their use is limited by toxicity, especially those that contain heavy metals [4, 13].

2. Carbon nanomaterials

Carbon is one of the most abundant elements in the Earth, a chemical element whose electronic configuration in its ground state is $1s^2\ 2s^2\ 2p^2$, and it forms a maximum of two covalent bonds. Although, in chemical compounds, due to the 2s electrons being rearranged toward the empty 2p orbital, it gives rise to four hybrid orbitals, and its configuration changes and therefore, it is capable of forming four covalent bonds depending on its chemical valence. The three existing hybridizations (sp, sp^2, and sp^3) can be distinguished according to the number of p orbitals mixed with the s orbital. It can be only single bonds (type σ) in sp^3 hybridization, double bonds (one σ and one π) in sp^2, and triple bonds (one σ and two π) in sp, according to hybridization de angle bond. The above explains the formation of a great diversity of compounds or geometric arrangements that the carbon atom has; this variety of spatial conformations is known as allotropes. Therefore, each allotrope has unique physicochemical properties depending on the hybridizations existing in its arrangement [14, 15].

Carbon nanomaterials were discovered in 1980, taking into account the discovery of C60 fullerene as a pillar in the development of these nanomaterials and the development of more sophisticated synthesis methods. Since then, important advances have been made in different fields such as agriculture, electronic devices, and the food industry, and also in biomedical applications [16–18]. These systems are highly hydrophobic and good thermal and electrical conductors. Furthermore, something that stands out is the harshness, which allows the integrity of the formulation to be preserved unlike other systems such as liposomes. It is also worth noting that in various studies they have presented low toxicity and good biocompatibility [19].

Among the great variety of carbon particles, various allotropes can be distinguished such as carbon dots, nanohorns, nanoribbons, and fullerenes. The following sections describe the structure, some physicochemical properties, and important interactions in biological systems, as well as some advances in drug delivery systems related to carbon nanotube structures, graphite and its derivatives such as graphene and graphene oxide, and carbon nanospheres, which are the most representative systems within carbon nanomaterials.

2.1 Graphene

Graphene is a two-dimensional allotrope of carbon in which carbon atoms are arranged in a single-atom-thick sheet packed in a hexagonal-shaped crystal plane having C-C (carbon-carbon) atoms with sp^2 hybridization which allows the formation of various allotropes. It comes from the exfoliation of graphite and, being made up of sp^2 carbon rings, it serves as a good electrical conductor and is hydrophobic [20].

Among graphene derivatives, graphene oxide (GO) is widely studied for biomedical applications (due to its hydrophilicity) as a drug, imaging agent delivery, and drug delivery system. GO is a highly oxidized form of graphene consisting of a single-atom carbon sheet with carboxyl and hydroxyl groups at the periphery, in addition to epoxide groups at the basal surface. The carboxyl groups provide a negatively charged surface and colloidal stability, while the basal planes containing functional groups provide a π-π interaction for the adsorption of molecules. Moreover, the oxygen groups on the surface turn into hydrophilic nanoparticles; thus, the stability in water or polyelectrolyte solutions increases [15]. The presence of reactive functional groups and localized π electrons, which promote π-π bonding with aromatic compounds, make graphene a suitable candidate as a drug delivery system. Besides, the hydroxyl groups are highly reactive to different kinds of molecules, such as polymers, increasing the biocompatibility [21].

There are many methods for the synthesis of GO, one of them is the micromechanical exfoliation of highly oriented pyrolytic graphite which is based on overcoming van der Waals forces of attraction between the graphene layers. Through the self-lubricating ability of graphite in the lateral direction, lateral force can also be exerted to promote relative motion between two layers of graphite. One system used is cutting graphite with a diamond wedge mounted on an ultrasonic oscillation system aligned with respect to a highly oriented pyrolytic graphite mount, obtaining thin layers of graphene [22].

Hummer's method is a method based on the chemical exfoliation of graphite, and the graphite powder is subjected to an oxidation process with $NaNO_3$ and $KMnO_4$. In an acidic medium of H_2SO_4, which integrates functional groups from these oxidizing agents into the double-bonded aromatic groups. Subsequently, a sonication process is carried out to exfoliate the graphite oxide and obtain individual GO sheets. This process can be modified and optimized to obtain biomedical grade GO from a centrifugation purification process at different temperatures until obtaining a neutral pH 6 [23].

The biocompatibility results of GO depend on various factors such as shape, size, and functional groups of the surface and charge. It is also important to consider the method used for its synthesis; for example, the Hummers method of chemical exfoliation can leave traces of oxidizing agents, enhancing the cytotoxic effect of the material. For this, it is suggested that the methodology be modified with a series of washes, greater control in the shape of the GO sheet through sonication processes, and that the material used is free of pyrogens and endotoxins such as lipopolysaccharides to obtain a material with high purity and biomedical grade. It is also known that materials smaller than 100 nm are better for drug delivery systems and cellular internalization [23, 24].

It has been shown that the affinity of GO for membrane lipids favors the cleavage mainly of cholesterol and the formation of oxidized cholesterol species in neutrophil membranes, in addition to finding a relationship between the size of GO sheets and the induction of neutrophil extracellular traps [25]. In addition, the internalization mechanisms of GO into the cytoplasm through endocytosis are mediated by clathrin/caveolin phagocytosis and pinocytosis, and exit the cell through the pathways of lysosome secretion, vesicle-related secretion, and non-vesicle-related secretion [26]. Once internalized, it causes lipid peroxidation and activation of reactive oxygen species, which generates a loss of mitochondrial membrane potential and, therefore, an increase in pro-apoptotic conditions through the activation of the epidermal growth factor receptor/protein kinase B (EGFR/AKT) signaling pathway. In addition to the above, damage to deoxyribonucleic acid (DNA) is caused by covalent binding to it and by the effect of reactive oxygen species. These effects were enhanced when the particle size decreased [27].

In *in vivo* systems, it has been seen to have effects on diverse organs, mainly due to the infiltration of macrophages. GO is mainly captured by the reticuloendothelial system and accumulated in the liver for subsequent excretion through the gastrointestinal system. In this accumulation, alterations in the morphology of hepatocytes have been noted, as well as morphological alterations in the kidney in murine models [28]. Macrophage infiltrates can trigger inflammatory processes in the lungs and damage neurotransmitter levels in the brain. A GO deposition was also observed in the head region that triggered damage to thermosensing neurons in *Caenorhabditis elegans* models [29].

One way to improve the biocompatibility of GO is functionalization. Surface modification allows for improving dispersibility and solubility, and changes the surface charge and particle size, but above all, biocompatibility. One way to do this is through the use of polymers, whether natural or synthetic. Among the most used, we can find polyvinyl alcohol, polyethylene glycol (PEG), polylactic acid, polyglycolic acid or polyacrylic acid, carboxymethyl cellulose, among others [30]. According to some researchers, functionalization with PEG decreases the capture of GO through reticuloendothelial system and decreases the formation of the corona protein in blood, which increases circulation time [24, 30]. In cytotoxicity assays of fibroblasts exposed to GO-PEG, cytocompatibility is shown, as well as in cells from cancer lines such as MCF-7. Furthermore, the biodistribution tests in mice were practiced and revealed a homogeneous organ distribution. Over time, a higher concentration was observed in the kidney and liver, which indicates a process of elimination of the product, which demonstrates that biocompatibility increases once the GO is functionalized [31].

The use of GO as a drug delivery system is extensive, and a wide range of drugs have been used. Initially, chemotherapy drugs in use, for example, doxorubicin (DOX), are taken into consideration in order to reduce adverse effects. Abdelhalim et al. created a system based on GO-DOX, and a cytotoxicity was observed toward the A549 cell line, significantly superior to individual DOX [32]. A more complex system formed by GO, carboxymethyl cellulose, and curcumin as a drug showed high inhibition of 76 and 81% against Saos2 and MCF7 cell lines *in vitro*; this is due to the addition of folic acid, which increases the specificity and selectivity by having a pharmacology target in the design of the release system [33]. The development of these systems can contemplate macromolecules such as nucleic acids or proteins. An example of this is the development of gene delivery such as siRNA. Imani et al. used GO as a nanocarrier, phospholipid-based amphiphilic polymer, and cell-penetrating peptide. It also included silencing siRNA designed for MCF-7 cells, where it was observed capable of inducing cell death in MCF-7 cancer cells compared to non-treated cells [34].

2.2 Carbon nanospheres

Hydrogenation of the carbon graphite structure can cause changes in the bonds of the carbon atoms (sp^2-sp^3), distortion of the carbon rings, increased interlayer distance, and structural disorder in the material, creating sites porous in the material. One of the many forms that come from these structural arrangements in carbon bonds is nanospheres, based on the formation of concentric graphitic layers, they are included in structures ranging from 50 nm to 1 μm in diameter [35].

In principle, the surface of nanospheres does not present oxygen and hydrogen functional groups, making it a hydrophobic structure, and it is only soluble in certain non-polar solvents. Changes in certain parameters in its synthesis or subsequent functionalization processes can modify said surface and generate oxygen and hydrogen functional groups, which increase its hydrophilicity, which is an important characteristic for its biomedical application. The nanospheres produced by the chemical vapor deposition (CVD) technique can be modified during the synthesis by adding ethanol to the precursor solution, resulting in obtaining nanospheres with oxygen and hydrogen functional groups on the surface [36].

Nanospheres synthesis is carried out through the CVD technique, which is based on promoting chemical reactions in a gaseous state from the pyrolysis reaction of a precursor. This can be of different origin; mainly hexane, methane, toluene, or acetylene are used. The reactions are carried out within a quartz reaction chamber at high temperatures and specific pressure and reaction time parameters. It is known that the change of parameters such as temperature, inert gas used, and pressure, as well as the type of precursor, generates different sizes of nanospheres, so it is essential to optimize the parameters according to the desired characteristics and parameters such as graphitization index and presence of argon or hydrogen [36, 37].

Carbon nanostructures typically below 200 nm could be easily internalized into cells by intracellular endocytosis and have therefore been successfully extended to biomedical and pharmaceutical applications, such as the delivery of drugs, genes, proteins, and targeting agents [38]. This use is more attractive if they are functionalized with polymers that provide the system with low cytotoxicity and the large pore volume of carbon nanospheres for safe compatibility with the human body and high drug loading, respectively [39].

Studies carried out in human colon carcinoma cells (Caco-2) have shown that carbon nanospheres of 500–800 nm are not cytotoxic at concentrations between 10 and 800 μg/mL and that they improve the solubility and delivery of non-cytotoxic polar drugs. On the other hand, studies carried out on fibroblasts of carbon materials indicate good biocompatibility results, as well as no hemolysis is observed, and it has a low percentage of protein binding, which is specified to be albumin and fibrinogen and is associated with the increase of the surface area of the material. An important fact to highlight is that this biocompatibility is closely related to the surface functional groups, since pristine materials contain a positive surface charge and generate membrane destruction, as well as activation of the complement system, unlike negatively charged molecules thanks to hydroxyl and carboxyl groups on the surface, which do not show any cytotoxic effect and do not present hemolysis [40]. Furthermore, as mentioned, it is always recommended to wrap the nanoparticles by using polymers such as PEG, which has been shown to have the ability to reduce interactions with the reticuloendothelial system and increases the system's solubility [41].

This opens up the panorama of how to create the design of a release system with nanospheres, since there are few systems developed using nanospheres synthesized by CVD, despite the fact that it is an appropriate synthesis method, since there are no traces of reagents or impurities. Kapri et al. created grass-derived nanospheres functionalized by polyethylenimine and folic acid. The drug under test was functionalization maintain therapeutic efficacy of the drug DOX to the targeted folate receptor overexpressed human cervical cancer cells (HeLa) and human breast cancer cells MDA-MB-231 through receptor-mediated endocytosis, whereas deficient normal cells human embryonic kidney 293 exhibit substantially lower endocytosis under identical conditions [42].

2.3 Carbon nanotubes

Carbon nanotubes were discovered in 1991.They consist of a rolled graphene sheet, either with open or closed ends. According to their type of synthesis, these nanotubes can be single-walled (SWNT) or multiple-walled (MWNT), which consist of two or more concentric graphite sheets around an initial tube. Typically, the distance between each sheet is 0.340 nm. It is important to emphasize that for it to be considered a nanotube, length/width ratios of 1000:1 nm are usually considered and they tend to have measurements widths 0.4–3 nm and lengths between 20 and 1000 nm. Due to carbon nanotubes are hollow, these materials have a high surface area, which allows them to be ideal for nanocarriers or drug delivery. They also have excellent photoluminescent activity and thermal conductivity, which places these materials in the development of phototherapies [14, 43].

Among the most used synthesis methods, CVD stands out; the difference with respect to the synthesis of nanospheres lies in the use of metal catalysts. On the other hand, the laser ablation method, where from the vaporization of graphite in a furnace with controlled temperature, pressures and an inert environment (mainly argon), the degradation of graphite begins and at the same time, it is obtained in the area of catalyst deposit (generally nickel or cobalt) the growth of the nanotubes. In addition to these methods, the initial method by which they were discovered is the arch discharged, which consists of a chamber at high temperatures, where an electric discharge is carried out in the presence of metal catalysts, with the discharge they reach high temperatures for the graphite sublimation, going from solid to plasma. In this process, carbon atoms are deposited in colder areas and rearranged in a graphitic form that finally generates carbon nanotubes [44, 45].

To generate a drug delivery system, it is necessary to functionalize the materials, since nanotubes are characterized by being highly hydrophobic and are only soluble in certain organic solvents. The most used strategies to follow are the use of polymers, or surfactants, and even the use of biomolecules such as amino acids. It can also be subjected to chemical oxidation or addition of amino groups [44].

Nanotubes can, due to their hydrophobic nature, have an affinity for membrane lipids, which generate membrane disruption or internalization of some short-sized components. Thus, various toxic effects have been observed such as the production of reactive oxygen species, lipid peroxidation, and elevation of mitochondrial succinate dehydrogenase activity, which translate into mitochondrial stress and activation of apoptosis signaling pathways. Furthermore, in distribution studies in murine models, a strong bioaccumulation has been found in the lungs and liver, related to the activity of macrophages from the reticuloendothelial system, which due to their shape is difficult to metabolize. On the other hand, carbon nanotubes can be metabolized by neutrophil myeloperoxidases. The inflammation in neurons activates necrosis tumor factor-α (TNFα), interleukin-1β (IL-1β), and interleukin-6 (IL-6) pathways [46].

It is known that functionalization allows cytotoxicity to be reduced; for example, Ahmadi et al. describe PEGylated and polysorbate-coated nanotubes administered in BALB/mice did not show signs of acute intoxication, although an increase in antioxidant activity was observed in liver tissue [47].

Some formulations of drug delivery systems from carbon nanotubes have been successful. Yu et al. prepared chitosan to the nanotubes to improve biocompatibility and added paclitaxel, a well-known chemotherapeutic, and observed that at pH 5.5, the drug was released better in cell culture A549. It also demonstrates a combination of the formation of reactive oxygen species, reflected in a cytotoxic effect [48]. On the other hand, Hassan et al. designed MWNT as a co-administer of immunomodulators such as cytosine-phosphate-guanine oligodeoxynucleotide and anti-CD40 Ig, with ovalbumin; it was demonstrated that co-administration increased the response in T cells in C57BL/6 mice, as well as inhibiting the growth of OVA-expressing B16F10 melanoma cells in subcutaneous or lung pseudo-metastatic tumor models, with respect to administration without MWNT [49].

Due to the structures being capable of being easily internalized into cells, they have been used for gene delivery with molecules such as siRNA, miRNA, or plasmids. Cao et al. oxidized SWNT and functionalized them with polyethylenimine, betaine, penetrating peptide (BR2) as a target molecule, siRNA for gene silencing, and DOX as a chemotherapeutic drug. It was observed that there was a significantly reduced tumor volume in A549 cell-bearing nude mice, in addition to the fact that there is no damage to non-tumor tissue [50].

3. Conclusion

Carbon nanomaterials have various important characteristics depending on the way in which they are structured, which offer an important range of possibilities in biomedical applications. One branch of this is the development of drug delivery systems, whose importance lies in improving the quality of life of patients by reducing the adverse effects of current therapies during and after treatment. To this end, exploring new materials with desirable characteristics for the development of these systems, such as large surface area, surface charge, rigidity, and ease of functionalization or drug loading, is of utmost importance. Although it has been observed that on the one hand they have important cytotoxic characteristics, it is desirable to review this information to propose improvements in functionalization and explore new synthesis techniques that allow us to increase their biocompatibility.

Acknowledgements

Luis A. De Sales Palestina thanks Consejo Nacional de Ciencia y Tecnología (CONACyT) for his doctoral fellowship.

Conflict of interest

The authors declare no conflict of interest.

Notes/thanks/other declarations

The authors do not have other declarations.

Author details

Luis Alberto De Sales Palestina, Sergio Alberto Bernal Chávez*
and Jessica Rosaura Campos Delgado
University of the Americas Puebla, Puebla, Mexico

*Address all correspondence to: sergio.bernal@udlap.mx

References

[1] Bayda S, Adeel M, Tuccinardi T, Cordani M, Rizzolio F. The history of nanoscience and nanotechnology: From chemical–physical applications to nanomedicine. Molecules;**25**(1):112. Available from: https://pmc.ncbi.nlm.nih.gov/articles/PMC6982820/

[2] Abid N, Khan AM, Shujait S, Chaudhary K, Ikram M, Imran M, et al. Synthesis of nanomaterials using various top-down and bottom-up approaches, influencing factors, advantages, and disadvantages: A review. Advances in Colloid and Interface Science. 2022;**300**. Available from: https://pubmed.ncbi.nlm.nih.gov/34979471/

[3] Farokhzad OC, Langer R. Impact of nanotechnology on drug delivery. ACS Nano. 2009;**3**(1):16-20. Available from: www.acsnano.org

[4] Mitchell MJ, Billingsley MM, Haley RM, Wechsler ME, Peppas NA, Langer R. Engineering precision nanoparticles for drug delivery. Nature Reviews. Drug Discovery. 2020;**20**(2):101-124. Available from: https://www.nature.com/articles/s41573-020-0090-8

[5] Diez-Pascual AM, Rahdar A. Functional nanomaterials in biomedicine: Current uses and potential applications. ChemMedChem. 2022;**17**(16). Available from: https://pubmed.ncbi.nlm.nih.gov/35729066/

[6] Idris AO, Akanji SP, Orimolade BO, Olorundare FOG, Azizi S, Mamba B, et al. Using nanomaterials as excellent immobilisation layer for biosensor design. Biosensors. 2023;**13**(2). Available from: https://pubmed.ncbi.nlm.nih.gov/36831958/

[7] Tiwari G, Tiwari R, Sriwastawa B, Bhati L, Pandey S, Pandey P, et al. Drug delivery systems: An updated review. International Journal of Pharmaceutical Investigation. 2012;**2**(1):2. Available from: https://pmc.ncbi.nlm.nih.gov/articles/PMC3465154/

[8] Li C, Wang J, Wang Y, Gao H, Wei G, Huang Y, et al. Recent progress in drug delivery. Acta Pharmaceutica Sinica B. 2019;**9**(6):1145-1162. Available from: https://pubmed.ncbi.nlm.nih.gov/31867161/

[9] García-Pinel B, Porras-Alcalá C, Ortega-Rodríguez A, Sarabia F, Prados J, Melguizo C, et al. Lipid-based nanoparticles: Application and recent advances in cancer treatment. Nanomaterials. Available from: www.mdpi.com/journal/nanomaterials

[10] Seidu TA, Kutoka PT, Asante DO, Farooq MA, Alolga RN, Bo W. Functionalization of nanoparticulate drug delivery systems and its influence in cancer therapy. Pharmaceutics. 2022;**14**(5). Available from: https://pmc.ncbi.nlm.nih.gov/articles/PMC9145684/

[11] Zielinska A, Carreiró F, Oliveira AM, Neves A, Pires B, Nagasamy Venkatesh D, et al. Polymeric nanoparticles: Production, characterization, toxicology and ecotoxicology. Molecules. 2020;**25**(16). Available from: https://pmc.ncbi.nlm.nih.gov/articles/PMC7464532/

[12] Begines B, Ortiz T, Pérez-Aranda M, Martínez G, Merinero M, Argüelles-Arias F, et al. Polymeric nanoparticles for drug delivery: Recent developments and future prospects. Nanomaterials. Available from: www.mdpi.com/journal/nanomaterials

[13] Luther DC, Huang R, Jeon T, Zhang X, Lee Y-W, Nagaraj H, et al. Delivery of drugs, proteins, and nucleic acids using inorganic nanoparticles. Advanced Drug Delivery Reviews. 2020. DOI: 10.1016/j.addr.2020.06.020

[14] Mohajeri M, Behnam B. Biomedical applications of carbon nanomaterials: Drug and gene delivery potentials. Journal of Cellular Physiology. 2018. Available from: https://www.researchgate.net/publication/326853418

[15] Rahman M, Akhter S, Ahmad MZ, Ahmad J, Addo RT, Ahmad FJ, et al. Emerging advances in cancer nanotheranostics with graphene nanocomposites: Opportunities and challenges. Nanomedicine (London, England). 2015;**10**(15):2405-2422. Available from: https://pubmed.ncbi.nlm.nih.gov/26252175/

[16] Sharma M. Understanding the mechanism of toxicity of carbon nanoparticles in humans in the new millennium: A systemic review. Indian Journal Of Occupational And Environmental Medicine. 2010;**14**(1):3. Available from: https://pmc.ncbi.nlm.nih.gov/articles/PMC2923423/

[17] Mukherjee A, Majumdar S, Servin AD, Pagano L, Dhankher OP, White JC. Carbon nanomaterials in agriculture: A critical review. Frontiers in Plant Science. 2016;7:168246

[18] Ozcakir G. Applications of nanomaterials in food industry: A review. Materials Proceedings. 2023;**14**(1):22. Available from: https://www.mdpi.com/2673-4605/14/1/22/htm

[19] Tiwari SK, Kumar V, Huczko A, Oraon R, De Adhikari A, Nayak GC. Magical allotropes of carbon: Prospects and applications. Critical Reviews in Solid State and Materials Sciences. 2016;**41**(4):257-317. Available from: https://www.tandfonline.com/doi/abs/10.1080/10408436.2015.1127206

[20] Jampilek J, Kralova K. Advances in drug delivery nanosystems using graphene-based materials and carbon nanotubes. Materials (Basel). 2021;**14**(5):1-39. Available from: https://pmc.ncbi.nlm.nih.gov/articles/PMC7956197/

[21] Oliveira AML, Machado M, Silva GA, Bitoque DB, Ferreira JT, Pinto LA, et al. Graphene oxide thin films with drug delivery function. Nanomaterials. 2022;**12**(7). Available from: https://pmc.ncbi.nlm.nih.gov/articles/PMC9000550/

[22] Yi M, Shen Z. A review on mechanical exfoliation for the scalable production of graphene. Journal of Materials Chemistry A. 2015;**3**(22):11700-11715. Available from: https://pubs.rsc.org/en/content/articlehtml/2015/ta/c5ta00252d

[23] Rodrigues AF, Newman L, Lozano N, Mukherjee SP, Fadeel B, Bussy C, et al. A blueprint for the synthesis and characterisation of thin graphene oxide with controlled lateral dimensions for biomedicine. 2D Materials. 2018;**5**(3):035020. Available from: https://iopscience.iop.org/article/10.1088/2053-1583/aac05c

[24] Liao C, Li Y, Tjong SC. Graphene nanomaterials: Synthesis, biocompatibility, and cytotoxicity. International Journal of Molecular Sciences. 2018;**19**(11). Available from: https://pmc.ncbi.nlm.nih.gov/articles/PMC6274822/

[25] Mukherjee SP, Lazzaretto B, Hultenby K, Newman L, Rodrigues AF, Lozano N, et al. Graphene oxide elicits membrane lipid changes and neutrophil

extracellular trap formation. Chem. 2018;**4**(2):334-358

[26] Piperno A, Scala A, Mazzaglia A, Neri G, Pennisi R, Sciortino MT, et al. Cellular signaling pathways activated by functional graphene nanomaterials. International Journal of Molecular Sciences. 2018;**19**(11). Available from: https://pmc.ncbi.nlm.nih.gov/articles/PMC6274994/

[27] Gurunathan S, Kang MH, Jeyaraj M, Kim JH. Differential cytotoxicity of different sizes of graphene oxide nanoparticles in Leydig (TM3) and Sertoli (TM4) cells. Nanomaterials. 2019;**9**(2):139. Available from: https://www.mdpi.com/2079-4991/9/2/139/htm

[28] Rhazouani A, Gamrani H, El Achaby M, Aziz K, Gebrati L, Uddin MS, et al. Synthesis and toxicity of graphene oxide nanoparticles: A literature review of In vitro and In vivo studies. BioMed Research International. 2021;**2021**. Available from: https://pubmed.ncbi.nlm.nih.gov/34222470/

[29] Kim M, Eom HJ, Choi I, Hong J, Choi J. Graphene oxide-induced neurotoxicity on neurotransmitters, AFD neurons and locomotive behavior in *Caenorhabditis elegans*. Neurotoxicology. 2020;**77**:30-39

[30] Ghosh S, Chatterjee K. Poly(ethylene glycol) functionalized graphene oxide in tissue engineering: A review on recent advances. International Journal of Nanomedicine. 2020;**15**:5991. Available from: https://pmc.ncbi.nlm.nih.gov/articles/PMC7656781/

[31] Guo Z, Chakraborty S, Monikh A, Varsou D-D, Chetwynd AJ, Afantitis A, et al. Surface functionalization of graphene-based materials: Biological behavior, toxicology, and safe-by-design aspects. Advanced Biology. 2021;**5**(9):2100637. Available from: https://onlinelibrary.wiley.com/doi/full/10.1002/adbi.202100637

[32] Abdelhalim AOE, Ageev SV, Petrov AV, Meshcheriakov AA, Luttsev MD, Vasina LV, et al. Graphene oxide conjugated with doxorubicin: Synthesis, bioactivity, and biosafety. Journal of Molecular Liquids. 2022;**359**:119156

[33] Sahne F, Mohammadi M, Najafpour GD. Single-layer assembly of multifunctional carboxymethylcellulose on graphene oxide nanoparticles for improving in vivo curcumin delivery into tumor cells. ACS Biomaterials Science & Engineering. 2019;**5**(5):2595-2609. Available from: https://pubmed.ncbi.nlm.nih.gov/33405765/

[34] Imani R, Shao W, Taherkhani S, Emami SH, Prakash S, Faghihi S. Dual-functionalized graphene oxide for enhanced siRNA delivery to breast cancer cells. Colloids and Surfaces. B, Biointerfaces. 2016;**147**:315-325

[35] Jin YZ, Gao C, Hsu WK, Zhu Y, Huczko A, Bystrzejewski M, et al. Large-scale synthesis and characterization of carbon spheres prepared by direct pyrolysis of hydrocarbons. Carbon N Y. 2005;**43**(9):1944-1953

[36] Panickar R, Sobhan CB, Chakravorti S. Chemical vapor deposition synthesis of carbon spheres: Effects of temperature and hydrogen. Vacuum. 2020;**172**:109108

[37] Yong G, Zhang X, She W, Zhang B. Easy synthesis of hydrogenated amorphous carbon from benzene. Carbon N Y. 2013;**55**:369-371

[38] Jeskey J, Chen Y, Kim S, Xia Y. EDTA-assisted synthesis of nitrogen-doped carbon nanospheres with uniform

sizes for photonic and electrocatalytic applications. Chemistry of Materials. 2023;**35**(7):3024-3032. Available from: https://pubs.acs.org/doi/full/10.1021/acs.chemmater.3c00341

[39] Zhao P, Wang L, Sun C, Jiang T, Zhang J, Zhang Q, et al. Uniform mesoporous carbon as a carrier for poorly water soluble drug and its cytotoxicity study. European Journal of Pharmaceutics and Biopharmaceutics. 2012;**80**(3):535-543. Available from: https://pubmed.ncbi.nlm.nih.gov/22193360/

[40] Misra SK, Chang HH, Mukherjee P, Tiwari S, Ohoka A, Pan D. Regulating biocompatibility of carbon spheres via defined nanoscale chemistry and a careful selection of surface functionalities. Scientific Reports. 2015;**5**(1):1-13. Available from: https://www.nature.com/articles/srep14986

[41] Pasut G, Veronese FM. State of the art in PEGylation: The great versatility achieved after forty years of research. Journal of Controlled Release. 2012;**161**(2):461-472. Available from: https://pubmed.ncbi.nlm.nih.gov/22094104/

[42] Kapri S, Majee R, Bhattacharyya S. Chemical modifications of porous carbon nanospheres obtained from ubiquitous precursors for targeted drug delivery and live cell imaging. ACS Sustainable Chemistry & Engineering. 2018;**6**(7):8503-8514. Available from: https://pubs.acs.org/doi/abs/10.1021/acssuschemeng.8b00785

[43] Ray SK, Chakrabarti M, Kiseleva R, Vertegel A. Carbon nanomaterials for drug delivery and cancer therapy. Journal of Nanoscience and Nanotechnology. 2015;**15**:5501-5511. Available from: www.aspbs.com/jnn

[44] Karimi M, Solati N, Amiri M, Mirshekari H, Mohamed E, Taheri M, et al. Carbon nanotubes part I: Preparation of a novel and versatile drug-delivery vehicle. Expert Opinion on Drug Delivery. 2015;**12**(7):1071. Available from: https://pmc.ncbi.nlm.nih.gov/articles/PMC4475469/

[45] Chrzanowska J, Hoffman J, Małolepszy A, Mazurkiewicz M, Kowalewski TA, Szymanski Z, et al. Synthesis of carbon nanotubes by the laser ablation method: Effect of laser wavelength. Physica Status Solidi B. Available from: www.pss-b.com

[46] Witkowska M, Florek E, Mrówczyński R. Assessment of pristine carbon nanotubes toxicity in rodent models. International Journal of Molecular Sciences. 2022;**23**(23). Available from: https://pmc.ncbi.nlm.nih.gov/articles/PMC9739793/

[47] Ahmadi H, Ramezani M, Yazdian-Robati R, Behnam B, Razavi Azarkhiavi K, Hashem Nia A, et al. Acute toxicity of functionalized single wall carbon nanotubes: A biochemical, histopathologic and proteomics approach. Chemico-Biological Interactions. 2017;**275**:196-209. Available from: https://pubmed.ncbi.nlm.nih.gov/28807745/

[48] Yu B, Tan L, Zheng R, Tan H, Zheng L. Targeted delivery and controlled release of Paclitaxel for the treatment of lung cancer using single-walled carbon nanotubes. Materials Science & Engineering. C, Materials for Biological Applications. 2016;**68**:579-584. Available from: https://pubmed.ncbi.nlm.nih.gov/27524057/

[49] Hassan HAFM, Smyth L, Wang JTW, Costa PM, Ratnasothy K, Diebold SS, et al. Dual stimulation of antigen presenting cells using carbon

nanotube-based vaccine delivery system for cancer immunotherapy. Biomaterials. 2016;**104**:310-322. Available from: https://pubmed.ncbi.nlm.nih.gov/27475727/

[50] Zakaria AB, Picaud F, Rattier T, Pudlo M, Saviot L, Chassagnon R, et al. Nanovectorization of TRAIL with single wall carbon nanotubes enhances tumor cell killing. Nano Letters. 2015;**15**(2):891-895. Available from: https://pubs.acs.org/doi/abs/10.1021/nl503565t

Chapter

Identifying Specific Brain Regions When a Perturbation in the Wrist Is Applied: A First Step before System Modeling

Martín Durán, Rocío Salazar and Gibran Etcheverry

Abstract

In the modeling of the cortical brain activity elicited by an external stimulation, the first required stage is to identify and define the brain regions that are activated. The objective of the present work is to propose a methodology to identify these regions. To achieve this goal, a public database was used. For this, 10 subjects were involved in the experiment, in which a mechanical perturbation (input) is applied to the wrist of each subject and the electroencephalography signal is recorded (output). The brain response was identified based on Independent Component Analysis and selecting the component with the maximal signal to noise ratio. The output response was established as the cortical response located in the left side of the brain, since the perturbation was applied to the right wrist of each subject, and it was identified by visualizing the spatial distribution of each independent component.

Keywords: brain response, electroencephalogram, identification of the cortical response, Independent Component Analysis, external wrist manipulation

1. Introduction

The human brain has been the object of neuroscientific research for so many years. However, there exists a wide set of processes that cannot be explained yet. The main factor of this challenge is the complexity of the human brain and the interaction between different regions; that is, the human brain is an extraordinary and vital organ and, moreover, it deviates from the relationship between body and brain size that applies to other primates [1].

On the one hand, some efforts have been developed to understand brain activity when performing specific tasks. For instance, Tongtong Li et al. [2] concluded that larger processing capacity is related to low levels of brain activity, while larger information arrival rate increases it.

Moreover, the advances in brain-computer interfaces (BCIs are systems that work as a communication channel between a human brain and a computer, an embedded system, or a mechatronic device) have accelerated in the last years thanks to the development of new signal processing methods, data acquisition devices, and specific software for neurologic applications.

The development of BCIs has allowed the extraction of information from brain activity when specific tasks are performed. For instance, in Mexico, Vera et al. proposed in 2017 a recording, processing, and classification scheme for a spelling BCI application [3].

In this work, a dataset from the University of Technology of Netherland [4] was used to analyze the brain activity due to afferent processes, where seven different mechanical signals perturb the wrist of 10 healthy subjects. The brain activity provoked by those perturbations was measured via EEG and the cortical response was identified by applying Independent Component Analysis (ICA), and the component with maximal signal to noise ratio (SNR) was taken. There are two main contributions in this work. The first contribution is the new way to gather information coming from EEG signals in matrix that take into account one single perturbation, which assures that the response belongs exclusively to that perturbation and makes the process of cortical response identification faster, which is crucial in BCI systems. The second contribution is the analysis in the time and frequency domains to reject artifacts, two simple methods that are useful in real-time applications.

2. Related work

Electroencephalographical (EEG) signals are essential resources to analyze brain activity. To understand underlying brain processes, blind source separation is a powerful tool. In this sense, one of the most common techniques is the analysis of independent components. For instance, Melnik et al. [5] used ICA to extract motor-related, sensory-related, and sensorimotor-related independent components, which they called sensorimotor ICs. The experiment that they designed was about performing categorization-response tasks. They concluded that some functional modules are participating in both motor and sensory actions.

An important work that intends to find out a link between cortical region activity and motor tasks is that of Yang et al. [6]. They provoked the flexion of the right-handed subject's wrist that changed the length of these muscles. When they measured the cortical activity, they found that 4 areas of the brain were activated: primary sensorimotor areas, prefrontal areas, supplementary motor areas, and the posterior parietal cortex.

Vlaar et al. [4] applied a high-pass filter at a cutoff frequency of 1 Hz for signals found in the same dataset used in this project. They applied these filters to attenuate low frequencies due to blood pressure or breathing. In the two other works [6, 7], a zero phase shift bandpass filter was applied to attenuate noise signals coming from electromagnetic interference with high frequency and inner signals with low frequencies. They also used a notch filter [6, 7], or an ideal filter [4] at a frequency of 50 Hz in order to attenuate the line signal (European line frequency). Finally, a downsampling was applied to the filtered signals with resampled frequency of 256 Hz [4] or 512 Hz [6, 7] in order to reduce computational complexity.

3. Fundamental concepts

3.1 Sensorimotor system

In this work, the sensing elements are the afferent fibers (fibers that send information from limbs to the brain) that convert the mechanical angular movement at the level of the wrist joint into electrical signals sent via the fibers $A\alpha$, $A\beta$, $A\delta$, and C [8].

A key system involved in the sensing process is the proprioceptive system, which consists of muscle spindles, Golgi tendon organs, joint capsules, and stretch-sensitive free endings. Mechanoreceptors are able to send information via the central nervous system [8]. In other words, the mechanoreceptors are receptors for converting mechanical events into neural signals [9].

When muscles are stretched, changes in both the muscle length and the rate of change in muscle length are sensed via the muscles' spindles. Groups Ia afferent and II afferent correspond to the fibers Aα and Aβ, respectively. Group Ia afferent transmits the information of the length or change in the muscle length, and group II afferent only sends the muscle length information [8].

3.2 Independent Component Analysis

Regarding biomedical signals, an artifact is a spurious signal that can be caused by internal or external events, such as blinking eyes, movements of muscles, and heart activity among others. Several methodologies have been proposed to preprocess the signal and remove artifacts. ICA is a widely used method, considering that it requires extensive computational resources because it requires large datasets to process the information and then obtain such components. Hence, ICA method is a time-consuming process.

Electrical brain activity can be captured at the scalp of the subjects via EEG recordings. Processing these signals, it is possible to analyze them and extract important information as the sources of brain activity, as shown in **Figure 1**.

Eq. (1) shows the relationship between the independent components and the EEG signals.

$$x = As \tag{1}$$

In Eq. (1), x represents the observed signals. It is a matrix of dimensions $r \times c$, where r represents the number of channels and c is the number of samples at each channel. A is the mixing matrix. It is used for plotting scalp maps of independent components. Finally, s contains the independent components.

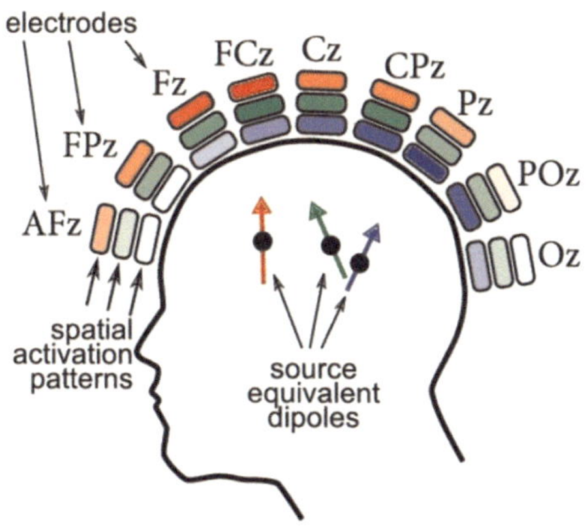

Figure 1.
Cortical response stage via Independent Component Analysis. Image taken from [10].

4. Materials

4.1 Informatic resources

The present results have been developed on Python 3.8, by using Dell Inspiron 15, 5000 Series, with Processor Intel(R) Core(TM) i7, 2.40GHz, and 16GB RAM, over the Operating System Ubuntu 18.04. The stages of preprocessing and selection of cortical sources require great capacity and computing time. That is why an agreement with the National Supercomputer Laboratory of Southeast Mexico was made (more information).

4.2 Database and experiment

The Delft University of Technology of Netherlands shared a dataset built with data collected from 10 healthy right-handed subjects (six men and four women). The dataset was originally introduced in [4, 8].

In this experiment, a person is seated and his arm is fixed to a robot, as is shown in the left image of **Figure 2**. The perturbation input is a rotational movement on his wrist joint. This perturbation consists of 7 different inputs (realizations). Three different realizations are shown in the color blue, red, and green, in which shadowed regions were removed to decrease transition effects. Each input is a multisine signal with frequencies 1, 3, 5, 7, 9, 11, 13, 15, 19, and 23 Hz and random phase, two input realizations are shown at the top right of **Figure 2**. The first three frequencies have the highest amplitude, and then it descends 20 dB at each frequency; this frequency content is visualized at the bottom right of **Figure 2**. Each realization has 210 periods of 1 s, which means that, for each subject, the experiment had a time duration of 7 × 210 = 1470 seconds. Brain activity is measured via EEG, where 126 channels were registered at a sampling frequency of 2048 Hz.

5. Method proposal

As it can be seen in **Figure 3**, from row signals of the database (126 EEG channels), a preprocessing stage was performed to reduce noise coming from internal and external factors, such as blood pressure and electrical line, respectively. After the filtering

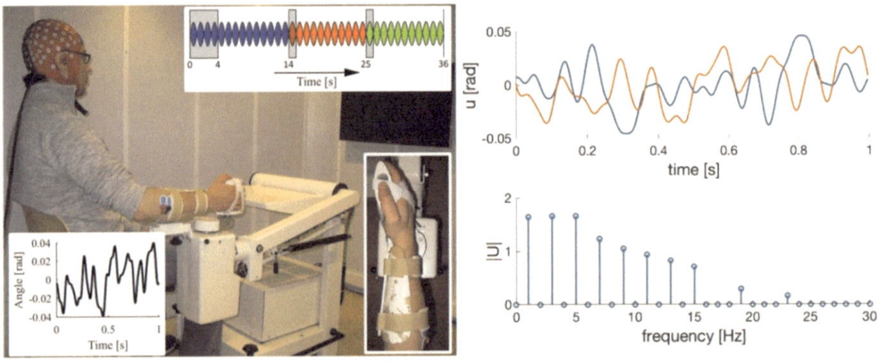

Figure 2.
Image taken from benchmark "Cortical responses evoked by wrist joint manipulation" [4, 8].

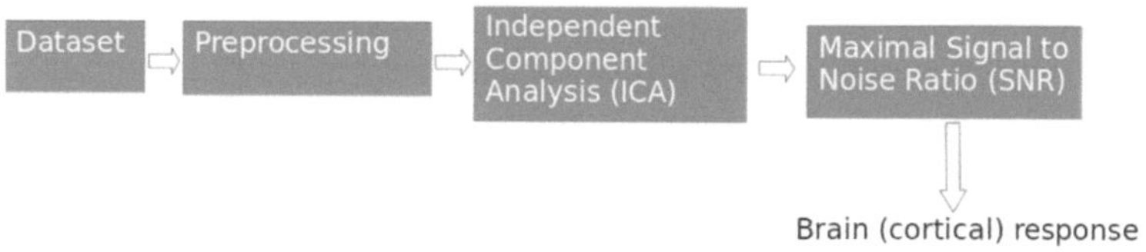

Figure 3.
General scheme of the proposal to identify the cortical response caused by external mechanical perturbations.

process, 125 independent components were extracted via ICA. A downsampling was applied to the independent components with the resampled frequency of 256 Hz to reduce computational complexity. For each independent component, the SNR was calculated. The component with the maximal SNR is chosen as output, that is, the cortical activity of the brain due to a mechanical external perturbation in the wrist.

As we want to obtain a model that explains the relation between the input and the output, it would be ideal that no noise affects both signals. That is why the independent component with the maximal SNR is chosen as output. In other words, to decide which IC represents the brain activity triggered by the mechanical perturbation, the first criterion to take into account is the maximal SNR for each IC. The SNR is calculated by obtaining the mean of each independent component divided by its averaged variance [4].

5.1 Preprocessing.

As signals are contaminated by noise, a zero phase shift bandpass filter was applied with cutoff frequencies at 1 Hz and 100 Hz to attenuate low frequencies due to blood pressure or breathing and high frequencies coming from electromagnetic interference of external environment. We also applied a notch filter at a frequency of 50 Hz to attenuate the line signal.

B. Identifying the cortical response.

SInce there are 126 channels, 120 periods, and 7 realizations and the sample frequency is 2048 Hz, a matrix x is built $\left(x \in R^{126x245760}\right)$. The number of rows of matrix x is equal to the number of channels, and the number of columns is equal to the number of periods times the sampling frequency. For each subject, seven matrices are built according to the number of realizations.

The objective of applying ICA is to obtain the independent components that can be established into a new matrix s, as shown in Eq. (1). For this project, x represents the EEG recordings matrix $\left(x \in R^{126x3010560}\right)$, A contains the weights' matrix $\left(A \in R^{126x125}\right)$, and s the independent components' matrix $\left(s \in R^{125x3010560}\right)$. Finally, the independent components' matrix s is segmented by periods and by realizations, where a tensor IC is built with dimensions $7 \times 210 \times 2048 \times 125$. Hence, $IC[M, P, n, i] \in R^{7x210x2048x125}$), where M is the realizations $\{1, 2, \ldots, 7\}$, P is the periods $\{1, 2, \ldots, 210\}$, n represents the samples $\{1, 2, \ldots, 2048\}$, and i is the independent components $\{1, 2, \ldots, 125\}$.

The independent component with the maximal SNR is chosen as the cortical response; that is, the output is the independent component with the lowest standard deviation. We applied a similar method as Vlaar et al. [4] to calculate the SNR of each component (see Algorithm 1). The main difference with our method is that in this project, ICA process is performed for each realization; that is, for each subject, there are seven ICA processes, instead of just one as Vlaar's method. This new way to gather information accelerates the process of cortical response identification, which is a valuable contribution when this project is implemented in a BCI system. From Algorithm 1, T is the total number or periods, $\overline{IC}_i(n)$ is the first moment of the ith independent component with $i \in [1, 125]$ $(i \in Z)$, and r is the index of the independent component with maximal SNR at each realization. The cortical response is represented as $y_M(n)$, which is provoked by the corresponding multisine perturbation $u_M(n)$.

Algorithm 1: SNR Algorithm

Initialization

$T = 210$

for $M = 1{:}7$ **do**

$\quad \overline{IC}_{M,i}(n) = \frac{1}{T}\sum_{P=1}^{210} IC[M,P,n,i]$

$\quad E_{M,i}(n) = \overline{IC}^2_{M,i}(n)$

$\quad \Sigma_{M,i}(n) = \frac{1}{T-1}\sum_{P=1}^{210}\left(IC[M,P,n,i] - \overline{IC}_{M,i}(n)\right)^2$

$\quad SNR_M(i) = \frac{\sum_{n=1}^{2048} E_{M,i}(n)}{\sum_{n=1}^{2048} \Sigma_{M,i}(n)}$

$\quad r = max\,(SNR_M(i))$

$\quad y_M(n) = \overline{IC}_{M,r}$

end

5.2 Artifacts' rejection

After applying ICA, we choose the five independent components with the maximal SNR, and we look into its associated matrix (the corresponding column of the weighted matrix that gives us the contribution of each channel). In this process, some artifacts, such as blinking eyes, could appear because the amplitude is greater than the cortical response. For that reason, an analysis in the time and frequency domains is proposed to correctly identify the brain response caused by the external mechanical perturbation.

5.3 System identification proposal

As we already know that delays in the sensorimotor system are important in our modeling, the proposed architecture contemplates two delays of the input signal. The basic structure of the cortical response model consists of two delays of the input and then multiplication of both delays as follows:

$$y2 = a_{1,1}y1_{\tau_1} + a_{1,2}y1_{\tau_2} + a_{1,3}y1^2_{\tau_1} + a_{1,4}y1_{\tau_1}y1_{\tau_2} + a_{1,5}y1^2_{\tau_2} \quad (2)$$

where $y1 = u(n)$ is the input signal, $y1_{\tau_1} = y1(n-\tau_1)$ and $y1_{\tau_2} = y1(n-\tau_2)$ are two delays of input signal, and $a_{s,k}$ (s = 1, k = 1, 2, 3, 4, 5) are the parameters to estimate, where s represents the number of stages (for Eq. (2), there is one stage) and k represents number of parameters per stage.

τ_1 and τ_2 are delays of the system set to 16 ms and 32 ms, respectively, according to the delays of groups Ia afferent and II afferent.

A block representation of Eq. (2) can be seen in **Figure 4**. We can see that the input $y1(n)$ corresponds to the averaged input signal $u(n)$, which is the external mechanical perturbation. This signal has two delays $y1(n-t1)$ and $y1(n-t2)$.

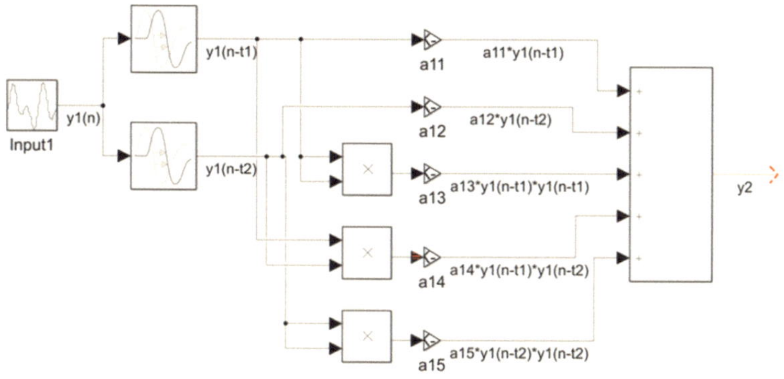

Figure 4.
Basic structure of the cortical response model.

From the basic structure of the cortical response modeled in Eq. (2) and shown in **Figure 4**, the output $y2$ is the input of another basic structure, so this cascade structure generates new harmonics at each stage, apparently the same as the sensorimotor system.

From Eq. (2), it is possible to generalize the basic structure of the cortical response model to any stage s as shown in Eq. (3).

$$(y)_{s+1} = a_{s,1}\left(y_{\tau_1}\right)_s + a_{s,2}\left(y_{\tau_2}\right)_s + a_{s,3}\left(y_{\tau_1}^2\right)_s + a_{s,4}\left(y_{\tau_1}\right)_s\left(y_{\tau_2}\right)_s + a_{s,5}\left(y_{\tau_2}\right)_s \tag{3}$$

where $(y)_s$ is the input signal at stage s, $\left(y_{\tau_i}\right)_s$ is the delayed signal $y(n-\tau_i)$ at stage s, $(i = 1, 2)$, and $a_{s,k}$ is the parameter a_k at stage s, $(k = 1, 2, 3, 4, 5)$.

For the proposal structure, $s = 1, 2, 3, 4$, and $(y)_1$ is the angular perturbation (for $s \leq 0, (y)_s = 0$).

All parameters to be estimated can be arranged into a matrix, as shown in Eq. (4):

$$A = \begin{bmatrix} a_{1,1} & a_{1,2} & a_{1.3} & a_{1,4} & a_{1,5} \\ a_{2,1} & a_{2,2} & a_{2.3} & a_{2,4} & a_{2,5} \\ a_{3,1} & a_{3,2} & a_{3.3} & a_{3,4} & a_{3,5} \\ a_{4,1} & a_{4,2} & a_{4.3} & a_{4,4} & a_{4,5} \end{bmatrix} \tag{4}$$

As seen in Eq. (4), the rows of matrix A correspond to the model stages; that is, each row represents the basic structure represented in **Figure 4**. The columns of matrix A correspond to the weights of signals $\left[y_{row}(n-\tau1), y_{row}(n-\tau2), y_{row}^2(n-\tau1), y_{row}(n-\tau1) * y_{row}(n-\tau2), y_{row}^2(n-\tau2)\right]$, respectively, from left to right.

6. Results and discussion

For each subject and each realization, five independent components with maximal SNR were calculated. The scalp maps (obtained from the matrix A of Eq. (1)) of these independent components are organized in a new matrix, where each row represents the number of realization and each column represents the number of independent components with maximal SNR, decreasing from left to right.

Figure 5 shows the scalp map for subject 06, and it highlights the components, which are most likely to be related to eye and muscle artifacts, by shading them in yellow color (location: [1,4], [1,5], [2,4], [2,5], [3,4], [5,3], and [7,3]). Those green boxes represent the scalp maps where it is likely to be the cortical response due to the external perturbation.

To separate cortical and noncortical responses, both the time domain and frequency domain of the independent components were considered. After analyzing the scalp map, the components in the time domain were taken into account. We obtained as a result other independent components that correspond to the noncortical response; therefore, we can reject them for further analysis.

Figure 6 highlights, in red color, four independent components that are not the desired cortical response (location: [1,4], [3,2], [5,2], and [7,3]). As we can see, these four components do not have a homogeneous behavior as most independent components have, which means that they do not follow a similar behavior at each period. In addition, components [3,2] and [5,2] have peaks that do not agree with the expected brain activity for the task that was performed. For that reason, they can be rejected for future analysis, as they do not represent the expected cortical response.

Finally, the component activity power spectrum is analyzed to separate cortical from the noncortical response. **Figure 7** shows the scalp maps and their response in the frequency domain for subject 06. According to this domain, there are 12 independent components that offer almost no information and a nonstructured scalp map (location: [1,2], [1,3], [1,5], [2,4], [3,2], [3,4], [5,2], [5,3], [5,4], [6,2], [6,5], and [7,5]). The other two independent components highlighted in red color (location: [1,4] and [7,3]) have similar frequency domain and scalp maps, they both have too much information in almost all range of frequencies, and the brain activity is focused in the right frontal area; hence, they can be considered as artifacts.

Figure 7 also shows the average (over all periods) of the frequencies of the independent component with maximal SNR. This signal has a descending amplitude as

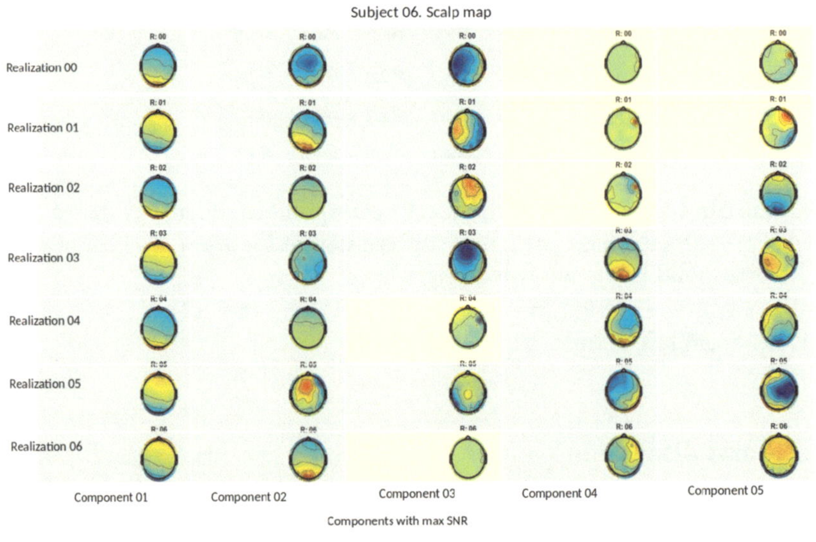

Figure 5.
Independent components' scalp maps for subject 06.

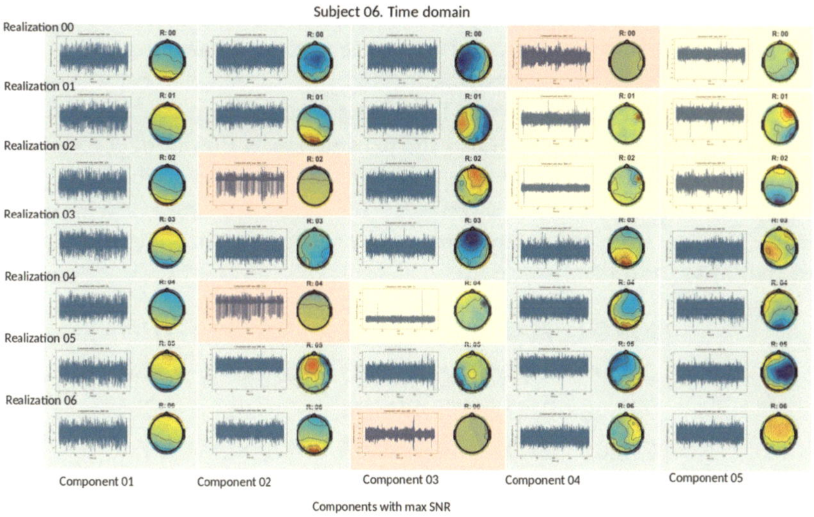

Figure 6.
Time domain of the five independent components with maximal SNR for subject 06.

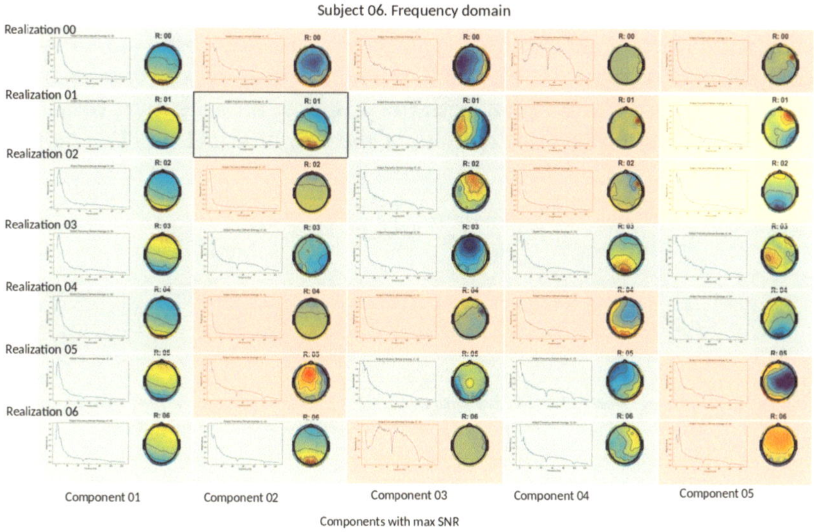

Figure 7.
Frequency domain of the five independent components with maximal SNR for subject 06.

frequency increases, which means that the central nervous system is generating harmonics from perturbation signal but with lower amplitude.

The input signal is the average (over the 210 periods) of the channel that contains the information about the angle that rotates the wrist of each subject. For subject 06, the output is the average (over the 210 periods) of the independent component with the maximal SNR, frequency response spike before 20 Hz, and contralateral activity in the scalp map (see **Figure 8b**).

In other words, the output signal in the frequency domain, corresponding with the cortical response, is the average of all normalized frequencies of the selected independent component for each period. In addition, the output signal in the time domain is the average of the selected independent component for each period as well. Both the averaged input signal and the averaged output response can be observed in **Figure 8a**.

No system identification process was applied yet. Instead, the parameters of the proposed model were found randomly by setting a range from 0 to 1000 to avoid quantification errors. Matrix A contains the parameters of this model (see Eqs. (3) and (4)), where it can be seen that the values of each row are equal, which means that each stage is the same and has the same behavior as the next one. However, these values could deviate from the subsystems of the sensorimotor system, which is more complex than that (add reference).

$$A = \begin{bmatrix} 298.00 & 2111.92 & -1728.93 & -570.81 & -457.48 \\ 298.00 & 2111.92 & -1728.93 & -570.81 & -457.48 \\ 298.00 & 2111.92 & -1728.93 & -570.81 & -457.48 \\ 298.00 & 2111.92 & -1728.93 & -570.81 & -457.48 \end{bmatrix}$$

By applying these parameters in Eqs. (3) and (4), the frequency domain of the estimated output was calculated. **Figure 9** plots the frequency domain of the perturbation input (blue signal), and the orange signal represents the frequency domain of the proposed model output $(y)_5$ (see Eq. (3)). It is important to mention that the objective of this work was not modeling but the methodology to identify the source to model, that is, to identify the cortical response as a previous step to model it.

By comparing both signals in **Figure 9**, the blue signal, which contains frequencies 1, 3, 5, 7, 9, 11, 13, 15, 19, and 23 Hz, with the proposed model output (orange signal), we can observe that the frequency domain of the output signal contains more

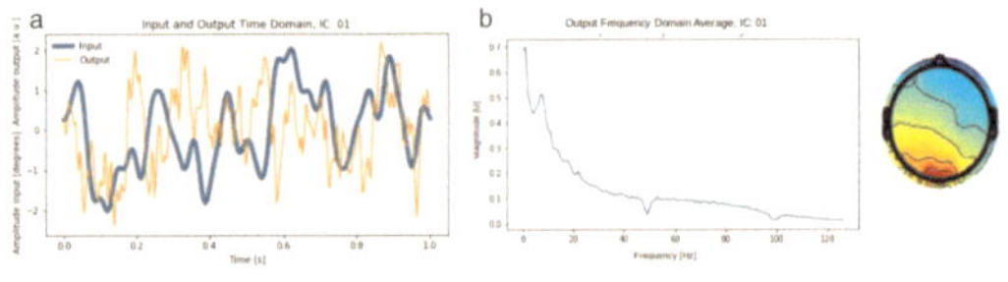

Figure 8.
a: Time domain of the averaged perturbation input signal (blue) and the averaged output signal (orange). b: Frequency domain of the averaged response and scalp map of the output signal for subject 06.

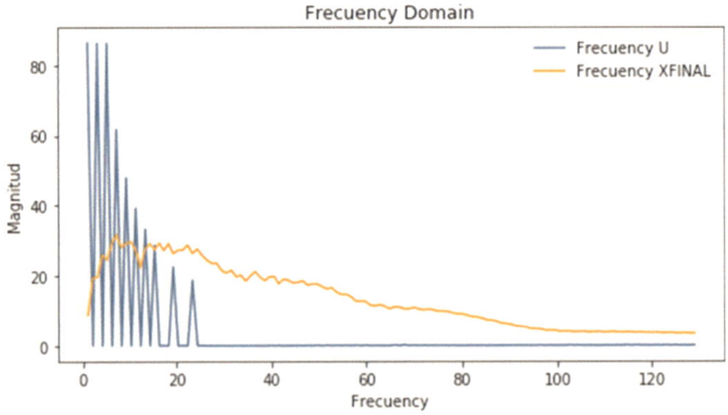

Figure 9.
Frequency domain of the perturbation input (blue signal) and frequency domain of the proposed model output $(y)_5$ *(orange signal).*

frequencies than the original input, which means that our proposed model is able to generate new harmonics from the input signal.

7. Conclusions and ongoing work

The cortical response caused by mechanical external perturbations was identified by choosing the independent component with maximal SNR and a contralateral response. This component represented the expected response because, when plotting its corresponding topology, it was confirmed that brain activity was focused on the left side, which meant a contralateral cortical response since the perturbation was applied at the right wrist of each subject.

A new cascade quadratic architecture was presented, and it can be a good approach to model the cortical response due to wrist joint perturbations because it generates new harmonics from the original perturbation input, and adding more stages generates more harmonics. This behavior is similar to the sensorimotor system, which we are trying to model. However, it is not enough to support the assertion that the proposed architecture may be a good approximation, since the parameters of each stage could probably deviate from physiological parameters of the sensorimotor system.

The number of stages of the proposed model needs to be analyzed. In order to establish the minimum number of stages that represent the parameters of the sensorimotor system for each subject, the physiology of the human brain must be considered. A recommended option could be five stages, so we have a square matrix with a full range, in which an orthogonal basis can be built from the model parameters.

No correlation can be established between the calculated coefficients of matrix A and the physiological system yet since these were found randomly. It can be speculated that the parameters are repeated because each stage of the model would correspond to a part of the sensorimotor system. However, further analysis is needed. It is essential to find out what is the link between the system identification method proposal and the physiological functions of the sensorimotor system to understand this type of process in the human brain in a way that some practical application could be developed in the future.

Acknowledgements

The authors thankfully acknowledge computer resources, technical advice, and support provided by the Laboratorio Nacional de Supercómputo del Sureste de México (LNS), a member of the Consejo Nacional de Humanidades, Ciencias y Tecnologías (CONAHCYT) national laboratories, with project No. 202101014C.

Author details

Martín Durán[1*], Rocío Salazar[1] and Gibran Etcheverry[2]

1 Universidad de las Américas Puebla, Cholula, Puebla, Mexico

2 External Researcher, Querétaro, Mexico

*Address all correspondence to: martin.duranss@udlap.mx

References

[1] Herculano-Houzel S. The human brain in numbers: A linearly scaled-up primate brain. Frontiers in Human Neuroscience. 2009;**31**(3)

[2] Li T, Zheng Y, Wang Z, Zhu DC, Ren J, Liu T, et al. Brain information processing capacity modeling. Scientific Reports. 2022;**12**(1):2174

[3] Vera SRV, Abud-Figueroa MA, Vásquez GR, Juárez-Martínez U, Peláez-Camarena SG, Gutiérrez JAM, et al. Desarrollo de una BCI utilizando el potencial p300 y la diadema Mindwave®. Research in **Computing Science**. 2017;**140**:151-164

[4] Vlaar MP, Birpoutsoukis G, Lataire J, Schoukens M, Schouten AC, Schoukens J, et al. Modeling the nonlinear cortical response in EEG evoked by wrist joint manipulation. IEEE Transactions on Neural Systems and Rehabilitation Engineering. 2018;**26**(1): 205-215

[5] Melnik A, Hairston WD, Ferris DP, König P. EEG correlates of sensorimotor processing: Independent components involved in sensory and motor processing. Scientific Reports. 2017;7(1):4461

[6] Yang Y, Solis-Escalante T, Van de Ruit M, Van der Helm FC, Schouten AC. Nonlinear coupling between cortical oscillations and muscle activity during isotonic wrist flexion. Frontiers in Computational Neuroscience. 2016;**10**:126

[7] Tian R, Yang Y, van der Helm FC, Dewald JP. A novel approach for modeling neural responses to joint perturbations using the NARMAX method and a hierarchical neural network. Frontiers in Computational Neuroscience. 2018;**12**:96

[8] Vlaar M. Characterizing cortical responses evoked by robotic joint manipulation after stroke [PhD thesis]. The Netherlands: Delft University of Technology; 2017

[9] Riemann BL, Lephart SM. The sensorimotor system, part i: The physiologic basis of functional joint stability. Journal of Athletic Training. 2002;**37**(1):71

[10] Van Vliet M, Chumerin N, De Deyne S, Wiersema JR, Fias W, Storms G, et al. Single-trial ERP component analysis using a spatiotemporal LCMV beamformer. IEEE Transactions on Biomedical Engineering. 2015;**63**(1):55-66

Chapter

The Application of Game Theory in Water Management: An Adaptative Approach

Miguel Angel Salomón-Vera and Carlos Patiño-Gomez

Abstract

Currently, one of the biggest problems facing both developing and developed countries in the water sector is the deterioration of water management mechanisms. This is mainly due to a series of factors such as: the competition of consumptive uses for the resource, the uses and customs of native peoples and the lack of water culture, which affect the creation of adequate legal instruments to support these mechanisms, thus affecting decision making by state and federal governments. A clear example of this can be observed in northern Mexico through social conflicts such as the Water War at La Boquilla Dam in 2020. This conflict largely arose due to a lack of communication among the different stakeholders involved and a reduced participation in decision-making by all actors, leading to inadequate regulation. The main objective of this work is to demonstrate how game theory emerges as a tool capable of improving decision-making in water management. This is achieved by presenting successful cases where it has helped address such issues and enhance the quality of life for present society.

Keywords: game theory, transboundary basins, water management, Conchos Basin, climate change

1. Introduction

In 2019, the World Economic Forum acknowledged in its Global Risks Report the importance of water resources as crucial for the development of societies due to their influence on health, population growth, both national and international politics, and the economies of nations. This highlighted that every human activity is closely related to water [1]. Indeed, water is vital for the development and subsistence of all beings, and it plays a fundamental role in the proper functioning of ecosystems and the acquisition of various resources. The relationship between humans and the environment has always been contradictory, and this contradiction has intensified in recent years, particularly due to the emergence of highly complex meteorological phenomena that have wreaked havoc at all levels of society. This has led to various issues, both of engineering and social nature. An example of these phenomena is drought, which is a cyclical phenomenon that affects climatic conditions and can have significant consequences on water availability and agriculture, among other aspects [2].

One of the most impactful effects of drought on societal development is water scarcity. In recent years, due to an increase in the frequency of drought, this issue has become common worldwide, affecting both developed and developing countries. Water scarcity has significant economic repercussions, as confirmed by the 2021 report from the Bank of Mexico. According to this report, 24.3% of the country's Gross Domestic Product (GDP) was generated in municipalities unaffected by drought, while 54.4% of the GDP was produced in drought-affected areas. Additionally, 42.9% were in municipalities in a state of emergency, equivalent to approximately 1.02 billion pesos each year [3]. Among the most affected industries are manufacturing and agriculture, with impact percentages of 0.39% and 0.8%, respectively, about the annual variation of the country's Gross Domestic Product (GDP). Additionally, 35.2% of manufacturing companies are affected by water scarcity [3]. Another problem that water scarcity poses for societies is the growing dependence on lower-quality water sources to meet the increasing demand for this resource [4]. Finally, one of the major issues that northern Mexico has faced due to the constant and increasingly prolonged periods of scarcity is social conflicts surrounding water. These conflicts have arisen due to an uneven distribution of resource, further exacerbated by a lack of communication and widespread ignorance about water issues among various societal actors. This situation has led to confrontations between civil society and the government, as seen in 2020 when farmers in the state of Chihuahua occupied the La Boquilla dam with the intention of "preventing the diversion of water for international treaty purposes." This conflict resulted in the loss of life and significant economic damage to the region [5]. These conflicts have often been shaped, in part, by the influence of the International Treaty signed in 1944, which focuses on the use and distribution of the international waters of the Colorado, Tijuana, and Bravo rivers. Although this treaty was conceived to resolve conflicts and ensure the equitable allocation of this vital resource, considering the needs of both sides of the border, it has also brought about various challenges in water availability in the border regions, resulting in significant impacts on agricultural production and economic development in the area [6].

Given all the issues, the need for a tool that provides a broad conceptual framework to maximize informed decision-making has become a fundamental objective. It is crucial to address these challenges, aiming to enhance the quality of life for society and improve efficiency in the use of water across various consumptive purposes. The main objective of this chapter is to analyze game theory as a tool capable of creating a suitable analytical framework to promote informed decision-making in water management, fostering cooperation among the various stakeholders.

1.1 Integrated water resource management (IWRM)

The Global Water Partnership (GWP) defines integrated water resource management (IWRM) as a process that drives the development and coordination of interconnected water, land, and other resources, to optimize economic and social well-being fairly, without compromising the sustainability of essential ecosystems. Water management seeks to protect the environment by promoting economic growth and sustainable development across all sectors of society. This approach is supported by democratic participation in decision-making, aiming to enhance human health and systems worldwide [7].

In other words, water management is based on a set of policies, roles, and responsibilities through which the state, key economic sectors (primarily agricultural, industrial, and energy), and civil society in general must collaborate to control and manage this essential resource [8].

1.1.1 Integral water resource management in Mexico

To address integrated water management in Mexico, we must go back to the 1930s during the presidency of Lázaro Cárdenas, when the importance of the nationalization of various resources and the economic protectionism of various environmental resources began to be emphasized. This laid the groundwork for the promotion of regulations, such as the creation of the National Water Commission (CONAGUA) in 1989 and the National Water Law of 1992. However, it was not until the early twenty-first century that the federal government declared the issue of water as a matter of national interest [9]. This declaration was in response to the increasing effects of climate change, which have been evident over the past few decades. A clear example of this is observed in northern Mexico, where the average annual temperature has increased between 1.2 and 2.4°C from 2000 to 2020 [8].

To thoroughly understand the integrated water management in Mexico, it is crucial to examine the hydrological functioning of the country. Rivers and streams form a hydrographic network spanning approximately 633,000 kilometers in length, which are divided into 757 basins, further categorized into 13 hydrological-administrative regions (RHA). Of all the hydrological-administrative regions in Mexico, the largest corresponds to the Rio Bravo region, with an area of 390,440 km^2, in contrast to the Waters of the Valley of Mexico region, which has an area of 18,229 km^2 [10].

Additionally, Mexico features eight transboundary basins, three of which are shared with the United States of America (Bravo, Colorado, and Tijuana), four with Guatemala (Grijalva-Usumacinta, Suchiate, Coatán, and Candelaria), and one with Belize and Guatemala (Río Hondo). It is of utmost importance to highlight that the water volume in transboundary basins with the United States is shared by the provisions of the "Water Treaty" of 1944. Regarding groundwater, Mexico possesses 653 aquifers, representing approximately 39.2% of the total volume allocated for consumptive uses in 2018 [11].

Of these basins, 652 are yet to be overused (**Figure 1**). Among them, 542 overexploited aquifers are identified (**Figure 2**) [12].

Once the country's water situation is understood, another essential factor to consider for proper water resource management is evaluating our overall water balance. According to CONAGUA, Mexico receives an annual volume of 1.449×10^{12} m^3 of precipitation, of which 70% returns to the atmosphere through evapotranspiration. Additionally, the country receives approximately 4.8×10^{10} m^3 through imports from rivers on the northern and southern borders and exports 4.3×10^8 m^3 annually from the Rio Grande to the United States. Thus, the average natural availability in the country is 4.516×10^{11} m^3 of water per year, with around 80% corresponding to national surface runoff and the remaining 20% contributing to aquifer recharge [13].

1.1.1.1 Consumptive uses in Mexico

The previously mentioned volumes help meet the demands of various consumptive uses, which are classified into three sectors: agriculture, public supply, and industry. However, we must understand that the quantity of water required by these sectors

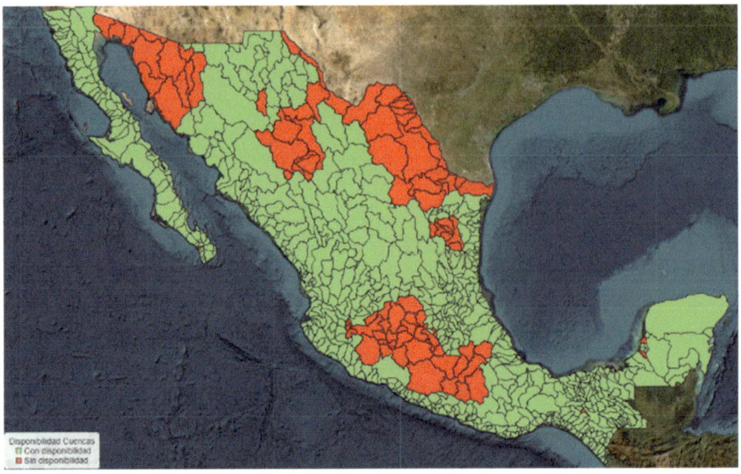

Figure 1.
Availability of basins in Mexico [12].

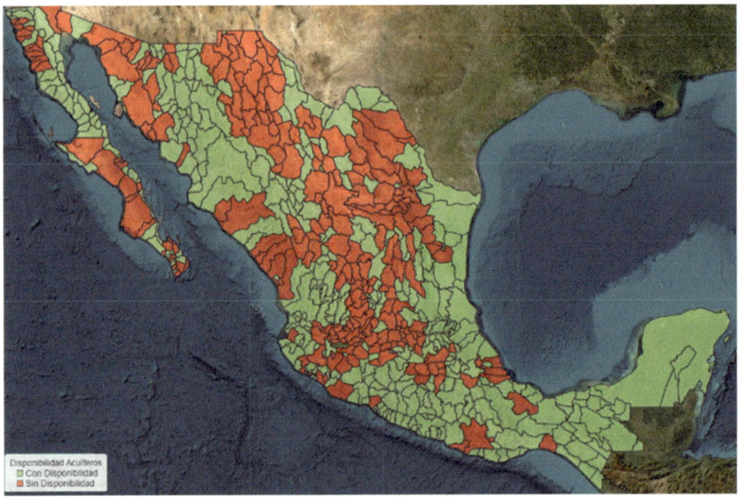

Figure 2.
Groundwater availability in Mexico [12].

is closely related to environmental, socioeconomic, and population characteristics. According to the Mexican Constitution, water is owned by the nation, so users from different sectors need concession or allocation permits, which by law will determine the volume of water and its designated use [14]. Nevertheless, these concessions have been increasing, often without considering the decrease in precipitation in recent years, as evidenced in 2017 when the volume granted for these consumptive uses was 21% higher than that recorded in 2001.

The increase in these concessions is primarily attributed to population growth in the main metropolitan areas of the country. This phenomenon is exemplified by examining the behavior of social migration, especially in rural areas, where it decreased from 57.3% in 1950 to 23.0% in 2015. This has led to an increase in

DOI: http://dx.doi.org/10.5772/intechopen.115472

industrial production and the demand for agricultural products, which, at times, due to insufficient monitoring or inadequate monitoring of water bodies, results in an overallocation of water resources. This scenario poses a significant challenge in water management [14].

1.1.1.2 Water challenges in Mexico

With all that has been mentioned previously, it can be observed that the water challenges in Mexico are numerous and primarily stem from the way decisions are made, the increasing competition among stakeholders for resources, and the theme of social inequality experienced in Mexico. This, especially in some more remote regions, leads to an uneven distribution of the resource in terms of time and space. Moreover, this is further compounded by an inefficient use of water, the lack of updating of many equipment, and the lack of maintenance of conduits, in addition to the growing number of bodies of water increasingly affected by pollution.

All these factors primarily lead to longer periods of water scarcity. These periods have a direct impact on the health of the population, as society is exposed to diseases such as cholera, hepatitis, and typhoid fever. Furthermore, the lack of water affects personal hygiene, which can increase the risk of respiratory and skin diseases [15].

Likewise, periods of water scarcity mainly affect the agricultural sector, as many farmers in these areas lack the necessary infrastructure and often depend solely on rainwater to irrigate their crops. Insufficient rainfall and reduced water in rivers and aquifers have led to a decrease in agricultural production and economic losses for farmers. However, the population continues to grow, leading to the need to export many agricultural products from other countries, resulting in a decline in the quality of life for farmers.

Water scarcity also has an impact on the country's economy. The lack of water can limit industrial production, electricity generation, and tourism, potentially reducing economic growth [15].

In other words, Mexico's problems can be summarized in the overexploitation and overallocation of basins and aquifers with larger populations and economic activity. There is a lack of awareness of water conservation, as around 70% of bodies of water are contaminated. Water security is compromised in many regions, and climate change is impacting various sectors through water. The number of conflicts and competition for water is on the rise, involving all three levels of government and water users. It also manifests in transboundary basins and aquifers, often exacerbated by a lack of communication between levels [1].

Similarly, one of the significant problems Mexico faces concerning water is education about this resource. This is because many Mexican customs involve water, leading to considerable waste due to these practices. Additionally, considering that most of the hydraulic infrastructure was built before the mid-twentieth century and has reached or will soon reach the end of its operational life, it becomes imperative to move toward an even more efficient use of the resource.

On the legal front, Mexico's legal framework regarding water resources is outdated. A clear example of this is the National Water Law, which serves as the fundamental basis for the country's water regulations. However, there are no laws that can adapt to the diverse needs of the country. This results in a lack of technological investment in the water infrastructure of the states.

1.1.1.3 The challenges in water management in Mexico

The water issue, as we have mentioned throughout the previous subtopics, will pose significant challenges to the proper management of this resource in Mexico. This is primarily because this problem has only recently begun to receive the attention it deserves. One of the main challenges facing Mexico in the realm of water is to position this resource as a matter of national security, guiding the creation of appropriate public policies. Likewise, the goal is to ensure the human right to water and sanitation, achieve Sustainable Development Goal 6, and pursue the decentralization of water management in the country [1].

1.2 Study area

The study area of this research project is the Conchos River Basin. This basin originates in the pine and oak forests of the eastern slope of the Sierra Madre Occidental and finally flows into the Rio Grande in the border area of Ojinaga. The basin is in the southeastern portion of the state of Chihuahua and encompasses 41 out of its 67 municipalities, as well as two municipalities in the northern part of the state of Durango. The approximate surface area of the basin is 74,371.79 km^2 [16].

The area presents another significant geographical feature as it is in a transitional zone from the monsoonal region to the plains of the USA. The basin consists of five sub-basins identified as "High," "Middle," "Low," "Río Florido," and "Río San Pedro." Each sub-basin exhibits a hydrological and socioeconomic dynamic by its geohydrological and climatic characteristics [17].

1.2.1 Hydrological and climatological aspects

The main natural runoff is the Conchos River, which originates in the foothills of the Sierra Madre Occidental with precipitation from the Pacific Ocean. It is supplied by a significant amount of perennial runoff during the rainy season and, for most of the year, by three tributaries: the Florido River, the San Pedro River, and the Chuvíscar River. The basin has experienced water stress in recent years due to increased demand from the agricultural, domestic, and industrial sectors, a situation exacerbated by the occurrence of drought events [16].

The National Commission for the Knowledge and Use of Biodiversity (CONABIO) identifies five types of climates in the basin: warm semi-arid, arid temperate, very warm semi-arid, very arid temperate, and semi-arid temperate with summer rains and medium humidity. The annual precipitation ranges between 500 and 800 mm, with an average annual temperature of 5–12°C [17].

1.2.2 Social issues

The importance of this basin lies not only at an economic level for the state of Chihuahua through its four irrigation districts but also in that this basin is subject to the regulations of the 1944 international treaty on boundaries and waters between Mexico and the United States. In addition to this, the basin provides a volume of water to supply a significant number of cities in the state of Chihuahua, considering that it spans 42 municipalities.

2. Methodology

2.1 Game theory

A mathematical way of understanding game theory is the study of competition or cooperation between several parties. There are two important areas in game theory: cooperative and non-cooperative game theory.

2.1.1 Cooperative game theory

Cooperative game theory is about fair splitting of the profits, and this theory involves a group of players, a set of coalitions, and a payoff distribution that assigns to each coalition the profit that this coalition obtains when the players of this coalition cooperate.

The idea involved in cooperative games is that given that the set of players has decided to cooperate, we want to split the total profit of the alliance in a way so that each player obtains a fair amount. This splitting is called a solution of the game.

In mathematical language, a cooperative game on $N=\{1,\ldots,n\}$ is a function $v:2^N \rightarrow R$, that assigns a real number to every subset (coalition) of N. This represents the profit of each coalition. A game is called super-additive if $v(A\cup B)\geq v(A)+v(B)$ whenever A and B are disjoint. This means that players are willing to cooperate.

A solution of the game is a function $x:N\rightarrow R$. This represents the splitting of the profit. The solution of the game is said to be efficient if $x(1)+\ldots+x(n)=v(N)$. If $S=\{i_1,\ldots,i_k\}$ let us denote by $x(S)=x(i_1)+\ldots+x(i_k)$. The core of the game is the set $C(v)=\{x|x(S)\geq v(S)\}$ for every *S* subset of *N* and *x* is efficient. One way of stating that a solution is fair is that it is inside the core.

A value is a function that assigns a solution to every game. The most important value is the Shapley value, which satisfy the properties of efficiency, symmetry, linearity, and null player,

One can prove that this characterizes the value. See more in Peters [18].

2.1.2 Non-cooperative game theory

Non-cooperative game theory is about strategies. Non-cooperative game theory involves a group of players, which represents the parties involved (persons, firms, unions, municipalities, etc..), a set of strategies for each player, which represents the actions that a player can decide to apply and a payoff function, which assigns a value to each player given that the players have chosen a particular strategy to apply.

The idea involved in non-cooperative games is that each player wants to maximize his payoff, so he must choose the strategy that helps him to achieve this goal. The problem becomes interesting by the fact that every player is trying to increase their own payoff, so it is not at all obvious what is the best strategy to play. Moreover, important results about the behavior of players, such as the existence of equilibrium points, describe what happens in real life problems. One of the most studied equilibriums is the Nash equilibrium, which gave Nash the Nobel prize [19].

In mathematical language, according to Peters, an n-person strategic game is a $2n+1$ tuple, $G=(N,S_1,\ldots,S_n,u_1,\ldots,u_n)$, where $N=\{1,\ldots,n\}$ is the set of players. For $i\in N$, S_i is the strategy set of player i and $u_i:S_1\times\ldots\times S_n\rightarrow R$ is the payoff function of player i. This means that for every strategy $(S_1,\ldots,S_n)$, we have that $U_i(S_1,\ldots,S_n)$ is the payoff of player i [18].

A Nash equilibrium is a strategy $S_1^*,\ldots,S_n^*$ such that for every player i, we have that $u_i\left(S_1^*,\ldots,S_n^*\right) \geq u_1(S_1^*,\ldots,S_{\{i-1\}}^*,S_i,S^*_{\{i+1\}}\ldots,S_n^*$, meaning each player does not have the motivation to change its strategy.

2.2 Game theory and water management

In Mckinnay and Teasly Cooperative Game Theory is used to analyze Transboundary River Basins in the Syr Darya Basin. In this analysis, the players involved are three countries, namely Kyrgyzstan, Uzbekistan, and Kazakhstan where the river flows. The chapter models the river basin by considering the allocation of water to energy and agricultural production within 1 year. The basin model is then used to calculate the payoff of each of the coalitions involving the three players. Each coalition is carefully analyzed in the scenario that they cooperate with. The chapter computes the core of the game as well as the Shapley value which is inside the core [20].

In Parrachiano, Dinar, and Patrone a survey of several applications of cooperative game theory in water resource problems is written. The paper deals with several water uses. As part of giving many examples of solutions to cooperative games (seen as allocation problems), the paper suggests that cooperation in the use of water is possible under some conditions. The main theme of the paper is to provide a solid basis for finding possible and stable cost-sharing arrangements [21].

In the study by Fu et al., a comparison of transboundary water resources allocation models based on game theory and multi-objective optimization is done. Concretely, the article studies the Huaihe River basin that contains Henan Province, Anhui Province, and Jiangsu province. The article studies the transboundary water resources allocation based on what is called the asymmetric Nash-Harsanyi Leader Follower game. The model involves a leader and multiple followers. The model can be divided into two stages. The first stage involves allocating public water, while the second involves followers allocating water resources after the decisions of the watershed management agency. A Multi-Objective Optimization Method is also discussed and compared with the game theory model [22].

In Frisvold and Caswell, transboundary water management is studied in the Game theory for projects on the US–Mexico border. The work with water management is a bargaining game. The Nash solution (i.e., the one that maximizes the product of the profits) is analyzed, as this solution is pareto efficient and guarantees that the standard is achieved at least cost. They also investigate environmental grant development as a strategic game and model the Nash solution again [23].

3. Discussion and conclusions

The Conchos River Basin in the state of Chihuahua represents a significant challenge for water management in various aspects. This issue arises due to the growing competition for water resources, both surface and underground, among the various involved sectors. The increasing water stress on water bodies is further exacerbated by the growing effects of climate change, leading to a crisis in decision-making not only in the study basin, but also throughout the entire Mexican territory.

The lack of availability studies has hindered obtaining precise readings of the volumes available in water bodies. Additionally, the lack of environmental awareness and resource care has led to episodes of violence and social discontent, such as the Water War in 2020 due to the international treaty, and more recently, the Three States

DOI: http://dx.doi.org/10.5772/intechopen.115472

War between Chihuahua, Nuevo León, and Tamaulipas over the volume of water flowing into La Amistad Dam, corresponding to 76% of the international treaty and originating from the study basin.

These conflicts reflect a deterioration in decision-making, which is necessary to ensure an adequate quality of life for the inhabitants of the area. Faced with this problem, game theory emerges as a tool that allows for a broad analytical framework by modeling various strategies that each actor must follow, whether collaborating or not. This could facilitate the creation of water markets that maximize profits through the efficient use of the resource without exceeding what is necessary.

However, to achieve the balance that game theory allows, it is necessary to consider how decisions are made in Mexico, as well as the traditions and customs of indigenous people that also influence the formulation of public policies. It has been demonstrated that this can be addressed by integrating other sub-models, such as positive mathematical programming, which would support the viability of proposals through economic analysis.

Acknowledgements

It is a pleasure to express my gratitude to CONAHCYT and the Universidad de las Américas Puebla for their valuable support during the development of my doctoral studies. I give special recognition to Dr. Carlos Patiño Gómez—who, unfortunately, passed away last December—for his mentorship and friendship. Additionally, I want to highlight and especially thank Dr. Gerardo Arizmendi from the Department of Actuarial Science, Physics, and Mathematics at this university for his invaluable guidance, direction, and continuous support. His contribution is essential to my academic and professional development. I also appreciate the unparalleled support of Dr. Benito Corona, who serves as my thesis advisor. Without their support, this research project would not have been possible.

Conflict of interest

The authors declare that they have no competing interests.

Author details

Miguel Angel Salomón-Vera[1,2]* and Carlos Patiño-Gomez[1,2]

1 Department of Civil and Environmental Engineering, Universidad de las Américas Puebla, San Andrés Cholula, Puebla, Mexico

2 Department of Actuarial Science, Physics, and Mathematics, Universidad de las Américas Puebla, San Andrés Cholula, Puebla, Mexico

*Address all correspondence to: miguel.salomonva@udlap.mx

References

[1] Arreguín-Cortés FI, López-Pérez M, Cervantes-Jaimes CE. Water challenges in Mexico. Tecnologia y Ciencias del Agua. 2020;**11**(2):341-371

[2] Carrillo González G, Constantino R. El Manejo del Recurso Hídrico, ¿Escasez o un Modelo de Gestión Inadecuado en México [Internet]. 2009. Available from: https://www.researchgate.net/publication/349914790

[3] Banco De Mexico. Sequía en México y su Potencial Impacto en la Actividad Económica. 2022

[4] Jiménez Cisneros B, Torregrosa y Armentia ML, Aboites Aguilar L. El Agua En México: Cauces Y Encauces. Academia Mexicana de Ciencias; 2010

[5] Financiero El. La 'guerra del agua' termina en Chihuahua: Entregarán presa La Boquilla [Internet]. 2021. Available from: https://www.elfinanciero.com.mx/estados/2021/10/22/la-guerra-del-agua-termina-en-chihuahua-entregaran-presa-la-boquilla/ [Accessed: March 1, 2023]

[6] Comisión Internacional de Límites y Aguas [CILA]. Tratado Entre El Gobierno De Los Estados Unidos Mexicanos Y El Gobierno De Los Estados Unidos De America De La Distribucion De Las Aguas Internacionales De Los Rios Colorado, Tijuana Y Bravo, Desde Fort Quitman. Texas: Hasta El Golfo De Mexico; 1944

[7] GWP. What is IWRM? [Internet]. 2011. Available from: https://www.gwp.org/en/GWP-CEE/about/why/what-is-iwrm/ [Accessed: August 16, 2023]

[8] IMCO. Aguas En México,¿Escasez O Mala Gestión? 2023

[9] Rolland L, Cárdenas YV. La Gestión Del Agua En México. Polis. 2010;**6**

[10] Breña Puyol AF, Naranjo JAB. Problemática Del Recurso Agua En Grandes Ciudades: Zona Metropolitana Del Valle De México. 2009

[11] Estadísticas del Agua en México. 2019

[12] CONAGUA. Sistema Nacional de Información del Agua [Internet]. 2024. Available from: https://sinav30.conagua.gob.mx:8080/SINA/?opcion=cuencas [Accessed: January 29, 2024]

[13] Informe del Medio Ambiente [Internet]. Available from: https://apps1.semarnat.gob.mx:8443/dgeia/informe18/tema/cap6.html [Accessed: January 29, 2024]

[14] CONAGUA. Atlas del Agua en México [Internet]. 2018. Available from: www.gob.mx/conagua

[15] JCarlos V, V. El Agua en México. Comisión Nacional del Agua; 2006

[16] Cuenca Alta Del Río Conchos [Internet]. Available from: http://www.conabio.gob.mx/conocimiento/regionalizacion/doctos/rhp_039.html [Accessed: April 17, 2022]

[17] Montero Martínez MJ, Ibáñez Hernández ÓF. La Cuenca Del Río Conchos: Una Mirada Desde Las Ciencias Ante El Cambio Climático. 2017. 267

[18] Peters H. Game Theory a Multi-Leveled Approach. 1st ed2008

[19] Gibbons R. Game Theory For Applied Economists. 1992

[20] Mckinney DC, Teasley RL. Cooperative Game Theory for

Transboundary River Basins: The Syr Darya Basin

[21] Parrachino I, Dinar A, Patrone F, Zara S, Marchiori C, Sgobbi A. Cooperative Game Theory and Its Application to Natural, Environmental and Water Resource Issue [Internet]. 2006. Available from: http://econ.worldbank.org

[22] Fu J, Zhong PA, Xu B, Zhu F, Chen J, Li J. Comparison of transboundary water resources allocation models based on game theory and multi-objective optimization. Water (Switzerland). 2021;**13**(10)

[23] Frisvold GB, Caswell MF. Transboundary water management game-theoretic lessons for projects on the US-Mexico border. Agricultural Economics. 2000;**24**

Chapter

Tissue Engineering Focused on Cardiac Regeneration

Mildred A. Sauce-Guevara, Miguel A. Mendez-Rojas, Sergio A. Bernal-Chavez and Emilio I. Alarcon

Abstract

In recent years, there has been an increasing interest in developing alternatives to restore damaged cardiac tissue. However, the problem of tissue biocompatibility has been a major obstacle. Tissue engineering offers unique methods of scaffold formation, which can easily overcome the biocompatibility problem by creating functional tissue substitutes to replace damaged heart tissue caused by various conditions. This work aims to summarize some of the most relevant techniques used in cardiac regenerative medicine and outline some of the most compelling future directions. The techniques of cell culture, 3D bioprinting, use of peptides, and nanomaterials are shown to be part of these innovative efforts aiming to regenerate damaged tissue. This area of study is a multidisciplinary approach that combines principles from the fields of materials, chemistry, engineering, and life sciences.

Keywords: nanomaterials, heart, tissue, regeneration, biomedical application

1. Introduction

Tissue engineering is a field that involves the combination of biologically active structures, cells, and molecules to create functional tissues. This is done with the aim of restoring, maintaining, or enhancing damaged tissues or organs. Cells are the basic building blocks of tissues, and tissues are the fundamental functional units of the body. Cell groups are responsible for manufacturing and secreting their own extracellular matrix, which provide structural support for the cell and act as a relay station for signaling molecules. These signals can trigger a chain of responses that activate cells within the organism. By understanding how individual cells respond to signals, interact with their environment, and organize themselves into tissues and organisms, various approaches have been developed for repairing damaged tissues [1].

To develop new tissue, researchers often begin by constructing scaffolds from various organic and inorganic sources. Once created, these scaffolds can be used to introduce different molecules such as cell's growth factors, proteins, and growth factors into the microenvironment, allowing damaged tissue to regenerate. Alternatively, cells can be removed from a donated or discarded organ during surgery and combined with the patient's own cells to create personalized organs that are not rejected by the immune system. Heart failure is a leading cause of death worldwide, accounting for over 30% of all reported deaths. In 2018, 17.5 million deaths were recorded, with 7.4 million of those due to ischemic heart disease. It is predicted that by 2030, these numbers will increase by 36%. In Mexico, the National Institute of Statistics and Geography (INEGI) reported 116,002 deaths due to heart failure in 2018, with 68% of those being caused by atherothrombotic ischemic heart disease [2]. The main cause of acute myocardial infarction (AMI) is the rupture of an atherosclerotic plaque, leading to thrombosis. This accounts for up to 79% of cases, regardless of clinical presentation [3]. AMI results in permanent loss of myocardial cells, leading to progressive pathological remodeling of the left ventricle and, ultimately, heart failure. Unfortunately, the heart muscle loses its regenerative capacity after birth and is unable to regenerate [4].

2. Techniques for cardiac regeneration

2.1 Cell culture used for fabrication of artificial human cardiac muscle

2.1.1 Skeletal myoblast

It is well known that the cardiac muscle does not express specific skeletal muscle proteins. However, studies have shown that skeletal muscle development has a very similar lineage to that of cardiac muscle. Researchers have discovered that there are different isoforms of myosin heavy chain (MHC), and both converge expressing a subclass I/b-MHC that is found in both skeletal muscle and cardiac muscle. Additionally, both tissues are regulated by troponins, and it is shown that troponin I, which is encoded by three different genes, is expressed in both [5–7].

Various studies have demonstrated that microribonucleic acids (micro-RNAs, miRs) play a significant role in post-transcriptional regulation of cardiac and skeletal myogenesis. For instance, miR-206 facilitates differentiation into skeletal muscle, while miR669 promotes cardiac differentiation as illustrated in **Figure 1**. Therefore, it has been reported that inhibition of miR-206 expression through muscle-specific microRNAs can promote differentiation into cardiac tissue [8, 9].

There are multiple benefits to using this type of cell for tissue regeneration, including a low risk of immune rejection and resistance to hypoxic conditions. On the downside, these cells can cause cardiac arrhythmias, which can lead to a low survival rate [10].

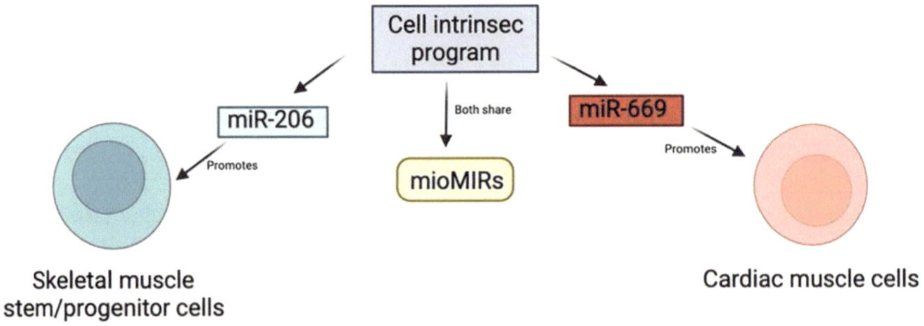

Figure 1.
MicroRNAs post-transcriptionally regulate myogenesis.

2.1.2 Mesenchymal stem cells

Mesenchymal stem cells are a type of stem cell that can be derived from the mesoderm and can be isolated from various sources such as bone marrow, umbilical cord, or adipose tissue. These cells possess proangiogenic and immunomodulatory properties, making them a promising field in the treatment of ischemic heart disease due to their ease of obtaining [11].

Berry et al. conducted a study on the treatment of ventricular injury caused by mesenchymal stem cell injection. The study found that after 8 weeks, the treatment did not regenerate cardiomyocytes, but it did reduce the rigidity of the scar and attenuated post-infarction remodeling [12]. This preservation of cardiac function is a positive outcome. However, the use of mesenchymal stem cells is still being studied due to a lack of conclusive evidence regarding their safety, particularly their profibrogenic potential [13].

2.1.3 Bone marrow cells

Bone marrow has two main types of stem cells: hematopoietic stem cells (HSCs) and stromal cells. Stromal cells include mesenchymal stem cells, multipotent stromal cells, and endothelial progenitors. Currently, Ixmyelocel-T is a multicellular therapy that is produced from a patient's own bone marrow. It selectively expands two key types of bone marrow mononuclear cells, which are CD90+ mesenchymal stem cells and CD45+ CD14+ autofluorescent activated macrophages [14, 15]. Henry et al. conducted a phase II trial with lIxmyelocel-T. The trial had a 1-year follow-up period and 61 patients participated. Patients were given an intramyocardial injection, and the results showed that major adverse cardiovascular events were reduced only in patients with ischemic-dilated cardiomyopathy [16]. Using these types of cells as a treatment offers several advantages. They can produce bioactive molecules that lead to no immunological conditions, have proangiogenic effects, and have been tested in clinical studies [16, 17].

2.1.4 Cardiomyocytes

Cardiomyocytes, also known as cardiac muscle cells, are a part of the cardiac muscle present in the walls of the heart. Their primary function is to pump blood by contracting and relaxing the cardiac ventricles [18]. Cardiomyocyte grafts derived from pluripotent stem cells have the advantage of remuscularizing infarcted myocardium in significant quantities. However, they may also lead to ventricular arrhythmias [19]. In a study, Fryer et al. transplanted around 750 million cryopreserved cardiomyocytes derived from human embryonic stem cells (hESC-CM) to macaque monkeys. The study showed an improvement of 10.6 ± 0.9% in the left ventricle after 1 month of transplantation with hESC-CM, as compared to 2.5 ± 0.8% in the controls. At 3 months, there was an additional 12.4% improvement in the treated group as compared to a decrease of 3.5% in the control group [20]. In another study, Liang, et al. isolated cells from the population of the cardiac side Sca1+/CD31- of mouse hearts (C57BL6/J). These cells were then injected into the heart of a mouse through an intramyocardial injection. The transplanted cells expressed specific cardiomyocyte markers and were capable of migrating to the damaged site from the non-ischemic area of the heart (**Figure 2**) [21].

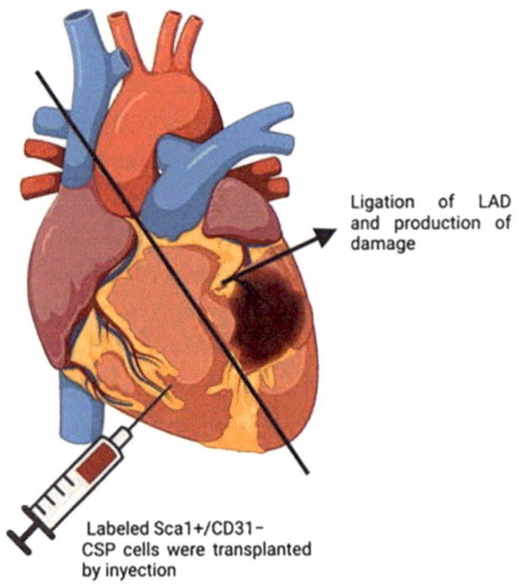

Figure 2.
Sca1+/CD31− cell administration process in the right ventricle (non-ischemic area) in mouse cardiac tissue.

One of the benefits of using cells that release extracellular vesicles for regenerative purposes is that they can enhance cell contraction and relaxation, and reduce tissue rejection. However, the main drawback is the ethical challenges associated with cultivating these cells in a responsible manner [22].

2.2 3D printing

Three-dimensional (3D) printing techniques are highly important in the field of tissue engineering, particularly for creating complex tissue structures or scaffolds, up to the point of generating complete organs. The complex structure of these tissues makes it necessary to generate them precisely and in a highly repeatable manner [23]. Scaffolds can be created through different methods, such as adding consecutive 2D layers of a material [24]. Bioprinting, on the other hand, generates 3D soft tissue scaffolds using biomaterials, living cells, and growth factors through layer-by-layer deposition of materials known as bioinks to create tissue-like structures [25].

2.2.1 Inkjet bioprinting

One type of bioprinting involves storing a combination of live cells and bioink in a chamber attached to the print head. This mixture helps in creating tissue constructs. This method is advantageous for its low cost and high cell viability. However, it has some drawbacks such as the clogging of the print head, uneven distribution of cells, and the inability to print viscous materials (**Figure 3**) [26, 27].

2.2.2 Laser-assisted bioprinting

This method of bioprinting involves using a layer of bioink, an energy-absorbing layer, a donor layer made of quartz or glass, and a collecting layer. During printing, the laser beam is focused on the energy-absorbing layer, which vaporizes and creates an air bubble between the bioink and donor layers, forming the structure.

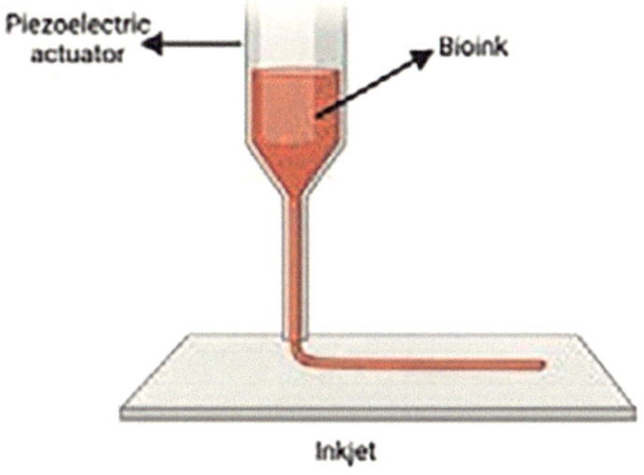

Figure 3.
Representation of the inkjet bioprinting.

This technique allows for the use of viscous materials with high cellular density and viability. However, it can be expensive to use (**Figure 4**) [26, 27].

2.2.3 Extrusion bioprinting

This technique involves forcing a liquid, which can either be a paste or a solution, through a syringe into a solution of controlled density. The materials are extruded as long strands to form various structures. This process has the advantage of being carried out in a high-temperature environment and is commonly used with hydrogels (**Figure 5**) [28].

2.2.4 Stereolithography

In this type of bioprinting, a laser beam is used to control the polymerization process of bioinks in a 2D layer. After each layer of material is deposited, a frame is formed, and a photosensitive hydrogel is exposed to ultraviolet or visible light. The process is repeated, layering the previous layer until the entire scaffold is completed.

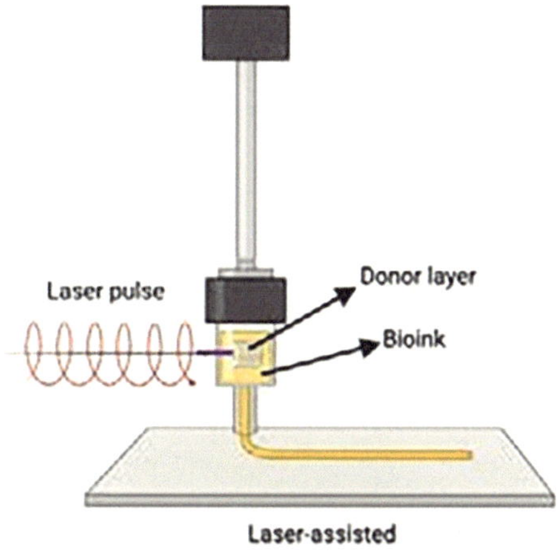

Figure 4.
Representation of the laser-assisted bioprinting.

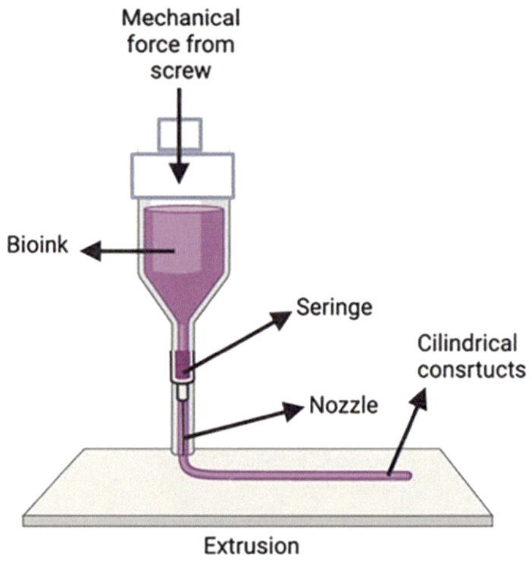

Figure 5.
Representation of the extrusion bioprinting.

Polyethylene glycol diacrylate (PEGDA) and gelatin methacryloyl (GelMA) are commonly used. The disadvantage is that it takes a long time to create the scaffold, but the advantage is that the process can produce detailed scaffolds (**Figure 6**) [29].

2.2.5 Applications

Gaetani et al. bioprinted a patch using human heart-derived progenitor cells (hCMPC) in a solution of hyaluronic acid and gelatin (HA/gel). The hCMPCs were printed on the HA/gel matrix (30 x 106 cells/ml) to create a biocomplex. This biocomplex was then transplanted into a mouse model of myocardial infarction (MI), resulting in repair and preservation of the cardiac performance. Additionally, hCMPC cells showed a temporary increase in cardiac and vascular differentiation markers during the 4-week follow-up period [30].

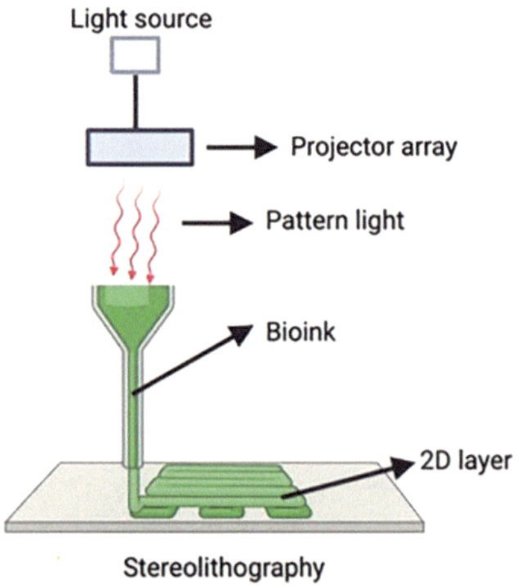

Figure 6.
Representation of the stereolithography bioprinting.

On the other hand, Duan et al. obtained hydrogels based on methacrylate hyaluronic acid (Me-HA) and methacrylate gelatin (Me-Gel) through 3D printing. They used these hydrogels to bioprint heart valve conduits containing encapsulated human aortic valve interstitial cells (HAVIC). The cells encapsulated within the bioprinted heart valves maintained high viability and remodeled the initial matrix by depositing collagen and glycosaminoglycans [31].

Kawai et al. manufactured engineered cardiac tissues (EHT) by bioprinting using human-induced pluripotent stem cells (hiPSCs). The EHT was composed of cardiomyocytes derived from hiPSCs, human umbilical vein endothelial cells, and human fibroblasts. The designed tissues were transplanted around the abdominal aorta and the inferior vena cava in mouse models, covered with the omentum, and the abdomen was then closed. After a period of time, the myocardial stria had vascularization, making it a potential treatment option for patients with congenital heart diseases [32].

2.3 Use of peptides

Peptides, a group of molecules consisting of short chains of amino acids containing 50 or fewer amino acids, have attracted significant attention in the medical field due to their remarkable potential in healing and tissue repair. Despite their small size, peptides perform a wide range of functions in the body, such as functioning as hormones, signaling molecules, neurotransmitters, and aiding in the regulation of several important biological processes [33].

Michal Shachar et al. developed different microstructures with porous materials that immobilized the RGD peptide. In this study, cardiac cells from neonatal rats were seeded in alginate scaffolds, and it was observed that the cellular adhesion in the matrix was promoted, cellular apoptosis was prevented, and the regeneration of damaged cardiac tissue was accelerated. Within just 6 days, the cardiomyocytes reorganized their myofibrils and reconstructed myofibers composed of multiple cardiomyocytes (**Figure 7**) [34].

Andukuri et al. developed a matrix made of electrospun polycaprolactone (ePCL) nanofibers and self-assembled Pas for cardiac repair. The ePCL provided a porous base structure, while the peptide YIGSR adhesive ligands and nitric oxide donors provided biofunctionality. When HUVEC and smooth muscle cells were grown on the matrix, they showed increased proliferation [35].

Uchikana et al. designed myoblast sheets that secrete the 7-amino acid SV peptide (Ser-Val-Val-Tyr-Gly-Leu-Arg), which has been shown to induce angiogenesis. The sheets were found to facilitate an improvement in long-term cardiac function since SV can induce the differentiation of fibroblasts into myofibroblasts through TGF-b-Smad signaling [36].

2.4 Use of nanomaterials

Nanomaterials are tiny particles that can be utilized to make superior biomaterials with specific physical, chemical, and biological traits. They are particularly valuable for developing biomedical technologies such as scaffolding, tissue regeneration, and controlled drug delivery [37]. Researching innovative methods for producing nanomaterials for tissue engineering applications, including the use of nanomaterials for soft tissue engineering, such as cardiac, neural, and cartilage tissue engineering, are some interesting areas of exploration available in the field.

Nanoparticles, which are materials with a size ranging from 1 to 100 nanometers, have become a focus of medical research for their potential use as therapeutic and

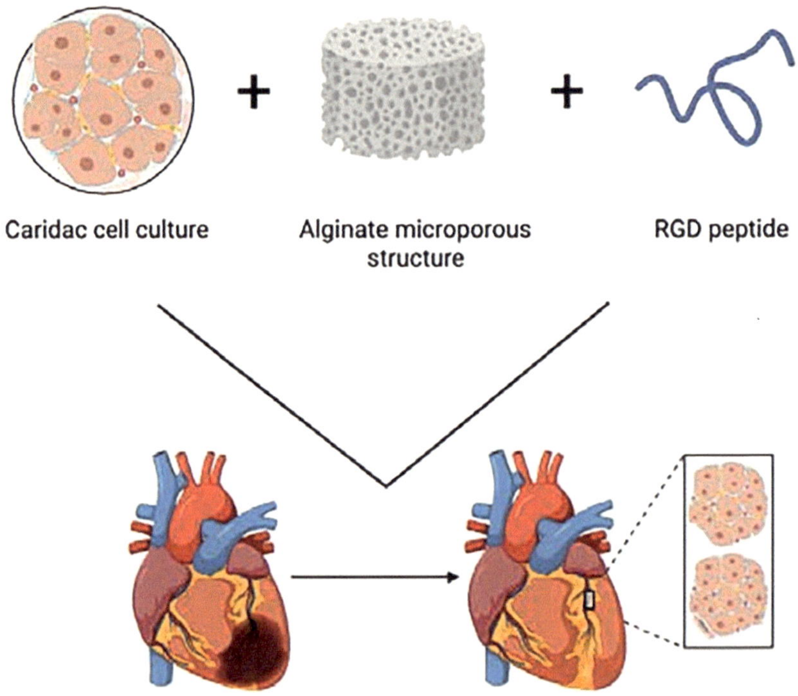

Figure 7.
Representation of the application of a microporous alginate structure combined with cardiac cells for regenerative purposes.

diagnostic tools. Their small size and high surface area-to-volume ratio make them attractive for medical applications, as they can be easily modified with biomolecules and can be taken up by cells [38].

2.4.1 Carbon nanomaterials

Carbon nanotubes were cultured with defunctionalized neonatal rat ventricular myopathy (NRVMs) cells to investigate their effect on cardiac myocytes. It was found that the nanotubes altered the cells' viability, proliferation, growth, maturation, and electrophysiological properties [39].

In another study by Hao, Tong et al., an injectable hydrogel made of fullerenol and alginate was developed. This hydrogel was found to effectively eliminate superoxide anion and hydroxyl radicals. Additionally, when tested on stem cells (BADSC) in the presence of H_2O_2, it was shown to have no cytotoxic effects and was able to suppress oxidative stress. The study also found that fullerenol can improve cardiomyogenic differentiation, decrease the level of reactive oxygen species in the MI zone, and induce angiogenesis, which can promote cardiac functional recovery (**Figure 8**) [40].

2.4.2 Metal nanomaterials

Metal nanoparticles (MNPs) are tiny materials with a one-dimensional shape and size ranging from 1 to 100 nm. One emerging strategy to improve and enhance the functionality of complex engineered tissues is to modify the scaffold surface using

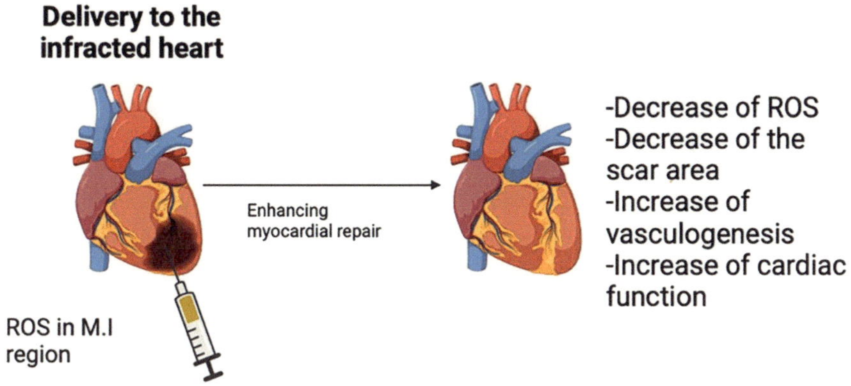

Figure 8.
Representation of the injectable hydrogel prepared based on fullerenol and alginate.

NPs. Nanoparticles imitate the nanometric size of ECM components present in physiological tissues [41].

Muñoz has developed a sprayable nanotherapeutic using nanogold grafted with a custom-designed multi-arm peptide. This therapy is applied to the *in situ* surface of an infarcted myocardial model. It has been demonstrated that after 1 week of exposure, cardiac function improves along with an enhancement in muscle contractility and electrical conductivity of the myocardium. The treatment remains at the site of application for 28 days without infiltrating off-target organs (**Figure 9**) [42].

Baei and colleagues conducted a study to examine the impact of gold nanoparticles (AuNPs) on chitosan thermosensitive hydrogels. These hydrogels were seeded with mesenchymal stem cells (MSCs). The results showed that the addition of AuNPs increased the conductivity of chitosan hydrogels and also increased gelation time in a concentration-dependent manner. There was no significant difference in the proliferation of MSCs over a period of 14 days. However, there was a notable increase in the expression of cardiac-specific markers Nkx-2.5 and α-MHC in cells grown on chitosan gels containing 1 wt% AuNPs compared to those without AuNPs [43].

In another study, magnetic iron oxide nanoparticles were used to guide tissue scaffold structure after implantation into a rat model. Zwi-Dantsis and colleagues

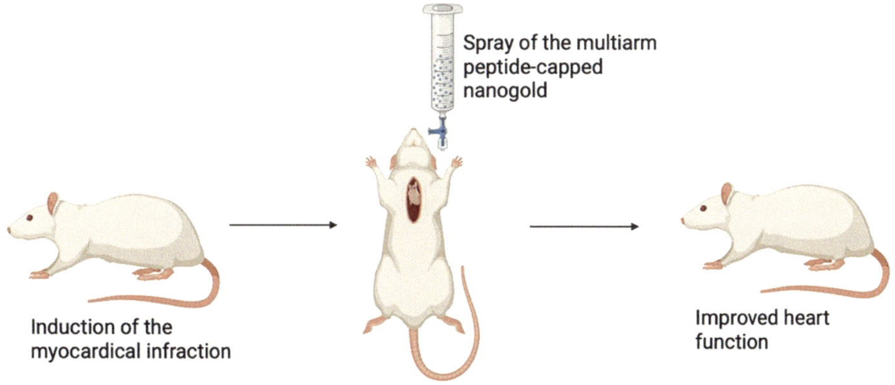

Figure 9.
Representation of the process of the sprayable nanotherapeutic.

demonstrated that magnetic iron oxide nanoparticles bound to antibodies directed at signal-regulatory protein alpha (SIRPA) could be utilized to manipulate human cardiomyocytes *in situ* with magnets [44].

3. Conclusions

In recent years, the development of various techniques to regenerate damaged tissues has gained significant importance. As research in this field advances, clinical investigation needs to consider all relevant properties, including the study of biological mechanisms and the evaluation of safety and long-term stability. Although the various techniques mentioned in this work show increased stability, conductivity, and heart function, it is crucial to ensure that these investigations can be applied in hospitals.

Acknowledgements

UDLAP and CONAHCYT are acknowledged for a Ph.D. scholarship to support MAS doctoral studies.

Conflict of interest

The authors declare that the research was conducted without any potential conflicts of interest arising from commercial or financial relationships.

Author details

Mildred A. Sauce-Guevara[1], Miguel A. Mendez-Rojas[1]*, Sergio A. Bernal-Chavez[1] and Emilio I. Alarcon[2,3]*

1 Department of Chemical and Biological Sciences, Universidad de las Américas Puebla, Cholula, Puebla, Mexico

2 Division of Cardiac Surgery, University of Ottawa Heart Institute, Ottawa, ON, Canada

3 Biochemistry, Microbiology and Immunology, University of Ottawa, Ottawa, ON, Canada

*Address all correspondence to: miguela.mendez@udlap.mx
and ealarcon@uottawa.ca

References

[1] NIH, National Institute of Biomedical Imaging and Bioengineering [Internet]. 2022. Available from: https://www.nibib.nih.gov/science-education/science-topics/tissue-engineering-and-regenerative-medicine [Accessed: November 03, 2023]

[2] Gómez Frödea CX, Díaz Echevarría A, Lara Moctezuma L, Maldonado Aparicio J, Rangel Paredes FV, Vázquez Ortiz LM. Infarto agudo del miocardio como causa de muerte. Revista de la Facultad de Medicina/Universidad Nacional Autónoma de México. 2021;**64**(1):49-59. Available from: https://www.medigraphic.com/pdfs/facmed/un 2021/un211h.pdf [Accessed: November 03, 2023]

[3] Dattoli-García CA, Jackson-Pedroza CN, Gallardo-Grajeda AL, Gopar-Nieto R, Araiza-Garaygordobil D, Arias-Mendoza A. Infarto agudo de miocardio: revisión sobre factores de riesgo, etiología, hallazgos angiográficos y desenlaces en pacientes jóvenes. Archivos de Cardiología de México. 2022;**91**(4):485-492. Available from: https://www.scielo.org.mx/scielo.php?pid=S1405 99402021000400485&script=sci_arttext [Accessed: November 03, 2023]

[4] Steinhauser ML, Lee RT. Regeneration of the heart. EMBO Molecular Medicine. 2011;**3**:701-712. DOI: 10.1002/emmm.201100175

[5] Sasse S, Brand NJ, Kyprianou P, Dhoot GK, Wade R, Arai M, Periasamy M, Yacoub MH, Barton PJ. Troponin I gene expression during human cardiac development and in end-stage heart failure. Circulation Research. 1993;**72**(5):932-938. DOI: 10.1161/01.res.72.5.932

[6] Clause KC, Tchao J, Powell MC, et al. Developing cardiac and skeletal muscle share fastskeletal myosin heavy chain and cardiac troponin-I expression. PLoS One. 2012;**7**:e40725

[7] Apple FS. Tissue specificity of cardiac troponin I, cardiac troponin T and creatine kinaseMB. Clinica Chimica Acta. 1999;**284**:151-159

[8] McCarthy JJ. MicroRNA-206: The skeletal muscle-specific myomiR. Biochimica et Biophysica Acta. 2008;**1779**:682-691

[9] Crippa S, Cassano M, Messina G, et al. miR669a and miR669q prevent skeletal muscle differentiation in postnatal cardiac progenitors. The Journal of Cell Biology. 2011;**193**:1197-1212

[10] Hassan N, Tchao J, Tobita K. Concise review: Skeletal muscle stem cells and cardiac lineage: Potential for heart repair. Stem Cells Translational Medicine. 2014;**3**:183-193. DOI: 10.5966/sctm.2013-0122

[11] Dabrowska S, Andrzejewska A, Janowski M, Lukomska B. Efectos inmunomoduladores y regenerativos de las células madre mesenquimales y vesículas extracelulares: perspectivas terapéuticas para enfermedades inflamatorias y degenerativas. Frontiers in Immunology. 2020;**11**:591065. DOI: 10.3389/fimmu.2020.591065

[12] Berry MF, Engler AJ, Woo YJ, Pirolli TJ, Bish LT, Jayasankar V, et al. La inyección de células madre mesenquimales después de un infarto de miocardio mejora la distensibilidad del miocardio. American Journal of Physiology-Heart and Circulatory Physiology. 2006;**290**(6):H2196-H2203. DOI: 10.1152/ajpheart.01017.200

[13] Lee SH. The advantages and limitations of mesenchymal stem cells in clinical application for treating human diseases. Osteoporos Sarcopenia. 2018;**4**:150. DOI: 10.1016/j.afos.2018.11.083

[14] Patel AN, Henry TD, Quyyumi AA, Schaer GL, Anderson RD, Toma C, et al. ixCELL-DCM Investigators. Ixmyelocel-T for patients with ischaemic heart failure: A prospective randomised double-blind trial. Lancet. 2016;**387**(10036):2412-2421. DOI: 10.1016/S0140-6736(16)30137-4

[15] Bartel RL, Cramer C, Ledford K, et al. La experiencia Aastrom. Stem Cell Research & Therapy. 2012;**3**:26. DOI: 10.1186/scrt117

[16] Henry TD, Traverse JH, Hammon BL, East CA, Bruckner B, Remmers AE, et al. Seguridad y eficacia de Ixmyelocel-T: una terapia multicelular autóloga ampliada en la miocardiopatía dilatada. Investigación de Circulación. 2014;**115**(8):730-737. DOI: 10.1161/circresaha.115.304

[17] Zhao E, Xu H, Wang L, Kryczek I, Wu K, Hu Y, et al. Bone marrow and the control of immunity. Cellular & Molecular Immunology. 2012;**9**:11-19. DOI: 10.1038/cmi.2011.47

[18] Bhattacharya S, Macdonald ST, Farthing CR. Molecular mechanisms controlling the coupled development of myocardium and coronary vasculature. Clinical Science (London, England). 2006;**111**(1):35-46

[19] Liu YW, Chen B, Yang X, Fugate JA, Kalucki FA, Futakuchi-Tsuchida A, et al. Células progenitoras cardíacas a partir de células madre: aprendizaje de la genética y los biomateriales. Celdas. 2019;**8**(12):1536. DOI: 10.3390/cells8121536

[20] Fryer BH, MacLellan WR, Thies RS, Murry CE. Human embryonic stem cell-derived cardiomyocytes restore function in infarcted hearts of non-human primates. Nature Biotechnology. 2018;**36**(9):597-605. DOI: 10.1038/nbt.4162

[21] Liang SX, Tan TYL, Gaudry L, y Chong, B. Diferenciación y migración de células de población del lado cardíaco Sca1+/CD31− en un modelo isquémico de miocardio murino. Revista Internacional de Cardiología. 2010;**138**(1):40-49. DOI: 10.1016/j.ijcard.2008.08.032

[22] Barreto S, Hamel L, Schiatti T, Yang Y, George V. Cardiac progenitor cells from stem cells: Learning from genetics and biomaterials. Cells. 2019;**8**(12):1536. DOI: 10.3390/cells8121536

[23] Zaszczyńska A, Moczulska-Heljak M, Gradys A, Sajkiewicz P. Advances in 3D printing for tissue engineering. Materials (Basel). 2021;**14**(12):3149. DOI: 10.3390/ma14123149

[24] Jain K, Shukla R, Yadav A. Ujjwal RR, Flora SJS 3D printing in nanomedicine development. Nanomaterials. 2021;**11**:420. DOI: 10.3390/nano11020420

[25] Wang C, Huang W, Zhou Y, He L, He Z, Chen Z, et al. 3D printing of bone tissue engineering scaffolds. Bioactives Mother. 2020;**5**:82-91. DOI: 10.1016/j.bioactmat.2020.01.004

[26] Xu T, Jin J, Gregory C, Hickman JJJJ, Boland T. Impresión por inyección de tinta de células de mamíferos viables. Biomateriales. 2005;**26**:93-99. DOI: 10.1016/j.biomaterials.2004.04.011

[27] Pepper ME, Seshadri V, Burg TC, Burg KJL. Groff RE Caracterización de los efectos de la sedimentación celular en la salida de la bioimpresora. Biofabricación. 2012;**4**:011001. DOI: 10.1088/1758-5082/4/1/011001

[28] Maher PS, Keatch RP, Donnelly K, Paxton JZ. Formed 3D bio-scaffolds using rapid prototyping technology. In: Vander Sloten J, Verdonck P, Nyssen M, Haueisen J, editors. Proceedings of the Fourth European Conference of the International Federation for Medical and Biological Engineering; Antwerp, Belgium. November 23 to 27, 2008. Berlin/Heidelberg, Germany: Springer; 2009. pp. 2200-2204

[29] Magalhães LSS, Santos FEP, Elias CDMV, Afewerki S, Sousa GF, Furtado AS, et al. Lobo AO printing 3D hydrogel structures using low-cost stereolithography technology. Journal of Function Biomatter. 2020;**11**:12. DOI: 10.3390/jfb11010012

[30] Gaetani R, Feyen DAM, Verhage V, Slaats R, Messina E, Christman KL, et al. Epicardial application of cardiac progenitor cells in a 3D-printed gelatin/hyaluronic acid patch preserves cardiac function after myocardial infarction. Biomaterials. 2015;**61**:339-*348*. DOI: 10.1016/j.biomaterials.2015.05.005

[31] Duan B, Kapetanovic E, Hockaday LA, Butcher JT. Conductos de válvulas de tres folletos impresos tridimensionales utilizando hidrogeles biológicos y células intersticiales de válvulas humanas. Acta Biomaterialia. 2014;**10**(5):1836-1846. DOI: 10.1016/j.actbio.2013.12.005

[32] Kawai Y, Tohyama S, Arai K, Tamura T, Soma Y, Fukuda K, et al. Tejido cardíaco tubular sin andamios diseñado a partir de células madre pluripotentes inducidas por humanos Utilizando la tecnología de impresión Bio-3D in vivo. Frontiers in Cardiovascular Medicine. 2021;**8**:806215. DOI: 10.3389/fcvm.2021.806215

[33] Nelson DL, Cox MM. Lehninger principles of biochemistry. 6th ed. New York, NY: W. H. Freeman; 2013

[34] Shachar M, Tsur-Gang O, Dvir T, Leor J, Cohen S. The effect of immobilized RGD peptide in alginate scaffolds on cardiac tissue engineering. Acta Biomaterialia. 2011;**7**(1):152-162. DOI: 10.1016/j.actbio.2010.07.034

[35] Andukuri A, Kushwaha M, Tambralli A, Anderson JM, Dean DR, Berry JL, et al. A hybrid biomimetic nanomatrix composed of electrospun polycaprolactone and bioactive peptide amphiphiles for cardiovascular implants. Acta Biomaterialia. 2011;7(1):225-233. DOI: 10.1016/j.actbio.2010.08.013

[36] Uchinaka A, Kawaguchi N, Hamada Y, Mori S, Miyagawa S, Saito A, et al. Transplantation of myoblast sheets that secrete the novel peptide SVVYGLR improves cardiac function in failing hearts. Cardiovascular Research. 2013;**99**(1):102-110. DOI: 10.1093/cvr/cvt088

[37] Kandhola G, Park S, Lim JW, Chivers C, Song YH, Chung JH, et al. Nanomaterial-based scaffolds for tissue engineering applications: A review on graphene, carbon nanotubes and nanocellulose. Tissue Engineering and Regenerative Medicine. 2023;**20**(3):411-433. DOI: 10.1007/s13770-023-00530-3. Epub 2023 Apr 15

[38] Gaharwar AK, Sant S, Hancock MJ, Hacking SA. Nanomaterials in Tissue Engineering: Fabrication and Applications. Cambridge, UK: Woodhead Publishing; 2013

[39] Valentina M, Giada C, Maria TF, Long Carlin S, Caldwell John H, Lorena Z, et al. Carbon nanotubes promote growth and spontaneous electrical activity in cultured cardiac myocytes. Nano Letters. 2012;**12**(4):1831-1838. DOI: 10.1021/nl204064s

[40] Hao T, Li J, Yao F, Dong D, Wang Y, Yang B, Wang C. Injectable fullerenol/

alginate hydrogel for suppression of oxidative stress damage in brown adipose-derived stem cells and cardiac repair. ACS Nano. 2017;**11**(6):5474-5488. DOI: 10.1021/acsnano.7b00221

[41] Cristallini C, Vitale E, Giachino C, Rastaldo R. Nanoengineering in cardiac regeneration: Looking Back and going forward. Nanomaterials. 2020;**10**(8):1587. DOI: 10.3390/nano10081587

[42] Muñoz M, Çimenci ÇE, Goel K, Comtois-Bona M, Hossain M, McTiernan CD, et al. Nanoengineered sprayable therapy for treating myocardial infarction. ACS Nano. 2022;**16**(3):3522-3537. DOI: 10.1021/acsnano.1c08890

[43] Baei P, Jalili-Firoozinezhad S, Rajabi-Zeleti S, Tafazzoli-Shadpour M, Baharvand H, Aghdami N. Electrically conductive gold nanoparticle-chitosan thermosensitive hydrogels for cardiac tissue engineering. Materials Science & Engineering. C, Materials for Biological Applications. 2016;**63**:131-141

[44] ZwiDantsis L, Wang B, Marijon C, Zonetti S, Ferrini A, Massi L, et al. Remote magnetic nanoparticle manipulation enables the dynamic patterning of cardiac tissues. Advanced Materials. 2020;**32**:1904598s

Chapter

ROC Curve Analysis of Molecular Docking for Drug Development

Ricardo Enrique Buendia Corona,
Lucila Isabel Castro Pastrana, Lenin Domínguez-Ramírez
and Paulina Cortés-Hernández

Abstract

The popularity of molecular docking as a drug repositioning tool is due to its perceived efficiency and scalability. Its ability to quickly screen thousands of compounds for binding affinities makes it a valuable asset for reducing both time and cost in drug development. However, the decision-making process following docking remains inconsistent across literature. Top-ranked pose is often assumed to be pharmacologically activity against the target. Such assumption, without validation for the specific target and docking model, can be erroneous and lead to misleading conclusions. Limitations of docking validation methods, such as re-docking co-crystallized ligands or ensemble docking, lie in their inability to discriminate between experimentally validated active compounds and inactive ones. This chapter explores the utility of receiver operating characteristic (ROC) curves as a statistical approximation for diagnosing performance of docking software. The generalities of docking procedures, its use for drug repositioning, and its inherent limitations are analyzed. ROC curves are explained, and a step-by-step pipeline is recapitulated. We emphasize that ROC curves diagnose docking model performance overall and do not directly select a single correct result. Evaluating predictive certainty of docking software across protein targets should be performed to ensure effective filtering of promising hits for successful drug repositioning.

Keywords: molecular docking, ROC curves, statistical validation, drug repositioning, drug development

1. Introduction

The term drug design (DD) describes the rational and systematic search for new compounds with biological activity against a specific molecular target or disease, supported by experimental information (**Figure 1A**) [1]. DD requires preclinical (computational, *in vitro/in vivo*) and clinical (toxicity/safety and effectivity) trials to identify molecules that produce a therapeutic effect and are safe for humans at therapeutic doses [2]. Compounds that pass these filters are called "hits" [3, 4]. DD is

IntechOpen

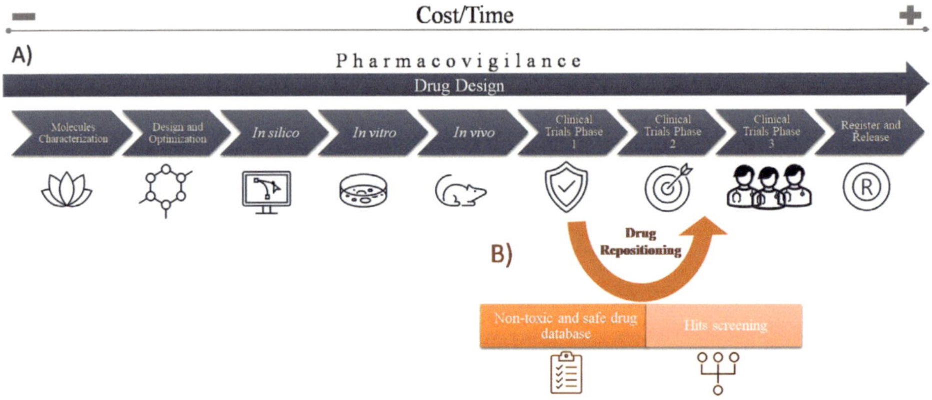

Figure 1.
Drug development process. (A) de novo *DD from molecule obtention and characterization from various natural or synthetic sources. (B) DR process. The intention is that human-safe compounds can take another path for another therapeutic target. The screening of thousands of drugs can filter and return some of them to the process of effectiveness and evaluation on large population scales in clinical trials 1 and 2. Pharmacovigilance is prevalent throughout the entire process.*

expensive in time and money, ranging around 15 years and a thousand million dollars per target [5]. Despite this huge effort, only about one in every thousand new compounds complete the preclinical phase to be tested in humans [6], and less than one in every five thousand is approved for commercial use [7].

To decrease long-term investments and high risks, an alternative effort, called drug repurposing (DR), has emerged. DR looks for new molecular targets for hits that already passed safety trials. In DR, hits are given a second chance against a different disease to the one they were originally designed for [8]. A well-defined, systematic DR pipeline has the potential to produce new treatments with less time and money (**Figure 1B**). However, initial DR success was notably the result of "applied pharmacovigilance" as exemplified by sildenafil (Viagra) that was famously repurposed after erections were reported as the side effect of a cardiovascular drug in development [9]. Pharmacovigilance (**Figure 1C**) continues to provide useful information for DR, as side effects can suggest new repositioning [10].

To accelerate DR, computational searches can be coupled to *in vitro* high-throughput robotic systems to test thousands of drugs against target cells [11]. While the procedures to grow eukaryotic cells take weeks, cost hundreds of dollars and typically test only dozens of compounds at a time; computational docking alone can explore thousands of compounds per day [12] and be conducted with free software [13]. "Target-based" docking requires a protein as a putative therapeutic target and a drug library to be screened against this target. An interesting and widely explored case of target-based docking was SARS-CoV-2. As of January 2024, 3933 structures of its viral proteins have been made public through the Research Collaboratory for Structural Bioinformatics Protein Data Bank (RCSB-PDB). Accessibility to the potential therapeutic targets led to hundreds of publications suggesting drug repurposing against SARS-CoV-2 using docking [14]. However, little feedback is available to answer whether these docking efforts produced any viable treatments, in part because the academic community that conducts and publishes docking efforts does not report a standard way of validating/scoring results against *in vitro* or clinical data.

A limitation of docking methods is that they only rank binding affinities without any other accuracy or pharmacological measure; that is, they do not intrinsically explore the likelihood of a compound with high affinity for the target to be a true drug. Where feasible, incorporating positive controls with established pharmacological activity enables discrimination between genuine and spurious positive outcomes identified via docking. This process can be facilitated by analyses like receiver operating characteristic (ROC) curves [12]. The aim of this chapter is to explain the fundamentals of molecular docking in its most popular implementations, as well as the fundamentals of ROC curves and how they can be applied to DR.

2. Molecular docking

Molecular docking is one of the *in silico* techniques that allow modeling of interactions between a receptor, usually a protein, and a ligand, usually a small molecule, to evaluate binding [15]. The recognition process of ligands and receptors is based on the coincidence of spatial and energy terms [16].

Molecular docking software uses a search algorithm that generates different conformations and orientations (poses) within a given binding site. Each pose is energetically evaluated in terms of Gibbs free energy using a scoring function. These functions are based on affinity energy and statistical thermodynamics, and recently, a few packages have introduced machine learning and deep learning [17]. Each orientation or pose is ranked based on its corresponding affinity energy. Lower energy values are interpreted as indicative of enhanced physiochemical favorability and increased likelihood of binding [18].

Recent reviews [19–21] mention the software in **Table 1** as the most frequently used in the scientific literature (**Table 1**). In these packages, both the search algorithms and the scoring functions depend on the developers and cannot easily be

Docking software	Hallmarks	Developer/License	O.S.	References
Autodock (v3 or v4)	Lamarckian genetic algorithm/semiempirical free energy force field	Center for Computational Structural Biology (CCSB)/GNU	Windows/MacOS/Linux	[22]
Autodock Vina	Monte-Carlo iterated search/semiempirical free energy force field	Center for Computational Structural Biology (CCSB)/GNU	Windows/MacOS/Linux	[23]
DOCK	Deterministic geometric algorithms/AMBER scoring function	UCSF/Academic free license	UCSF/Unix based	[24]
FRED	Systematic non-stochastic algorithm/Chemgauss4 scoring function	OpenEye/Eligible for academic free license	Windows/MacOS/Linux	[25]
GLIDE	Side point search/Empirical GlideScore	Shrödinger/Buy-into license	Windows/MacOS/Linux	[26]

GNU: GNU is Not Unix.
O.S.: Operative System. Names are listed by popular usage.
UCSF: University of California San Francisco.

Table 1.
Most used docking software.

changed by the user. Cross-platform in various operative systems and user-friendly graphical interfaces are also limited.

Docking methods can run thousands of compounds in a few days [27], providing a computationally accessible platform to rapidly test pharmacological hypotheses. Consequently, there has been a notable increase in the number of publications that combine molecular docking and drug repurposing during the last decade (**Figure 2**).

Docking does not have a gold standard considered as valid by the scientific community. Although there have been efforts to evaluate agreement across software packages [28], the docking methodology is not uniform across research groups. Considering a certain software better by its scoring function or by whether users need to pay a license is inaccurate and poorly supported by evidence.

We can explore two primary factors that have been identified as potential sources of bias in docking techniques.

- First, docking reliability depends on input. Thus, the starting material is an obvious bias that impacts the process. The selection of small molecule ligands and protein receptors can utilize curated experimental databases or involve *in silico* approaches, including homology modeling and artificial intelligence methods. The subsequent use of these candidates demands preliminary quality control measures, encompassing validation and/or optimization, to ensure their suitability for docking studies.

- Second, the docking process can be subdivided into *rigid*, where ligand and receptor lack free movement; *flexible ligand*, where the drug has torsion angles allowed in the search algorithm; and *flexible residues*, where both the ligand and the sidechains from the receptor cavity are allowed to move [29]. Employing the previously described docking subdivisions empowers researchers to refine their docking results, leading to the development of more accurate atomistic models.

The selection of specific docking methods is often contingent upon available computational capabilities and individual research group needs. However, even with pre-processed molecules and multiple docking simulations, accurately assessing the software's performance in identifying true positives solely based on affinity ranking

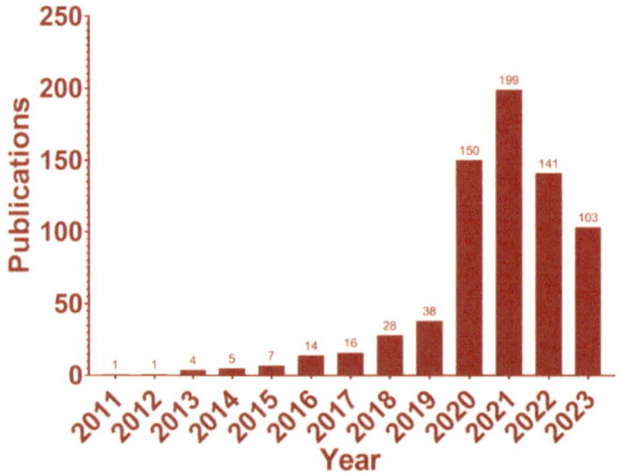

Figure 2.
Drug repositioning and molecular docking number of publications. Retrieved from PubMed under search "(Molecular Docking Simulation [MeSH Terms]) AND (Drug Repositioning [MeSH Terms])." Accessed on January 16, 2024.

remains challenging. Some authors propose validation by approaches such as the following:

- Co-crystallized ligands from protein structures can be re-docked so the user can verify that the software is finding the same pose [30].
- Another approach is the use of decoys as ligands, so the user can determine how well software is ranking true positives from random entries [31].
- Alternatively, consensus docking is another path that can help filter hit compounds, so the user can select hits that are better ranked in a variety of docking programs [32].

Although these approaches are highly accepted and used in the field, none of them intrinsically determine whether software can discriminate a true positive or a true negative from random input.

3. ROC curves

Decisions based on any diagnostic test or measured parameter ultimately depend on the statistical confidence of the technique itself. This is true for all dichotomous classification tasks and is not limited to docking techniques. ROC curves are a statistical analysis and visualization tool that diagnoses the performance of a test (in this case, a docking program, let us say Autodock Vina) and that validates decision-making when test results (in this case, the binding affinity results from the docking program, Autodock Vina) are compared to known true positives (TP) (in this case the binding affinity results of known pharmacologically active ligands obtained with the same docking program, Autodock Vina) [33]. The ROC curve can suggest a cut-off [34] value that distinguishes between true positives (TP) and true negatives (TN). In this way, the ROC process shares similarities with adjusting a radio bandwidth to maximize a desired signal (TP) while minimizing background noise (FP, or false positive). ROC curves are a way to process data to assess what portion of true positives can be identified correctly by the test (sensitivity = TP/TP + FN) as well as what portion of true negatives (specificity = TN/TN + FP, where FN are false negatives) are assigned [35]. Arithmetic between the estimated cost of TN, TP, FP, FN, and P (prevalence), see Eq. (1), is used to identify an optimal cut-off value that separates known positives from negatives:

$$S = \left(\frac{FP_c - TN_c}{FN_c - TP_c} \right) \left(1 - \frac{\mathrm{P}}{\mathrm{P}} \right) \quad (1)$$

Where the subindex c indicates this is the cost of each decision. In turn, the Area under the curve (AUC) in ROC curves is a measure of how well the model separates positives from negatives. Results with different docking programs can be compared (e.g., the same set of ligands tested with docking programs Autodock Vina compared to GLIDE) when a ROC curve is performed for each program, then comparing AUC, sensitivity, specificity, etc. In general, the program with the highest AUC is the best model.

To explain ROC curves, we can look at a simple medical example where biochemical data is used. Let us consider the case of blood glucose as a parameter that describes diabetic patients. Blood glucose values of known diabetic patients are labeled as positive. The blood glucose values from the population that we want to classify (test set or "challenge" population) will be compared to diabetic values to visualize how well they match the positives. A model will be created and evaluated with a confusion table, and the resulting accuracy (proportion of total correct predictions found by the model, including true positives and true negatives with the formula TP + TN/TP + TN + FP + FN), precision (the ability to find true positives among all the positives, with the formula TP/TP + FP), sensitivity, specificity, and AUC can be calculated and describes how well blood glucose separates diabetic and non-diabetic individuals. If blood glucose values are a good proxy for diabetes or its diagnosis, AUC will be a value above 0.7–0.8. A model as good as random identification will have an AUC of 0.5, and the curve will trace a diagonal. Any AUC below 0.5 points to a bad model that incorrectly assigns positives as negatives (**Figure 3B**). Furthermore, the ROC will provide the cut-off for the best blood glucose threshold that most discriminate between positive and negative results.

Similarly, for docking applications, ROC curves are generated using experimentally validated binding or inhibition results as "ground truth" for true positives (active drugs). Both known actives and the ligand library under investigation (challenge data) are processed through the docking program. Binding scores corresponding to confirmed actives are subsequently labeled as such for ROC analysis, while all the remaining results from the challenge set are designated as decoys. **Figure 3A** depicts a potential range of docking results, categorized as "optimal," "random," and "poor," that may be obtained with docking software depending on how good the model is, that is, how well the docking program finds "actives" (known active drugs) and separates them from "decoys" (non-binders in a data set). Under "optimal" conditions, the docking model is expected to prioritize known active ligands by assigning them

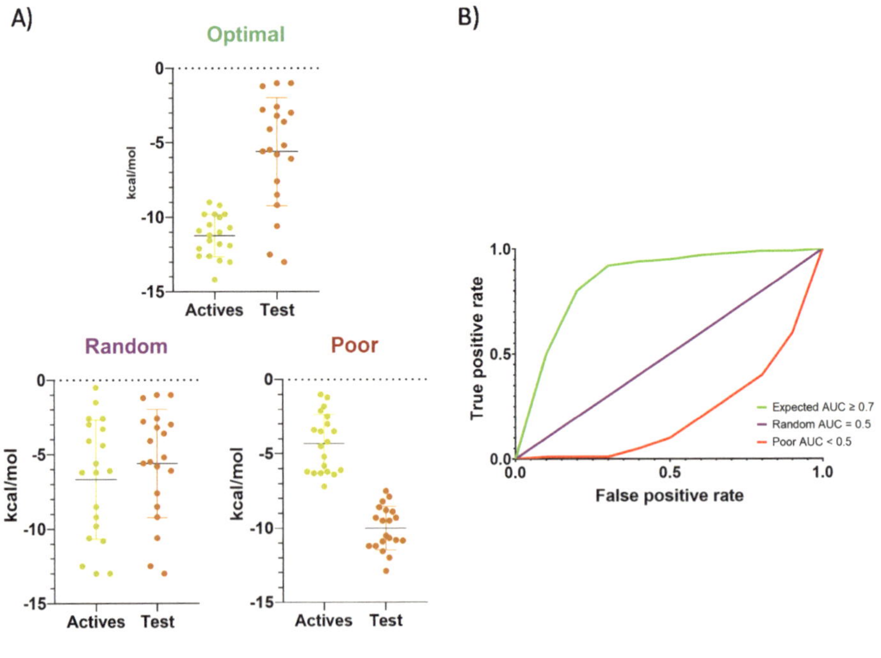

Figure 3.
Docking results and ROC representation. (A) Categorization of docking results. (B) ROC curves attempted for every categorization.

the lowest binding free energy values (kcal/mol). "Random" results indicate that the software ranks known active ligands in a non-specific, probabilistically distributed manner, indicating a lack of predictive power in the docking model. Notably, "poor" results may indicate a misclassification of the true positives used in the ROC. In the context of docking, an AUC above 0.7 indicates a good model and the cut-off would suggest the minimum affinity required to select positive results.

rDock, a docking software developed by GitLab® [36], made a pipeline publicly available to evaluate its results. Besides rDock, it requires R and the ROCR library [37]. This pipeline is illustrated in **Figure 4**:

The docking process with ROC analysis is summarized below:

- Download and prepare receptor structure. Usually, the Protein Data Bank Database (www.rcsb.org) is the best repository for experimentally solved 3D protein structures.

- Download or create a library of compounds known experimentally as actives for the receptor. One way to find this library is the UCSF ZINC database (zinc.docking.org) in the biological subset.

- Download a library of decoy compounds or test set. A useful and popular database is DUD-E (dude.docking.org).

- Run docking for both libraries to the receptor. While we highly recommend ADFR [38] for docking, this pipeline can be modified and applied for any docking suite.

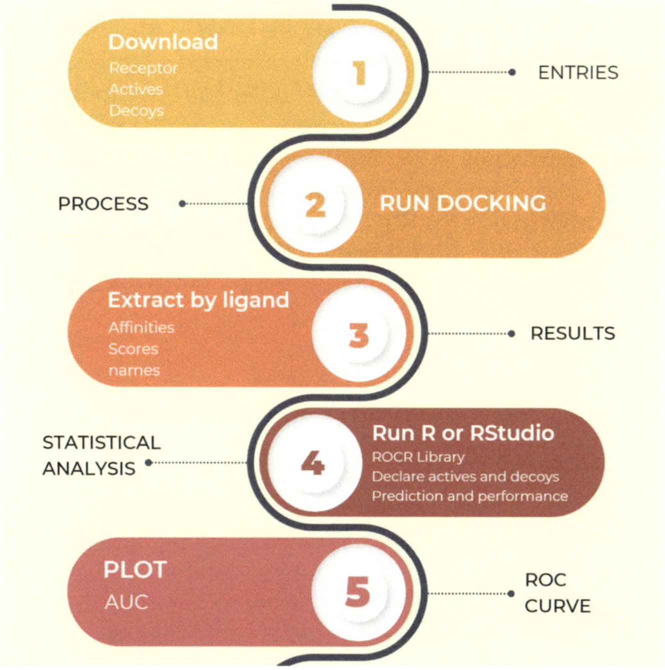

Figure 4.
Docking and ROC calculation process.

- Extract ligand names and affinities in a csv or text data file.
- Call the ROCR library in R or RStudio.
- Declare variables for actives and decoy results files.
- Run ROC prediction and performance commands.
- Plot results.

ROC curves are flexible in the information that can be used, and they require a "yes"or "no" decision and any quantitative value on a continuous scale. For example, Gnina [39] docking package produces affinity in kcal/mol as well as the logarithm of the affinity estimated through its Convolutional Neural Network (CNN) model and CNN pose score. These three parameters can be used for ROC calculation, independently or simultaneously. Since ROC allows for the comparison of the models, any user can easily compare AUC results and decide which parameters afford better results. Furthermore, any user can also use data such as ligand molecular weight and the number of hydrogen bonding donors.

It is important to recognize that ROC curve analysis does not directly enhance docking results. While it provides valuable statistical insights into the model's ability to rank results relative to positive controls, it does not offer actionable information for improving poor results. Consequently, their primary application lies in providing comparative data for different software versions or target-specific performance evaluations.

A recommended reading is a series of papers describing the development of UCSF DOCK [24, 40, 41]. These publications employ ROC curves to evaluate the version's ability to handle diverse protein and RNA targets, as well as its effectiveness in discriminating between active compounds and decoys from established databases. It is worth noting that each target shown in these studies has a different independent ROC curve, even if the libraries screened are the same. That is, the ROC curve is docking software and target-dependent.

It is not obvious from the general characteristics of the targets employed by the UCSF DOCK team what makes the ROC curve better or worse when using the same docking algorithm. Their attempts to improve UCSF DOCK performance by including pharmacophore-based similarity to their scoring function have not yielded consistent improvement so far [42].

4. Truth or randomness? Validating drug repurposing predictions with ROC curves

DR, specifically target-based, employs screening for a variety of compounds. One of the most used techniques to screen molecules is docking. Since neither a technique calibration nor a gold standard is available, authors tend to validate or "diagnose success" via re-docking and ensemble docking.

ROC curves of docking results are displayed in published papers of DR only when positive controls are available from wet-lab experiments. One outstanding example is the work from Llanos and collaborators in 2021 [43]. Llanos' et al. explore the weakness and strengths of some of the efforts to reposition drugs via docking against COVID-19. The authors analyze 168 published works that utilize Mpro from SARS-COV-2 as a therapeutic target. Authors emphasize that only three out of 168 publications performed a ROC analysis of the docking protocol. Llanos also performed ROC curves for their results with three docking packages: Vina, Autodock4-GPU, and Autodock 4 hydrated. All three protocols revealed either random or poor results in ROC curves. A protonated state of Mpro was also used with the intention of improving the model. Nonetheless, its ROC curve did not show promising results. Publications assume that an energetic ranking through docking results describes the inhibitory activity of a compound. Given the results evaluated by the ROC curves, these assumptions should be at least reconsidered and revised.

5. Conclusions

Docking is a very powerful tool for drug repurposing but must be used with caution. Based on the analysis exposed above, we put forward the following recommendations:

While docking is an accessible tool, a systematic method must be adopted and reproduced before selecting assumed positive results.

Decoy docking scores alone are not enough for a correct hit identification. Ideally, experimental data must be used to calibrate the certainty of docking results through the inclusion of true positives in the docking.

Many different docking packages must be tested against any given target to find the best model identified by its AUC and other ROC parameters; the most popular are not accurate for every target.

In vitro and *in vivo* tests remain a central part of drug development, so it is highly suggested to validate docking results with these techniques.

Acknowledgements

The authors thank Universidad de las Américas Puebla (UDLAP) and to the Consejo Nacional de Humanidades, Ciencias y Tecnologías (CONAHCYT, Mexico) for his doctoral scholarship. The authors would like to acknowledge the UDLAP Research and Postgraduate Department for supporting this publication. The authors also recognize the support given from Centro de Investigación Biomédica de Oriente (CIBIOR) for the development of this publication.

Conflict of interest

The authors declare no conflict of interest.

Author details

Ricardo Enrique Buendia Corona[1*], Lucila Isabel Castro Pastrana[1], Lenin Domínguez-Ramírez[2] and Paulina Cortés-Hernández[2]

1 Universidad de las Américas Puebla (UDLAP), Puebla, Mexico

2 Centro de Investigación Biomédica de Oriente (CIBIOR), Instituto Mexicano del Seguro Social (IMSS), Atlixco, Mexico

*Address all correspondence to: ricardo.buendiaca@udlap.mx

References

[1] Talevi A. Computer-aided drug design: An overview. Methods in Molecular Biology. 2018;**1762**:1-19

[2] Zhou SF, Zhong WZ. Drug design and discovery: Principles and applications. Molecules. 2017;**22**. DOI: 10.3390/molecules22020279

[3] Attene-Ramos MS, Austin CP, Xia M. High throughput screening. In: Wexler PBT-E of T, editor. Encyclopedia of Toxicology. 3rd ed. Oxford: Academic Press. pp. 916-917

[4] Saldívar-González F, Prieto-Martínez FD, Medina-Franco JL. Descubrimiento y desarrollo de fármacos: un enfoque computacional. Educacion Quimica. 2017;**28**:51-58

[5] Wouters OJ, McKee M, Luyten J. Estimated research and development investment needed to bring a new medicine to market, 2009-2018. JAMA. 2020;**323**:844

[6] Ashburn TT, Thor KB. Drug repositioning: Identifying and developing new uses for existing drugs. Nature Reviews. Drug Discovery. 2004;**3**:673-683

[7] Williams CT. Food and drug administration drug approval process: A history and overview. The Nursing Clinics of North America. 2016;**51**:1-11

[8] Pushpakom S, Iorio F, Eyers PA, et al. Drug repurposing: Progress, challenges and recommendations. Nature Reviews. Drug Discovery. 2018;**18**:41-58

[9] Osakwe O, Rizvi SAA. Introduction. In: Osakwe O, SAA R, editors. Social Aspects of Drug Discovery, Development and Commercialization. Boston: Academic Press. pp. xvii-xxx

[10] Liu Y, Liu Y, Fan R, et al. Pharmacovigilance-based drug repurposing: Searching for putative drugs with hypohidrosis or anhidrosis adverse events for use against hyperhidrosis. European Journal of Medical Research. 2023;**28**. DOI: 10.1186/s40001-023-01048-z

[11] Nunes-Alves C. New tricks for old drugs. Nature Reviews. Microbiology. 2015;**13**:68

[12] Deng W, Verlinde CLMJ. Evaluation of different virtual screening programs for docking in a charged binding pocket. Journal of Chemical Information and Modeling. 2008;**48**:2010-2020

[13] Pagadala NS, Syed K, Tuszynski J. Software for molecular docking: A review. Biophysical Reviews. 2017;**9**:91-102

[14] Hosseini F, Azin M, Ofoghi H, et al. Evaluation of drug repositioning by molecular docking of pharmaceutical resources to identification of potential SARS-CoV-2 viral inhibitors. In: Saxena SK, editor. Drug Repurposing—Molecular Aspects and Therapeutic Applications. Rijeka: IntechOpen; 2022. DOI: 10.5772/intechopen.101395

[15] Meng X-Y, Zhang H-X, Mezei M, et al. Molecular docking: A powerful approach for structure-based drug discovery. Current Computer-Aided Drug Design. 2011;**7**:146-157

[16] Lin X, Li X, Lin X. A review on applications of computational methods in drug screening and design. Molecules. 2020;**25**. DOI: 10.3390/molecules25061375

[17] Crampon K, Giorkallos A, Deldossi M, et al. Machine-learning

methods for ligand–protein molecular docking. Drug Discovery Today. 2022;**27**:151-164

[18] Salmaso V, Moro S. Bridging molecular docking to molecular dynamics in exploring ligand-protein recognition process: An overview. Frontiers in Pharmacology. 2018;**9**. DOI: 10.3389/fphar.2018.00923

[19] Timo GO, Reis RSSV d, Melo AF d, et al. Predictive power of in silico approach to evaluate chemicals against m. tuberculosis: A systematic review. Pharmaceuticals. 2019;**12**. DOI: 10.3390/ph12030135

[20] Zev S, Raz K, Schwartz R, et al. Benchmarking the ability of common docking programs to correctly reproduce and score binding modes in SARS-CoV-2 protease Mpro. Journal of Chemical Information and Modeling. 2021;**61**:2957-2966

[21] Sahoo RN, Pattanaik S, Pattnaik G, et al. Review on the use of molecular docking as the first line tool in drug discovery and development. Indian Journal of Pharmaceutical Sciences. 2022;**84**. DOI: 10.36468/pharmaceutical-sciences.1031

[22] Morris GM, Huey R, Lindstrom W, et al. AutoDock4 and AutoDockTools4: Automated docking with selective receptor flexibility. Journal of Computational Chemistry. 2009;**30**:2785-2791

[23] Trott O, Olson AJ. AutoDock Vina: Improving the speed and accuracy of docking with a new scoring function, efficient optimization, and multithreading. Journal of Computational Chemistry. 2010;**31**:455-461

[24] Allen WJ, Balius TE, Mukherjee S, et al. DOCK 6: Impact of new features and current docking performance. Journal of Computational Chemistry. 2015;**36**:1132-1156

[25] McGann M. FRED pose prediction and virtual screening accuracy. Journal of Chemical Information and Modeling. 2011;**51**:578-596

[26] Friesner RA, Banks JL, Murphy RB, et al. Glide: A new approach for rapid, accurate docking and scoring. 1. Method and assessment of docking accuracy. Journal of Medicinal Chemistry. 2004;**47**:1739-1749

[27] Agu PC, Afiukwa CA, Orji OU, et al. Molecular docking as a tool for the discovery of molecular targets of nutraceuticals in diseases management. Scientific Reports. 2023;**13**. DOI: 10.1038/s41598-023-40160-2

[28] Parks CD, Gaieb Z, Chiu M, et al. D3R grand challenge 4: Blind prediction of protein–ligand poses, affinity rankings, and relative binding free energies. Journal of Computer-Aided Molecular Design. 2020;**34**:99-119

[29] Azad I. Molecular docking in the study of ligand-protein recognition: An overview. In: Istifli ES, editor. Molecular Docking. Rijeka: IntechOpen; 2023. DOI: 10.5772/intechopen.106583

[30] Shivanika C, Deepak Kumar S, Ragunathan V, et al. Molecular docking, validation, dynamics simulations, and pharmacokinetic prediction of natural compounds against the SARS-CoV-2 main-protease. Journal of Biomolecular Structure & Dynamics. 2022;**40**:585-611

[31] Hevener KE, Zhao W, Ball DM, et al. Validation of molecular docking programs for virtual screening against dihydropteroate synthase. Journal of Chemical Information and Modeling. 2009;**49**:444-460

[32] Blanes-Mira C, Fernández-Aguado P, de Andrés-López J, et al. Comprehensive survey of consensus docking for high-throughput virtual screening. Molecules. 2023;**28**. DOI: 10.3390/molecules28010175

[33] Nahm FS. Receiver operating characteristic curve: Overview and practical use for clinicians. Korean Journal of Anesthesiology. 2022;**75**:25-36

[34] Christian T. An introduction to cutpointr. cran.r-project.org

[35] Park SH, Goo JM, Jo C-H. Receiver operating characteristic (ROC) curve: Practical review for radiologists. Korean Journal of Radiology. 2004;**5**:11

[36] Ruiz-Carmona S, Alvarez-Garcia D, Foloppe N, et al. rDock: A fast, versatile and open source program for docking ligands to proteins and nucleic acids. PLoS Computational Biology. 2014;**10**. DOI: 10.1371/journal.pcbi.1003571

[37] rDock. How to calculate ROC curves. CBDD Research Group Blog

[38] Ravindranath PA, Forli S, Goodsell DS, et al. AutoDockFR: Advances in protein-ligand docking with explicitly specified binding site flexibility. PLoS Computational Biology. 2015;**11**:1-28

[39] McNutt AT, Francoeur P, Aggarwal R, et al. GNINA 1.0: Molecular docking with deep learning. Journal of Cheminformatics. 2021;**13**. DOI: 10.1186/s13321-021-00522-2

[40] Balius TE, Tan YS, Chakrabarti M. DOCK 6: Incorporating hierarchical traversal through precomputed ligand conformations to enable large-scale docking. Journal of Computational Chemistry. 2024;**45**:47-63

[41] Brozell SR, Mukherjee S, Balius TE, et al. Evaluation of DOCK 6 as a pose generation and database enrichment tool. Journal of Computer-Aided Molecular Design. 2012;**26**:749-773

[42] Jiang L, Rizzo RC. Pharmacophore-based similarity scoring for DOCK. The Journal of Physical Chemistry. B. 2015;**119**:1083-1102

[43] Llanos MA, Gantner ME, Rodriguez S, et al. Strengths and weaknesses of docking simulations in the SARS-CoV-2 era: The main protease (Mpro) case study. Journal of Chemical Information and Modeling. 2021;**61**:3758-3770

Chapter

Use of Antimicrobial Agents in Reduced Concentrations for the Preservation of Partially Baked Bread Stored in Refrigeration

Ricardo Hilario Hernández Figueroa, María Eugenia Bárcenas Pozos, Fidel Vergara Balderas and Aurelio López-Malo

Abstract

Bread is a food that is an important part of the diet of the Mexican population and of the culinary traditions of our country. The disadvantage of this product is its shelf life, which is very short, mainly due to the aging process, which occurs only a few hours after making the bread. Aging occurs mainly due to water migration from the crumb to the crust and from the crust to the environment; the crumb's hardening occurs mainly due to the retrogradation of starch and the loss of water. For this reason, partially baked bread (PBB) and refrigeration storage have been studied as an alternative to increasing the shelf life of bakery products; however, the main problem with this form of preservation is the growth of molds on the surface of the bread. Therefore, this research evaluated the stability of partially baked bread added with low doses (<0.1%) of potassium sorbate or calcium propionate during refrigerated storage. The results confirm that adding either 0.05 or 0.075% potassium sorbate or 0.075% calcium propionate allows the PBB to be kept refrigerated with an acceptable microbial load and free of visible molds for at least 28 days of storage.

Keywords: bread, antifungal, shelf life, molds, baking goods

1. Introduction

Currently, the food industry depends on chemical preservatives to extend the shelf life and control the growth of spoilage fungi. Some food products depend greatly on antifungal chemical preservatives such as propionic acid or propionates to assure their safety and extend their shelf-life. However, consumers are demanding processed foods with free or reduced preservatives due to the increased awareness about the health hazards associated with some chemical additives [1, 2]. Specifically, bread has a very short shelf life, mainly due to staling phenomena that gradually occur just a few hours after making the bread. These phenomena mainly comprise two processes: the retrogradation of starch and the migration of water from the crumb to the crust and from the crust to the environment; thus, hardening occurs. For this reason, partially baked bread (PBB) and refrigeration storage have been studied as an alternative to increasing the shelf life of bakery products; however, the main problem with this form of preservation is the growth of molds on the surface of the bread [3, 4].

The storage and preservation of PBB are mainly done by freezing. However, this process has important disadvantages, such as high energy consumption and the high cost of maintaining the cold chain. Likewise, the formation and growth of ice crystals can directly damage the structure of the bread. For these reasons, various studies have been carried out that include the evaluation of refrigeration as an alternative method to freezing storage. Bárcenas and Rosell [3] demonstrated that the damage caused to partially baked bread by refrigeration (2°C) is insignificant compared to that caused by the same type of bread when stored under frozen conditions. Otherwise, Lainez et al. [4] demonstrated that one of the main problems presented by partially baked bread stored in refrigeration is the growth of molds on the surface of this product, which occurs within a period of 9 days when the bread is stored at 7°C. Therefore, this research evaluated the stability of partially baked bread added with low doses (<0.1%) of potassium sorbate or calcium propionate during refrigerated storage.

2. Methodology

2.1 Bread preparation

The bread was manufactured in the Universidad de las Américas Puebla (UDLAP) Food Engineering pilot plant, applying a direct dough process and using a basic formulation (flour, 61%; water, 36%; salt, 1%; sugar, 1% and yeast, 1%). The ingredients were mixed and kneaded using a mixer (Hobart, Troy, OH, USA); the dough was cut and formed manually; 180 g pieces of dough were formed into a baguette shape. Fermentation was carried out for 30 minutes at 35°C and 65% relative humidity using a Mini combo oven (Zuchelli Alpha, Trevenzuolo, Verona, Italia). The partial baking was then carried out at 180°C for 10 minutes, and then the bread cooled until it reached 20°C in the center of the crumb. The partially baked bread was packaged in polyethylene bags, placed in corrugated cardboard boxes, then stored in a refrigeration chamber at 4°C. To complete baking, the PBB was baked at 180°C for 10 minutes. The bread was then cooled to room temperature for 1 hour. To evaluate the effect of the selected antimicrobial agents (potassium sorbate and calcium propionate) at different concentrations (0.05 or 0.075%), these were added directly to the ingredients to form the dough [4].

2.2 Determination of pH and water activity a_w

The pH of the PPB samples was measured following the AACC method 02–52 by electrode immersion with a pre-calibrated pH meter (HI 2210 Hanna Instruments, Woonsocket, RI, USA) [5]. An AquaLab 4TEV Series equipment (Meter Food, Pullman, WA, USA) was used for aw analysis. All determinations were performed in triplicate.

2.3 Microbiological analysis

Serial dilutions of (10^{-1} and 10^{-2}) of refrigerated PBB were prepared with sterile peptone water. To count aerobic mesophilic bacteria (AMB), serial dilutions were poured on Petri with nutrient agar plates. The plates were incubated at 35°C, and after 24 hours of incubation, the AMB colonies were counted using a colony counter. Results are presented as colony-forming units per gram of sample (CFU/g). Similarly, pour-plating of 10^{-1}, and 10^{-2} dilutions of refrigerated PBB were done on potato dextrose agar plates to count molds and yeasts. The plates were incubated at 25°C, and after 5 days of incubation, mold and yeast colonies were counted using a colony counter. The results were presented as colony-forming units per gram of sample (CFU/g). In both cases, the analyses were carried out in triplicate for each dilution.

2.4 Direct observations of partially baked bread

The packaged bread pieces were observed daily until mold growth was visually detected on the surface of the control or the pieces of bread with antimicrobials and determine the antifungal activity of the reduced doses of the tested synthetic antimicrobials [5].

2.5 Crumb hardness measurement

After baking the PPB samples, the crumb hardness was determined by adapting the AACCI method 74-10a [6], using a TA-XT2 texturometer (Texture Technologies Co., NY, USA) in manual mode with a 36 mm diameter cylindrical probe. The probe's speed was 1.7 mm/s, and the distance traveled was 25%. A slice of the central part of the baguette, 25 mm thick, was used as a sample. The measurement was carried out in triplicate.

2.6 Determination of bread moisture content, specific volume, and width/height ratio

The PPB samples were completely baked and the bread moisture was measured following the AOAC 930.15 method [7]. Volume determination was measured by seed displacement according to AACC method 10–05 [6], and the specific volume of bread was expressed as the volume between the weight of bread (cm^3/g). An electronic Vernier was used to measure the width/height ratio, and both dimensions were measured in central slices of 2.5 cm thickness [4]. All determinations were performed in triplicate.

2.7 Statistical analysis

The results are presented as mean values and standard deviations. The obtained data were statistically analyzed using analysis of variance (ANOVA). Pair comparisons for mean values were performed with Tukey's test at $p < 0.05$, using the statistical software Minitab 20 software (Minitab LLC, State College, PA, USA).

3. Results

3.1 Bread moisture content, specific volume, and width/height ratio

Table 1 shows the results obtained for the bread moisture content; it is observed that the presence of antimicrobial agents does not affect the moisture of the PBB

Time (days)	Control	Potassium sorbate		Calcium propionate	
		0.05%	0.075%	0.05%	0.075%
0	38.53	38.74	39.28	38.50	37.77
7	37.77	37.88	38.51	39.22	38.21
14	38.18	38.18	38.21	39.52	38.89
21	ND	38.66	39.04	ND	38.58
28	ND	38.12	38.53	ND	38.93

ND: not determined due to the presence of mold on the surface.

Table 1.
Moisture content (wb%) of partially baked bread at different refrigerated storage times (−4°C).

(after baking) significantly ($p > 0.05$). On the other hand, in the five batches studied, the moisture content remained practically constant during storage, indicating no notable water losses throughout the packaging and refrigerated storage. Results are comparable with those obtained by Lainez [4].

The specific volume and width/height ratio of bread baked in two stages were obtained to verify that the baking process did not present any difference between the different batches. Table II shows the values obtained for these parameters from the four batches of bread added with the tested antimicrobials and from the bread without antimicrobials used as a control.

Although there is no significant difference ($p > 0.05$) between the specific volume of the pieces of bread added with antimicrobial agents and the control, the latter presented a larger total mean than the added pieces of bread (**Table 2**). This behavior agrees with what was reported by Carrillo [8], who found that the specific volume of bread baked in two stages added with 0.1% calcium propionate was lower than that obtained for bread that was not added with antimicrobials.

With respect to the width/height ratio, which is an indication of the proportionality in the shape of the bread, it is found that the bread with propionate (0.075%) had a significantly ($p < 0.05$) higher value for this parameter, when compared with the other samples (**Table 2**). This means that it presented a flatter shape. According to Carrillo [8], a high width/height ratio (1.7–1.9, for loaves of 180 g dough) indicates

Bread	Specific volume (cm^3/g)	Width/height ratio
Control	3.95 ± 0.12 a	1.43 ± 0.06 a
Potassium sorbate		
0.05%	3.87 ± 0.21 a	1.54 ± 0.14 a
0.075%	3.47 ± 0.15 a	1.38 ± 0.03 a
Calcium propionate		
0.05%	3.44 ± 0.02 a	1.44 ± 0.05 a
0.075%	3.91 ± 0.39 a	1.75 ± 0.06 b

For each analyzed parameter, samples with the same letter, there is no significant difference ($p > 0.05$).

Table 2.
Specific volume and width/height ratio of PPB added with antimicrobial agents after baking.

that the breads are compact and do not have a well-proportioned shape. In the case of the pieces of bread obtained in this study, only the product containing 0.075% calcium propionate (CaP) is in the abovementioned situation. The remaining samples had width/height ratio values within the range reported as adequate (1.3–1.6 for loaves of 180 g dough), being well-proportioned breads [3, 8, 9]. The higher value of the width/height ratio of the bread with 0.075% CaP does not necessarily indicate a severe deviation in the baking conditions (over-kneading, over-fermentation), especially if it is considered that the specific volume of this product was not significantly different than the others. Instead, this difference can be attributed to the variability of the baking process.

3.2 Water activity and pH of bread

The results obtained for aw and pH of the bread were 0.971 ± 0.01 and pH 5.44 ± 0.03, respectively. Both values are high; therefore, the stability of PPB against microbial growth is not sufficient to keep the product stored refrigerated for a long time, hence the need for other preservation factors, in this case, the addition of antimicrobial agents.

3.3 Bread crumb hardness

It is observed that in all batches of bread evaluated, the hardness increased considerably during the first days of storage (**Figure 1**). In the case of PBB with 0.05% potassium sorbate (KS) and 0.075% CaP, the hardness continued to increase as the storage time passed. The main cause of bread hardening is the retrogradation of starch, which occurs faster at temperatures around 4°C. The increase in crumb rigidity can be associated with a reduction in the water mobility of the system and, therefore, with greater difficulty for microbial growth [3].

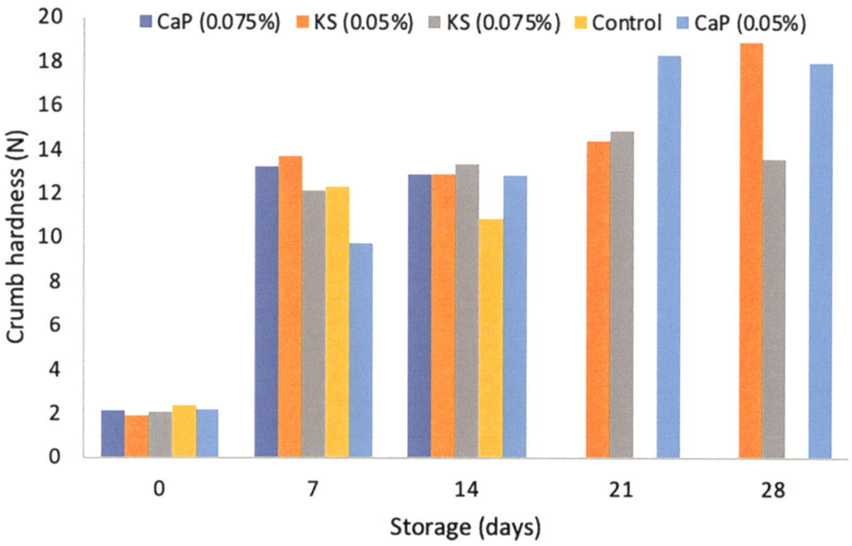

Figure 1.
Hardness (N) of PPB after baking at different refrigerated storage times (4°C). After 21 days, the crumb hardness was not determined in the control and CaP (0.05%) samples due to the appearance of molds on the surface.

3.4 Microbiological analysis

Only the batches of control bread and bread added with 0.05% CaP showed mold growth on the surface after 14 and 21 days of storage, respectively. In the rest of the products, no mold appearance was observed in the 28 days that the study lasted. **Figure 2** shows the appearance of mold growth on the surface of the PBB.

On the other hand, **Table 3** shows the data obtained from the evaluation of aerobic mesophilic bacteria (AMB) present in the PBB at different storage times. The 0.075% batches, potassium sorbate and CaP, had the lowest AMB counts after 28 days of storage.

In the case of molds and yeasts (**Tables 4** and **5**), the data obtained show that the batches added with 0.075%, both KS and CaP, presented the lowest growth levels. The Official Mexican Standard (NOM-147-SSA1-1996) establishes that the maximum limit of aerobic mesophilic bacteria, molds, and yeasts in white bread is 1000, 20, and 20 CFU/g, respectively.

After 28 days of refrigerated storage, both the PPB added with KS and the one added with 0.075% CaP presented microbial counts higher than those established by the standard as the maximum limit. However, it must be kept in mind that PPB is not consumed as such, before consumption, it is subjected to baking (180°C for 10 minutes), which significantly reduces the microbial load of the final product. This was demonstrated by Lainez et al. [4]. Therefore, it can be stated that using KS in concentrations of 0.05 and 0.075% or of CaP at 0.075% allows the PPB to be kept refrigerated, with an acceptable microbial load for at least 28 days.

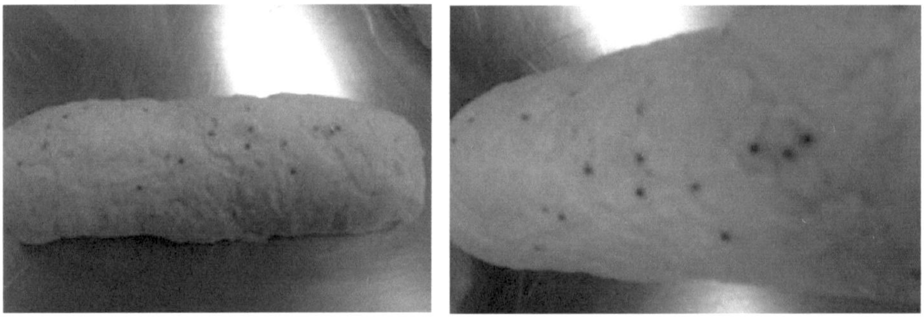

Figure 2.
Appearance of partially baked bread with mold growth on the surface.

Time (days)	Control	Potassium sorbate		Calcium propionate	
		0.05%	0.075%	0.05%	0.075
0	1.60×10^2	1.30×10^2	1.31×10^2	1.40×10^2	1.10×10^2
7	2.25×10^3	40	10	80	60
14	4.70×10^4	4.30×10^4	8.77×10^3	6.91×10^3	4.48×10^3
21	ND	1.57×10^4	6.91×10^3	ND	7.65×10^3
28	ND	4.67×10^4	8.59×10^3	ND	8.03×10^3

ND: not determined due to the presence of mold on the surface.

Table 3.
Effect of concentration and type of antimicrobial agent on the growth of aerobic mesophilic bacteria (CFU/g) in partially baked bread stored under refrigeration.

Time (days)	Control	Potassium sorbate		Calcium propionate	
		0.05%	0.075%	0.05%	0.075%
0	20	30	20	1.10 x 10^2	40
7	20	40	20	40	10
14	1.30 x 10^2	40	30	20	<10 (estimated)
21	ND	50	<10 (estimated)	ND	<10 (estimated)
28	ND	1.30 x 10^2	70	ND	<10 (estimated)

ND: not determined due to the presence of mold on the surface.

Table 4.
Effect of concentration and type of antimicrobial agent on mold growth (CFU/g) in partially baked bread stored under refrigeration.

Time (days)	Control	Potasium sorbate		Calcium propionate	
		0.05%	0.075%	0.05%	0.075%
0	1.20 x 10^2	40	40	1.00 x 10^2	80
7	2.74 x 10^3	10	<10 (estimated)	1.60 x 10^2	1.4 x 10^2
14	2.37 x 10^4	<10 (estimated)	<10 (estimated)	6.91 x 10^3	1.9 x 10^2
21	ND	<10 (estimated)	10	ND	3.3 x 10^2
28	ND	8.40 x 10^3	2.80 x 10^3	ND	7.4 x 10^2

ND: not determined due to the presence of mold on the surface.

Table 5.
Effect of concentration and type of antimicrobial agent on yeast growth (CFU/g) in partially baked bread stored under refrigeration.

4. Conclusion

According to the values of the specific volume and the width/height ratio, the baking processes to obtain the different batches of partially baked bread (PBB) were carried out homogeneously. The water activity and pH of the PBB are adequate to allow the growth of microorganisms. Refrigerated storage conditions do not affect the PBB's moisture content but favor the product's hardening. The use of potassium sorbate (KS) or calcium propionate (CaP) in concentrations less than 0.1% (0.05% or 0.075% for KS, 0.075% for CaP), allows PBB to be stored refrigerated (3–4°C) with an acceptable microbial load, for at least 28 days.

Acknowledgements

Author Hernández-Figueroa acknowledges financial support for his Ph.D. studies in Food Science from the National Council for Humanities, Science and Technology (CONAHCyT) and Universidad de las Américas Puebla (UDLAP).

Conflict of interest

The authors declare that they have no known competing financial interests or personal relationships that could have appeared to influence the work reported in this paper.

Author details

Ricardo Hilario Hernández Figueroa*, María Eugenia Bárcenas Pozos,
Fidel Vergara Balderas and Aurelio López-Malo
Departamento de Ingeniería Química, Alimentos y Ambiental, Universidad de las Américas Puebla, San Andrés Cholula, México

*Address all correspondence to: ricardoh.hernandezf@udlap.mx

References

[1] Muhialdin BJ, Algboory HL, Kadum H, Mohammed NK, Saari N, Hassan Z, et al. Antifungal activity determination for the peptides generated by lactobacillus plantarum TE10 against aspergillus flavus in maize seeds. Food Control. 2020;**109**:106898. DOI: 10.1016/j.foodcont.2019.106898

[2] Luz C, Saladino F, Luciano FB, Mañes J, Meca G. In vitro antifungal activity of bioactive peptides produced by lactobacillus plantarum against aspergillus parasiticus and *Penicillium* expansum. LWT—Food Science and Technology. 2017;**81**:128-135. DOI: 10.1016/j.lwt.2017.03.053

[3] Bárcenas ME, Rosell CM. Effect of HPMC addition on the microstructure, quality and aging of wheat bread. Food Hydrocolloids. 2005;**19**(6):1037-1043. DOI: 10.1016/j.foodhyd.2005.01.005

[4] Lainez E, Vergara F, Bárcenas ME. Quality and microbial stability of partially baked bread during refrigerated storage. Journal of Food Engineering. 2008;**89**(4):414-418. DOI: 10.1016/j.jfoodeng.2008.05.020

[5] Hernández-Figueroa RH, Mani-López E, López-Malo A. Antifungal activity of wheat-flour sourdough (type II) from two different lactobacillus in vitro and bread. Applied Food Research. 2023;**3**(2):100319. DOI: 10.1016/j.afres.2023.100319

[6] American Association of Cereal Chemists, editor. Approved Methods of the American Association of Cereal Chemists. 10th ed. St. Paul, MN: AACC; 2000

[7] Latimer GW, AOAC International, editors. Official Methods of Analysis of AOAC International. 21st ed. Vol. 3. Gaithersburg, MD: AOAC International; 2019

[8] Carrillo A. Evaluación Del Uso de Antimicrobianos Sobre La Estabilidad de Pan Parcialmente Horneado Almacenado En Refrigeración. Tesis de Licenciatura: Universidad de las Américas Puebla; 2007

[9] Bárcenas ME, Benedito C, Rosell CM. Use of hydrocolloids as bread improvers in interrupted baking process with frozen storage. Food Hydrocolloids. 2004;**18**(5):769-774. DOI: 10.1016/j.foodhyd.2003.12.003

Chapter

Mucilage and *Opuntia ficus-indica* Cladode Powder as Functional Ingredients in Amaranth Bars

Rocío Aparicio-Ortuño, J. Daniel Lozada-Ramírez and Ana E. Ortega-Regules

Abstract

The cladodes of *Opuntia ficus-indica*, commonly known as nopales, possess high nutritional value attributable to their content of dietary fiber, bioactive compounds, minerals, and protein. Consequently, they offer potential as an alternative in the formulation and development of functional foods, given the growing demand for foods that promote health and well-being, making functional foods one of the fastest growing sectors in the industry. Nopales, recognized economically as one of the most important crops, exhibit nutritional, functional, and therapeutic properties, with potential applications in the cosmetic, pharmaceutical, bioenergy industries, and ecological contexts. The objective of this study is to compare traditional amaranth bars with two formulations incorporating mucilage and cladode powder while reducing sugar content. Differences in water activity, moisture percentage, color, hardness, nutritional composition, and sensory acceptance of the bars were evaluated. The results revealed significant differences ($p < 0.05$) in the traditional amaranth bar, particularly in moisture content, color, hardness, and sensory evaluation. Additionally, it displayed a higher caloric content (215 kcal/100 g) and a lower contribution of dietary fiber (4.8%) compared to Formulation 1 (155 kcal/100 g and 14.43% fiber) and Formulation 2 (138 kcal/100 g and 15.56% fiber). Mucilage and nopal powder emerge as viable, functional ingredients in various food products, offering nutritional benefits and positively impacting human health by increasing antioxidant and dietary fiber intake.

Keywords: *Opuntia ficus-indica*, cladodes, mucilage, functional ingredients, dietary fiber

1. Introduction

The cladodes of *Opuntia ficus-indica*, commonly known as nopales, are predominantly cultivated in temperate, semi-arid, and dry tropical zones. Indigenous to America, they were initially distributed from Mexico to Spain and have since spread to Africa, Asia, Europe, and Oceania. In Mexico, the consumption of wild varieties of cladodes, also referred to as nopales, has a history dating back approximately 25,000 years [1]. Nopales have played a crucial role since pre-Hispanic times, serving as a significant component in various culinary preparations, agricultural practices, fodder, and soil erosion control, owing to their affordability and abundance [2]. *Opuntia ficus-indica* is a cactus species that adapts well to extreme climates, particularly in the semi-arid regions of central Mexico, which boasts the highest diversity of this cactus globally. The *Opuntia* spp. genus encompasses around 1500 cactus species, and the chemical composition of nopales varies depending on the species, maturity stage, environmental conditions, harvest season, and post-harvest treatment [3, 4]. As per the Mexican standard NMX-FF-068-SCFI-2006, nopal is defined as the young cladodes (tender shoots) of the plant belonging to the cactus family, encompassing the genera *Opuntia* spp. and *Nopalea* spp. Cladodes are flattened stem segments furnished with temporary reduced leaves, glochids (ahuates), and spines at specific points known as areoles. Cladodes are categorized by quality grade, size, and variety, with lengths ranging from 7 cm (Cambray) to 25.1 cm or more, in accordance with Mexican standards. In 2019, data from the Agri-Food and Fisheries Information Service (SIAP) reported that nationwide cultivation of prickly pear cactus covered 12,799 hectares, yielding 821,000 tons and averaging 71 tons per hectare. According to the Codex Alimentarius (2006), the prickly pear cactus is defined as the modified stem of commercial nopales obtained from various *Opuntia* species within the Cactaceae family. This product must be supplied fresh to consumers after conditioning and packaging, with cactus lengths varying from 9 to 30 cm.

Cladodes are composed of water, carbohydrates, proteins, lipids, vitamins, minerals, and bioactive compounds [5]. Bioactive compounds may reduce the risk of several diseases associated with oxidative stress, such as cardiovascular disease, cancer, diabetes, and neurodegenerative diseases [6]. Among the carbohydrates present in cladodes, dietary fiber is a crucial component, consisting of carbohydrate polymers and non-starchy polysaccharides that form the main structure of plant cell walls. These include cellulose, hemicelluloses, hemiglucans, and pectins, along with other polysaccharides from vegetables and algae, such as gums and mucilage. Additionally, non-digestible reserve polysaccharides such as inulin and resistant starch are included. The terms soluble dietary fiber and insoluble dietary fiber offer a straightforward classification based on their physiological properties. Many soluble fibers are viscous and form gels in the small intestine, while insoluble fibers, including cellulose and lignin, have a greater impact on the function of the large intestine [7].

In recent years, there has been a growing interest in formulating foods with dehydrated nopal due to its high levels of bioactive compounds, including dietary fiber, flavonoids, flavonols, carotenes, ascorbic acid, phytosterols, and chlorophyll [8]. The consumption of fiber is associated with various benefits, such as preventing diseases like diabetes, gastrointestinal disorders, and conditions linked to low dietary fiber intake. It can also contribute to the reduction of blood glucose values, anti-hyperlipidemic, and anti-hypercholesterolemic effects [9]. Moreover, dietary fiber can facilitate the transport of numerous bioactive compounds, such as polyphenols and carotenoids, linked to the dietary fiber matrix through the human gut. On average, 2.5% of insoluble dietary fiber consists of polyphenols [10]. Polyphenols are characterized by the presence of several phenolic groups, associated with complex chemical groups of high molecular weight. These compounds, derived from plant metabolism, are gaining interest due to their potential antioxidant capacity, contributing to health benefits such as inflammation prevention, cardiovascular regulation, neurodegenerative disease prevention, and demonstrated anticancer activity [11].

Mucilage, a component of soluble dietary fiber, is composed of arabinose, galactose, xylose, and rhamnose as neutral sugars, along with a small amount of galacturonic acid. Due to its viscosity and chemical composition, mucilage is considered a functional ingredient [5, 9, 12]. This biopolymer has found applications in the pharmaceutical, cosmetic, medical, and food industries due to its physicochemical and rheological properties, emulsifying capacity, and its role as an auxiliary in gel formation. It is also utilized in the production of edible films and coatings [13]. Similarly, powdered cladodes offer an alternative in the formulation and development of functional foods, responding to the growing consumer demand for foods promoting health and well-being [12].

On the other hand, amaranth (*Amaranthus viridis*), one of the oldest edible cereals found in Tehuacán, Puebla, Mexico, with records dating back to about 4000 B.C., possesses high nutritional value. It serves as a source of high-quality protein, unsaturated lipids, dietary fiber, vitamins, and minerals [14].

The term "functional foods" was initially coined in Japan in the 1980s. Recently, Granato et al. [15] defined functional foods as processed or natural foods that, when regularly consumed within a diet at certain concentrations, exhibit potentially positive effects on health beyond basic nutrition. In the consensus document on scientific concepts of functional foods in Europe (1999), it is emphasized that there is no universally accepted definition of functional foods. Instead, functional foods are regarded more as a concept than a precisely defined group of food products. In Europe, a food is deemed "functional" if it is adequately demonstrated to impact one or more objective functions of the body, surpassing basic nutritional effects. This impact should be relevant for improving health and well-being and reducing the risk of disease. It is essential for functional foods to maintain their status as regular foods and to demonstrate their effects at quantities typically consumed in the diet. Furthermore, a food can be considered functional for specific populations, defined, for instance, by age or genetic makeup. The United States of America has allowed claims of "disease risk reduction" in certain foods accepted by the Food and Drug Administration (FDA) since 1993, provided there is objective and scientific evidence establishing a correlation between nutrients or foods in the diet and certain diseases.

The objective of this study was to develop amaranth bars by substituting a portion of sugar with mucilage, serving as an adhesion and compaction agent, and enhancing the content of dietary fiber, both soluble and insoluble, using powdered *Opuntia ficus-indica* cladodes. This approach was adopted as amaranth has low natural fiber content. The impact of adding fresh mucilage and powdered cladodes on the physical, nutritional, and sensory properties of the bars was thoroughly evaluated.

2. Materials and methods

2.1 Plant material and sample preparation

Fresh *Opuntia ficus-indica* cladodes, harvested in Atlixco, Puebla, Mexico, measuring 20–30 cm in size and weighing 90–100 g, were utilized. The spineless cladodes underwent washing with potable water and disinfection using a 10% sodium hypochlorite solution [2]. Subsequently, they were sliced into 2 cm strips, dried in a convection oven at 45°C for 30 h (MAPSA, HDP-334, Mexico), and ground at

22 rpm/3 min (WARING, Mixer 7011 HS, USA). The resulting powder was packed and stored in airtight aluminum bags in a cool, dark place. Amaranth was sourced from the company Dul Cerel S.A. de C.V., while piloncillo and honey were acquired from the local market.

2.2 Extraction of nopal mucilage

The extraction process followed the method outlined by León-Martínez et al. [16] with certain modifications. Spineless cladodes were washed and disinfected before being cut into 1-cm squares. These were then placed in a stainless-steel container with distilled water at a ratio of 1:3 w/v and allowed to soak for 24 h at 25°C in the absence of light. Mucilage was separated from the solid mass through decantation and stored at 4°C for subsequent use.

2.3 Preparation of Amaranth bars

Various formulations were prepared (**Table 1**). The procedure involved melting piloncillo with honey and/or water or mucilage until a syrup of thick consistency was obtained. This syrup was then poured over the amaranth and mixed with powdered nopal until fully incorporated. Finally, the mixture was poured into a mold and left to rest for 15 min. The bars were stored in a dry container and kept in a cool place for subsequent analysis.

2.4 Physical properties

2.4.1 Water activity

The determination of water activity (aw) utilized an Aqualab 4TEV machine. A bar sample was placed in the tray, and the measurement was performed on the equipment. Three determinations were conducted for each formulation, and the results were expressed as the mean ± standard deviation.

2.4.2 Humidity percentage

Moisture content was determined following the AOAC 950.46 method [17]. Two grams of the sample were weighed in an aluminum dish at a constant weight,

Ingredient (g)	C	F1	F2
Amaranth	30.0	30.0	30.0
Piloncillo	25.0	25.0	25.0
Honey	25.0	13.6	11.9
Cladodes powder	—	10.0	10.0
Mucilage	—	20.0	30.0
Water	20.0	—	—

C: Control; F1: Formulation 1; F2: Formulation 2.

Table 1.
Formulation of amaranth bars.

introduced into an oven at 135 ± 2°C for 2 h, cooled in a desiccator for 20 min, and then reweighed. The percentage of moisture on a dry basis was calculated using Eq. (1):

$$\%Hbs = \frac{wi - wf}{wf} \times 100 \tag{1}$$

where w_i is the initial weight of the sample in grams, and w_f is the final weight in grams of the sample. Three determinations were performed for each formulation, and the results were expressed as the mean ± standard deviation.

2.4.3 Color parameters

The color parameters were determined using the technique described by Anchondo-Trejo et al. [18]. The bar was placed on the plate, and the color was assessed with a colorimeter (Konica Minolta CR-410, Tokyo, Japan) calibrated with white light. The recorded values were L^* (Luminosity), a^* (tendency toward green−, or red+), and b^* (tendency toward blue−, or yellow+). Three determinations were conducted for each formulation, and the results were expressed as the mean ± standard deviation.

2.4.4 Hardness

Hardness was measured with an EZ-SX texturometer (Shimadzu Corporation, Kyoto, Japan). The force was measured with the cutting device in the middle of the bar. The upper knife moved downwards at a speed of 1 mm/s and with a penetration of 5 mm, measuring the maximum force required to break the sample. Hardness values were expressed in Newton (N). Five determinations were made for each formulation, and the results were expressed as the mean ± standard deviation.

2.5 Nutritional analysis

Nutritional analysis utilized the tables of the Mexican equivalent food system [19]. Composition values of different ingredients used in amaranth bar preparation were considered for protein, lipids, available carbohydrates, dietary fiber, sodium, and total calorie content.

2.6 Sensory analysis

For sensory evaluation, an affective method assessed the response (preference or acceptance) provoked in potential consumers. Twenty-seven adults participated in the study, evaluating color, taste, texture, and general acceptance using a nine-point hedonic scale. Ratings ranged from 9 ("I like it very much") to 1 ("I dislike it very much"). The samples were placed on white plates identified with random three-digit numbers, and participants rinsed their mouths with water between each test.

2.7 Statistical analysis

All results were expressed as the mean ± standard deviation. The comparison of means was analyzed using analysis of variance (ANOVA), followed by Tukey's test to detect significant differences ($p \leq 0.05$) between samples. Minitab statistical software (version 19, Minitab Inc., USA) was employed for data analysis.

3. Results and discussion

3.1 Physical properties

3.1.1 Water activity

The water activity values ranged from 0.35 to 0.41 (**Table 2**). The control sample and the F1 sample exhibited significant differences, while the F2 sample did not show a significant difference compared to the other two samples ($p < 0.05$). The highest water activity value was observed in the F1 bar, possibly due to its higher honey content in comparison with the F2 bar. Various ingredients, including powdered cladodes, mucilage, honey, and piloncillo, can influence the final water activity due to their chemical structure and water-holding capacity.

3.1.2 Humidity percentage

The humidity percentage values (**Table 2**) ranged from 4.69 to 6.21%, showing significant differences ($p < 0.05$) among the three bars. The control bar had the lowest value (4.69%), coinciding with the lowest water activity value found. Similarly, the values of F1 and F2 (5.21%, 6.21% H, respectively) aligned with the observed water activity values. Dick et al. [6] reported moisture values in crackers formulated with mucilage and powdered prickly pear cactus in the range of 5.43–7.39%, consistent with the observed values.

3.1.3 Color

Significant differences ($p < 0.05$) were observed for color parameters (L^*, a^*, b^*) (**Table 2**). Regarding the L^* color parameter, a significant difference existed between F1 (69.26) and the control (65.14) and F2 (65.91), with no significant differences between F1 and F2. For the a^* parameter, all three bars showed significant differences, with the lowest value found in F1 (−1.02), possibly due to its higher mucilage content, imparting a greater tendency toward the green color. Regarding the b^* parameter, a significant difference was noted between the control (24.32) and F1 and F2 (19.09 and 17.34, respectively), indicating a trend toward yellow.

3.1.4 Hardness

Hardness values exhibited significant differences (**Table 2**) between the control (18.18 N) and F1 and F2 (2.99 N, 2.95 N, respectively). However, F1 and F2 did not show significant differences between them ($p < 0.05$). The control bar displayed greater hardness, suggesting that the addition of mucilage and cladodes powder influenced a decrease in bar hardness, along with the reduction in honey in these formulations. The F1 and F2 bars may be suitable for consumption by elderly individuals or those with dental problems, where the consumption of hard or very compact products is restricted.

3.2 Nutrition information

The data presented in **Table 3** reveals notable differences. In a 100 g portion, the control bar had a higher energy content (215.56 kcal) compared to the F1 and F2

Formulation	aw	% H	L^*	a^*	b^*	Hardness *N*
Control	0.35 ± 0.02 b	4.95 ± 0.13 c	65.14 ± 0.37 b	4.46 ± 0.30 a	24.32 ± 0.37 a	18.18 ± 4.93 a
1	0.41 ± 0.006 a	6.21 ± 0.05 a	69.26 ± 0.62 a	0.88 ± 0.24 b	19.09 ± 2.23 b	2.99 ± 1.95 b
2	0.38 ± 0.01 a,b	5.21 ± 0.07 b	65.91 ± 1.02 b	−1.02 ± 0.69 c	17.34 ± 0.73 b	2.95 ± 2.56 b

The values are the average of three replicas. Different letters in the same column are statistically different ($p \leq 0.05$).

Table 2.
Physical properties of amaranth bars, made with mucilage and cladodes powder.

	Control	Formula 1	Formula 2
Portion	(100 g)	(100 g)	(100 g)
Energy content (kcal)	215.56	155.72	138.41
Proteins (g)	16.00	17.81	17.81
Fats (g)	0.06	0.24	0.22
Saturated fats (g)	0.00	0.00	0.00
Cholesterol (mg)	0.00	0.00	0.00
Trans Fats (g)	0.00	0.00	0.00
Available carbohydrates(g)	52.87	36.58	32.44
Sugars (g)	0.00	0.81	0.75
Added sugars (g)	0.00	0.00	0.00
Dietary fiber (g)	4.80	14.43	15.56
Sodium (mg)	83.33	86.00	90.93

Table 3.
Nutritional information for amaranth bars.

bars (155.72 kcal and 138.41 kcal, respectively), directly correlating with the available carbohydrate content in the bars (52.87 g, 36.58 g, and 32.44 g, respectively). F2 exhibited the highest dietary fiber content (15.56 g) due to its higher mucilage content. Protein values were also higher for F1 and F2 relative to the control. Uriarte-Frías et al. [14] reported protein values in biscuits made with mushroom, amaranth, and prickly pear flour ranging from 16.37 to 18.21, consistent with the nutritional information values observed.

3.3 Sensory analysis

The sensory evaluation of different bar formulations, including the control, aimed to assess color, texture, flavor, and general acceptance, and the results are presented in **Table 4**. For the color attribute, a significant difference ($p < 0.05$) was noted between the control bar, scoring 8.19, and the F1 and F2 bars, scoring 6.96 and 7.15, respectively. Regarding the texture characteristic, the control bar exhibited a significant difference (7.81) compared to F1 (6.5) and F2 (6.69). In terms of taste, the control achieved a score of 7.88, while F1 scored 6.23 and F2 scored 6.77, showing significant differences ($p < 0.05$). Similarly, for the general acceptance of the bars, the control displayed a significant difference, scoring 8.11, compared to F1 with a score of 6.38 and F2 with a score of 6.88.

The observed values and significant differences indicate a preference for the control bar with a higher sugar content (50 g) among the evaluators (ranging from "I like it very much" to "I like it moderately"), in comparison with F1 with 38 g and F2 with 37 g of sugar (both rated as "I like it little"). Dick et al. [6] reported general acceptance values ranging from 5.73 to 7.38 for cracker cookies made with mucilage and prickly pear powder. Uriarte-Frías et al. [14] found values of 7.0–8.0 in biscuits formulated with mushroom, amaranth, and prickly pear flour. Castillo et al. [20] observed values of 5.76–7.06 in the general acceptance attribute for alfajores made with prickly pear flour. The values obtained align with those reported by these authors for products made with mucilage and/or powdered prickly pear cactus.

DOI: http://dx.doi.org/10.5772/intechopen.115472

Formula	Color	Texture	Taste	Overall acceptance
Control	8.19 ± 1.09 a	7.81 ± 1.35 a	7.88 ± 1.24 a	8.11 ± 1.07 a
1	6.96 ± 1.34 b	6.50 ± 1.67 b	6.23 ± 1.90 b	6.38 ± 1.52 b
2	7.15 ± 1.48 b	6.69 ± 1.59 b	6.77 ± 1.53 b	6.88 ± 1.30 b

The values are the average of 30 replicas. Different letters in the same column are statistically different ($p \leq 0.05$).

Table 4.
Sensory analysis of amaranth bars.

4. Conclusions

Amaranth bars with reduced sugar content were successfully developed by incorporating mucilage and powdered cactus, resulting in higher protein and fiber proportions compared to amaranth bars without these additions. The physical properties, including water activity (aw), color, moisture percentage, and hardness, exhibited significant differences compared to the control bar. However, sensory evaluation indicated a preference for the original bar, possibly attributed to its sweet taste, which is favored by many consumers. The potential applications of mucilage and prickly pear powder as functional ingredients in various food products were highlighted, offering benefits to human nutrition and health by enhancing antioxidant intake and dietary fiber. The obtained results serve as a valuable foundation for future research on prickly pear mucilage and prickly pear powder.

Acknowledgements

R.A.O. expresses gratitude to Universidad de las Américas Puebla (UDLAP) and Consejo Nacional de Humanidades, Ciencias y Tecnología (CONAHCYT) for the scholarship that supported her doctoral degree.

Conflict of interest

The authors declare no conflict of interest.

Author details

Rocío Aparicio-Ortuño[1], J. Daniel Lozada-Ramírez[2] and Ana E. Ortega-Regules[3*]

1 Department of Chemical, Food and Environmental Engineering, Universidad de las Américas Puebla, Puebla, Mexico

2 Department of Chemical and Biological Sciences, Universidad de las Américas Puebla, Puebla, Mexico

3 Department of Health Sciences, Universidad de las Américas Puebla, Puebla, Mexico

*Address all correspondence to: ana.ortega@udlap.mx

References

[1] Bautista-Justo M, Pineda Torres RI, Camarena-Aguilar E, et al. El Nopal fresco como fuente de fibra y calcio en panqués. Acta Universitaria. 2010;**20**:11-17. DOI: 10.15174/au.2010.62

[2] Hernández-Urbiola MI, Contreras-Padilla M, Pérez-Torrero E, et al. Study of nutritional composition of Nopal (*Opuntia ficus indica* cv. Redonda) at different maturity stages. The Open Nutrition Journal. 2010;**4**:11-16. DOI: 10.2174/1874288201004010011

[3] Astello-García MG, Cervantes I, Nair V, et al. Chemical composition and phenolic compounds profile of cladodes from Opuntia spp. cultivars with different domestication gradient. Journal of Food Composition and Analysis. 2015;**43**:119-130. DOI: 10.1016/j.jfca.2015.04.016

[4] Aruwa CE, Amoo SO, Kudanga T. Opuntia (Cactaceae) plant compounds, biological activities and prospects – A comprehensive review. Food Research International. 2018;**112**:328-344. DOI: 10.1016/j.foodres.2018.06.047

[5] del Socorro Santos Díaz M, Barba de la Rosa A-P, Héliès-Toussaint C, et al. Opuntia spp.: Characterization and benefits in chronic diseases. Oxidative Medicine and Cellular Longevity. 2017;**2017**:1-17. DOI: 10.1155/2017/8634249

[6] Dick M, Limberger C, Cruz Silveira Thys R, et al. Mucilage and cladode flour from cactus (*Opuntia monacantha*) as alternative ingredients in gluten-free crackers. Food Chemistry. 2020;**314**:126178. DOI: 10.1016/j.foodchem.2020.126178

[7] Gray J. Dietary fibre: Definition, analysis, physiology & health. In: ILSI Europe Concise Monograph Series. Brussels: ILSI Europe; 2006, 2006. p. 35

[8] Medina-Torres L, Vernon-Carter EJ, Gallegos-Infante JA, et al. Study of the antioxidant properties of extracts obtained from nopal cactus (*Opuntia ficus-indica*) cladodes after convective drying. Journal of the Science of Food and Agriculture. 2011;**91**(6):1001-1005. DOI: 10.1002/jsfa.4271

[9] Hernández-Urbiola MI, Pérez-Torrero E, Rodríguez-García ME. Chemical analysis of nutritional content of prickly pads (*Opuntia ficus indica*) at varied ages in an organic harvest. International Journal of Environmental Research and Public Health. 2011;**8**:1287-1295. DOI: 10.3390/ijerph8051287

[10] Bensadón S, Hervert-Hernández D, Sáyago-Ayerdi SG, Goñi I. By-products of *Opuntia ficus-indica* as a source of antioxidant dietary fiber. Plant Foods for Human Nutrition. 2010;**65**:210-216. DOI: 10.1007/s11130-010-0176-2

[11] El-Mostafa K, El Kharrassi Y, Badreddine A, et al. Nopal cactus (*Opuntia ficus-indica*) as a source of bioactive compounds for nutrition, health and disease. Molecules. 2014;**19**:14879-14901. DOI: 10.3390/molecules190914879

[12] Liguori G, Gentile C, Gaglio R, et al. Effect of addition of *Opuntia ficus-indica* mucilage on the biological leavening, physical, nutritional, antioxidant and sensory aspects of bread. Journal of Bioscience and Bioengineering. 2020;**129**:184-191. DOI: 10.1016/j.jbiosc.2019.08.009

[13] Quintero-García M, Gutiérrez-Cortez E, Bah M, et al. Comparative

analysis of the chemical composition and physicochemical properties of the mucilage extracted from fresh and dehydrated *Opuntia ficus indica* cladodes. Food. 2021;**10**:2137. DOI: 10.3390/foods10092137

[14] Uriarte-Frías G, Hernández-Ortega MM, Gutiérrez-Salmeán G, et al. Pre-Hispanic foods oyster mushroom (*Pleurotus ostreatus*), Nopal (*Opuntia ficus-indica*) and Amaranth (Amaranthus sp.) as new alternative ingredients for developing functional cookies. Journal of Fungi. 2021;7:911. DOI: 10.3390/jof7110911

[15] Granato D, Barba FJ, Bursać Kovačević D, et al. Functional foods: Product development, technological trends, efficacy testing, and safety. Annual Review of Food Science and Technology. 2020;**11**:93-118. DOI: 10.1146/annurev-food-032519-051708

[16] León-Martínez FM, Méndez-Lagunas LL, Rodríguez-Ramírez J. Spray drying of nopal mucilage (*Opuntia ficus-indica*): Effects on powder properties and characterization. Carbohydrate Polymers. 2010;**81**:864-870. DOI: 10.1016/j.carbpol.2010.03.061

[17] AOAC. Official Methods of Analysis Association of Official Analytical Chemist. EUA; 2000

[18] Anchondo-Trejo C, Loya-Carrasco JA, Galicia-García T, et al. Development of a third generation snack of Rice starch enriched with nopal flour (*Opuntia ficus indica*). Molecules. 2020;**26**:54. DOI: 10.3390/molecules26010054

[19] Pérez Lizaur AB, Palacios González B, Castro Becerra AL. Sistema mexicano de alimentos equivalentes. México: Fomento de Nutrición y Salud, A.C; 2008

[20] Castillo SF, Estrada L, Margalef MI, Tóffoli SL. Obtención de harina de nopal y formulación de alfajores de alto contenido en fibra. Diaeta (B. Aires). 2013;**31**(142):20-26

Chapter

Physical Properties of Bread Added with Cricket

Sheila Cortazar-Moya, Ricardo H. Hernández-Figueroa, Victor E. Vera-Santander, Aurelio López-Malo and Jocksan Ismael Morales-Camacho

Abstract

Recently, there has been an increasing interest in alternative and sustainable sources of foods that increase protein and add nutritional value to the diet. In this sense, edible insects are a sustainable and nutritious alternative source of protein. Due to its popularity, bread is often enriched with functional natural ingredients like insect flour, which can increase its nutritional value and change some techno-functional and physical properties that need to be evaluated. In this study, three formulations of bread: control (C) made of wheat flour and two mixtures substituting 5% (CB5) and 10% (CB10) of wheat flour (WF) for cricket flour (CrF) were analyzed. The addition of CrF significantly reduced the specific volume from 4.27 in control to 3.17 in CB5. The pH also differed between control (5.96), CB5 (5.75), and CB10 (5.82). The width/height ratio decreased in CB5 (3.17) and CB10 (3.51) compared to the control (4.27), and finally the crumb and crust color lost lightness compared to the control. The ΔE and the browning index of the bread (CB5 and CB10) were significantly different from the control. On the other hand, the hardness between the CB10 and the C significantly differs; the CrF decreased the hardness of CB10.

Keywords: bread, insect flour, hardness, color, crumb, crust

1. Introduction

It has been predicted that the world population will reach nearly 10 billion people in the next 30 years [1]. With the population growing at this speed and the demand for protein increasing, insects are being considered a cost-effective alternative [2, 3]. In this regard, edible insects are also promoted by the Food and Agriculture Organization as a source of food, and more than 2000 species of edible insects have been reported worldwide [4]. Their content of macronutrients can range around 35–61% of proteins, 15–40% of fats, and 3–10% of minerals [5]. Insects have been a part of the human diet for centuries and have recently gained

IntechOpen

interest in Western countries as a food source. The most consumed insects belong to the families of Coleoptera (beetles), Lepidoptera (caterpillars of butterflies and moths), Hymenoptera (bees, wasps, ants), Orthoptera (grasshoppers, locusts, crickets, termites), Hemiptera (cicadas, leaf and plant hoppers, true bugs, scale insects), Odonata (dragonflies) and Diptera (flies) [6]. The production of certain insects is also more sustainable than conventional protein sources in terms of water use and greenhouse gas emissions [7]. Crickets are particularly popular as they have high food and feed conversion efficiency and can be fed on industrial by-products and cereal grains [3].

Insects are known to be an excellent source of protein and have several sustainable benefits. However, one of the greatest challenges associated with the consumption of insects is their low acceptance in Western culture due to their appearance [3]. Therefore, it is crucial to explore alternative ways of presenting insects as food, as people would be more willing to eat them in a less visible form, in modified products that are indistinguishable from familiar ones [1]. Currently, many commercial food products are enriched with proteins derived from cereal grains such as rye or oats, or legumes [8]. However, insects are much richer in protein than beans, lentils, or soybeans. Insect flours have many advantages in food production, such as high protein content, low fat content, and a neutral taste and smell. Despite this, there are few reported applications of insect flours in food products. However, a recent proposal suggested using insect flours from mealworm larvae and silkworm pupae to replace 10% of lean pork in emulsion sausages. The result was an increase in the cooking yield and hardness of the sausages, which confirms the potential of insect flour as a new protein ingredient. Additionally, maize tortillas have been enriched with protein by adding larvae of the *Tenebrio molitor*, which increased the protein content by 2%. These findings highlight the immense potential of insect flour as a sustainable alternative protein source in food production [3].

The potential of certain insect species to be used as a food source has been explored, and incorporating insect powder into food products could provide additional nutrients and improve various techno-functional properties. One such food product that could benefit from this is bread [3]. Bread is a staple food that is consumed every day all around the world, and it is of great significance to consumers. Due to its popularity, bread is often enriched with functional products of natural origin [9]. Incorporating insect flour into bread recipes could significantly enhance its nutritional value and modify some techno-functional and physical properties that require further evaluation. Therefore, insect powder in bread formulation has the potential to offer a range of benefits to consumers and could help address some of the challenges associated with food security and nutrition [1].

In previous studies, wheat bread fortification with various edible insect flours, i.e., mealworm, buffalo worm, cricket, and the larvae of the black soldier fly improved the biological value of bread due to the high protein content characteristics [1]. Pauter et al. [10] applied 10% cricket powder to increase the protein content of muffins. In addition to nutritional value, cricket enrichment helps products have a unique aroma and improved texture, increased hardness, and improved consistency [11]. This research aimed to evaluate the changes in the physicochemical and quality properties of bread added with cricket (*Acheta domesticus*) flour in different percentages.

2. Methodology

2.1 Bread making

A standard formulation of bread with wheat flour (WF), as reported by Hernández-Figueroa et al. [12], was used as a control (C), including wheat flour (62%), salt (1%), compressed yeast (1%), and water (37%). Wheat flour was replaced in quantities of 5% (CB5) and 10% (CB10) based on the initial wheat flour weight of cricket flour (CrF). The CrF was bought in a supermarket in Puebla City. A Legacy HL200 mixer/kneader (Hobart, Troy, OH, USA) was used to mix and knead the ingredients. All the loaves of bread were fermented for 30 min at 40°C using a Mini combo oven (Zucchelli Alpha, Trevenzuolo, Verona, Italy); then, the dough was divided and formed into pieces (60 g) and baked at 200°C for 18 min in a Mini combo electric oven (Zucchelli Alpha, Trevenzuolo, Verona, Italy). Then, the bread was cooled at room temperature for 60 min and stored in sealed polyethylene bags for further use.

2.2 Determination of bread's physicochemical properties

Breads were evaluated by assessing physicochemical properties: pH, water activity (aw), specific volume, and weight/high ratio. The bread's pH measurement was carried out with a pH meter model HI2210 (Hanna Instruments, Woonsocket, RI, USA) following the 02–52 method of the AACCI [13]. The a_w was measured in a mixed sample from crumb and crust from the slides of bread in an AquaLab 4TEV Series equipment (Meter Food, Pullman, WA, USA); the analysis was performed in duplicate. Volume determination was measured by seed displacement according to AACC method 10–05 [13], and the specific volume of bread was expressed as the volume between the weight of bread (cm^3/g). An electronic Vernier was used to measure the width/height ratio, and both dimensions were measured in central slices of 2.5 cm thickness. All determinations were performed in duplicate. The measurements were carried out the day before baking.

2.3 Quality properties of the bread

The hardness was measured a day before baking the pieces of bread. A texture profile analysis (TPA) was carried out using the EZ Test EZ-LX texture meter (Shimadzu, Kyoto, Japan) controlled by Trapezium X Texture Pl software. A crumb sample from the central part of the bread was compressed twice with a 10 s gap between compresses along its axis with a 50 mm-diameter stainless steel disc at a speed of 1.7 $mm \cdot s^{-1}$ until 25% deformation was obtained [12]. The measurements were taken in duplicate. Hardness was recorded from the TPA measurement.

Crumb and crust color was measured using a chroma meter CR-410 (Konica Minolta Sensing Inc., Tokyo, Japan). Color differences were recorded in the CIE L*a*b* scale regarding lightness (L*) and color (a* – redness, b* – yellowness). Numerical values of a* and b* were converted into hue angle (h°) and Chroma (C*) that represent the hue and the saturation index (Eqs. (1) and (2)). Two measurements were made at different points in each crumb and crust. Additionally, the total color difference (ΔE) was calculated according to Eq. (3):

$$h^{\circ} = \left(\frac{\left(tan^{-1}\left(\frac{b*}{a*}\right)\right)}{(2\pi)} \right) 360 \quad (1)$$

$$\mathbf{C}^{*} = \sqrt{\left(\mathbf{a}^{*}\right)^{2} + \left(\mathbf{b}^{*}\right)^{2}} \quad (2)$$

Results were expressed as hue angle (h°) and Chroma (C*).

$$\Delta E = \sqrt{\Delta L^{2} + \Delta a^{2} + \Delta b^{2}} \quad (3)$$

where ΔL, Δa, and Δb are differences in the L*, a*, b* values between the reference sample and the test sample, respectively.

Browning Index (BI) Eqs. 4 and 5, was calculated using L*, a*, and b* values as follows.

$$x = \left(\frac{(\mathbf{a}* + \mathbf{1.75\,L})}{(\mathbf{5.645L}* + \mathbf{a}* - \mathbf{3.012b}*)} \right) \quad (4)$$

$$BI = \left(\frac{100(x - 0.31)}{0.17} \right) \quad (5)$$

2.4 Statistical analysis

Results are expressed as mean values and standard deviation and subjected to one-way ANOVA. Pair comparisons for mean values were performed with Tukey's test at $p < 0.05$, using the statistical software Minitab 18® (Minitab LLC, State College, PA, USA). Before statistical analysis, the data's normality and the homogeneity of variances were evaluated and subjected to two-way ANOVA and Tukey's comparison tests using a confidence level of 95% using Minitab 18® (Minitab Inc., USA). Before statistical analysis, the data's normality and the homogeneity of variance were evaluated.

3. Results and discussion

3.1 Determination of physicochemical properties of the bread

Bread was evaluated by assessing physicochemical properties (pH, a_w, specific volume, W/H ratio, and hardness). Zhao et al. [14] and Hernández-Figueroa et al. [15] reported pH values in white bread between 5.9 and 6.1, which makes it slightly acidic. This pH range may be produced due to the presence of organic acids produced by the fermentation process, which include lactic acid and acetic acid. It is known that the pH of bread dough can affect the texture and flavor of the bread; a pH that is too low (more acidic) can make the dough sticky and difficult to work with, leading to

Sample	pH	a_w	Specific volume (cm³/g)	W/H ratio	Hardness (N)
C	5.96 ± 0.02^a	0.95 ± 0.01^a	4.27 ± 0.02^a	2.05 ± 0.01^b	60.12 ± 5.14^a
CB5	5.75 ± 0.01^b	0.92 ± 0.01^b	3.17 ± 0.04^c	2.70 ± 0.16^a	36.08 ± 2.21^{ab}
CB10	5.82 ± 0.02^b	0.93 ± 0.01^{ab}	3.51 ± 0.08^b	2.27 ± 0.01^b	32.25 ± 3.8^b

Each value is presented as the mean ± SD (n = 2). Values followed by a different superscript letter in each column are significantly different (p < 0.05). C, control with wheat flour; CB5, bread with 5% of cricket flour; CB10, bread with 10% of cricket flour.

Table 1.
Physicochemical properties of the bread of wheat flour and cricket flour.

dense and heavy bread. A too-high pH (more alkaline) can prevent the dough from rising properly, resulting in dense, heavy bread with poor texture. A higher pH value (**Table 1**) was observed in the control bread (5.96); this value is significantly different ($p < 0.05$) from CB5 and CB10 (5.75, 5.82), respectively. In this sense, it has also been reported that similar pH values for bread when adding 10 and 30% of cricket flour (5.71 and 5.84), respectively [16]. This pH reduction could affect some physical characteristics such as the W/H ratio and the specific volume of the bread [12].

Table 1 displays the water activity (a_w) of the analyzed breads. The loaves of bread CB10 showed significant differences ($p < 0.05$) compared to the control. The loaves of bread CB10 presented a lower water activity (0.93). This could be due to the change in the amount of gluten and the interaction between the gluten and the proteins in the CrF which influenced moisture retention in bread [12]. Forming a correct gluten network and the swelling of starch granules reduce water vapor loss during product baking [17]. The replacement of a percentage of wheat flour for CrF in CB5 and CB10 contributed to a significant reduction in water activity. The physical properties of the different analyzed breads are shown in **Table 1**. Bread added with the CrF (CB5, CB10) presented a lower specific volume (3.17 and 3.51 cm^3/g), respectively. The bread CB5 presented the lowest specific volume, significantly different ($p < 0.05$) from the control and the bread CB10. The control bread had the highest specific volume (4.27 cm^3/g). The specific volume of bread significantly decreased in the presence of CrF content. The decrease in specific volume may be due to the interruption of gluten formation, resulting in the decrease of gas retention ability during proofing and baking. It has been reported that breadmaking is sensitive to the addition of non-gluten ingredients into wheat flour, which is related to the disturbance of gluten development and the reduction in gas production [18]. In this sense, the chemical composition of CrF may be responsible for interfering with the proper development of the dough, especially in bread made using 5% CrF [19]. The bread production process is sensitive to replacing wheat flour with non-starch additives by disrupting the development of the gluten network [1]. It has been reported that adding insect powder reduces the specific volume of bread [20].

Along with the specific volume, the width/height ratio is an important quality parameter in bakery products. Some studies showed that good-quality bread has a width/height ratio of 1.3–1.6 [17] and is directly related to the proper formation of gluten. In this sense, the control and CB10 were closer to those values (2.05, 2.27), respectively, with no significant difference between them (**Table 1**). Only CB5 had a higher W/H ratio (2.70); this means that CB5 was wider and less high than CB10 and the control (**Figure 1d–f**).

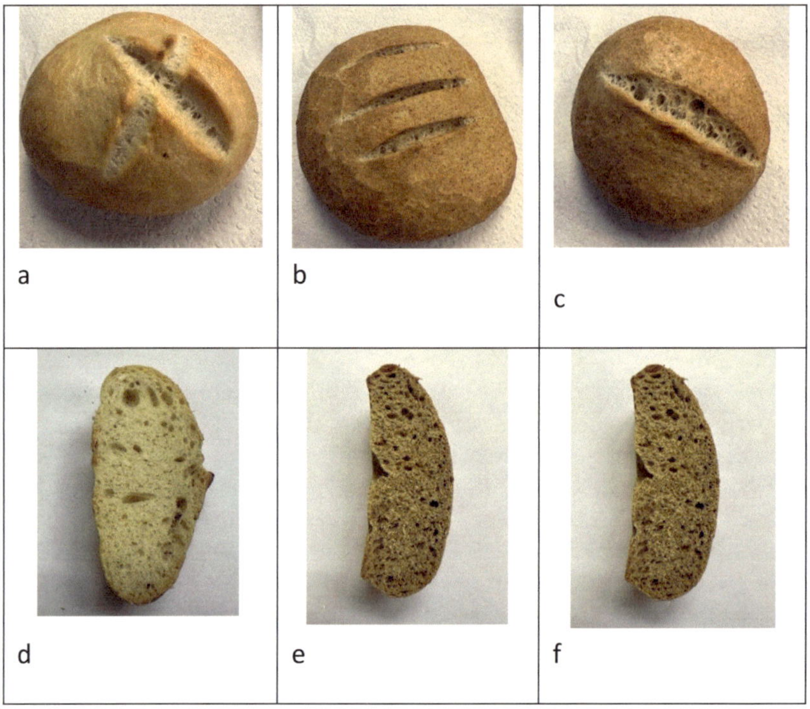

Figure 1.
Images (a–c), crust appearance of the breads: (a) control, (b) CB5 substituted 5% wheat flour with cricket flour, (c) CB10 substituted 10% wheat flour with cricket flour; images (d–f), the cross-section of bread slices: (d) Control, (e), CB5 (f) CB10.

3.2 Quality properties of the bread

The texture of bread is determined by its qualitative characteristics, particularly its mechanical properties. These properties provide information on the product's durability and how acceptable it is to consumers [21]. Among the most frequently used parameters to assess bread texture is its hardness, which is affected by several factors such as the type of additives, dough handling, and specific weight of the bread crumb, among others. The statistical analysis showed that the hardness of C and CB10 were significantly different (60.12 N and 32.25 N), respectively (**Table 1**). The bread with 10% CrF (CB10) was the one with the lowest hardness. The higher hardness in the control could be related to its higher adsorption of water, which determines the greater hardness. In the study conducted by Mafu et al. [11] with bread enriched with cricket powder, they reported a hardness of 33.69 N in the bread enriched with 10%; the addition of protein components could act on various gluten proteins directly or through some reaction by promoting the formation of additional interprotein bonds that favor a less hardness in CB10. In the study conducted by Kowalski et al. [1], the bread with mealworm and cricket flour had significantly lower hardness than wheat bread. Finally, similar results were obtained by Roncolini et al. [9], who also observed a decrease in bread hardness with a 5 and 10% share of mealworm flour.

Color is an important feature of baked products because, together with texture and volume, they influence consumer acceptance [3]. The color attributes measured

DOI: http://dx.doi.org/10.5772/intechopen.115472

	Crust			Crumb		
Parameter	**C**	**CB5**	**CB10**	**C**	**CB5**	**CB10**
L*	71.77 ± 0.14[a]	54.68 ± 0.10[c]	62.18 ± 0.86[b]	76.54 ± 2.33[a]	58.54 ± 0.19[b]	63.86 ± 0.38[b]
a*	4.90 ± 0.28[c]	8.25 ± 0.14[a]	6.25 ± 0.04[b]	−0.56 ± 0.011[c]	3.19 ± 0.02[a]	1.81 ± 0.02[b]
b*	30.87 ± 0.28[a]	25.89 ± 0.19[b]	25.65 ± 0.43[b]	16.79 ± 0.36[b]	18.10 ± 0.11[a]	17.45 ± 0.12[ab]
BI	59.76 ± 1.14	72.28 ± 2.37	59.36 ± 0.15	23.65 ± 1.51	40.25 ± 0.14	33.31 ± 0.52
HUE	80.97 ± 0.44	72.06 ± 0.78	76.31 ± 0.11	91.61 ± 0.42	80.00 ± 0.02	84.09 ± 0.02
Chroma	31.26 ± 0.33	26.80 ± 0.66	26.39 ± 0.43	16.80 ± 0.35	18.38 ± 0.11	17.54 ± 0.12
ΔE	—	18.11	11.00	—	18.43	12.91

Each value is presented as the mean ± SD (n = 3). Values followed by a different superscript letter in each row corresponding to crust and crumb are significantly different ($p < 0.05$). C, control with wheat flour; CB5, bread with 5% of cricket flour; CB10, bread with 10% of cricket flour; BI, browning index.

Table 2.
Properties related to the color of the crust and crumb of the bread with cricket flour (CrF) and the control (C).

in C, CB5, and CB10 were lightness (L*), redness (+a*), and yellowness (+b*). The chroma (the saturation of color) is the radius, i.e., the distance of the spectrum from the origin. Spectra with no chroma (white, gray, black) will appear at the center (the origin), and those with high chroma will appear toward the outside. Hue (H°), the type of color, corresponds to the angle clockwise from the positive y-axis. Red colors will have small values (i.e., be close to 0°), while yellow, green, and blue colors will be near 90, 180, and 270°, respectively [22]. In particular, the crust and crumb of bread with 5 or 10 percent of CrF had lower lightness (L*). The hue angle (h°) decreased in the crust, indicating that it was closer to red-brown. The Chroma (C*) in the crust decreased in CB5 (26.80) and CB10 (26.39) with respect to the C (31.26). On the contrary, in the crumb, C* increased in CB5 (18.38) and CB10 (17.54) with respect to C (16.80). This means that the crumb and crust of bread with CrF appear to be darker, tending to brown and assuming a more vivid color compared with the control bread.

The color determination of the crust and crumb of the bread is shown in **Table 2**. There was a significant difference ($p < 0.05$) in the lightness between the crust in C (71.77), CB5 (54.68), and CB10 (62.18) being the C lighter in color. Also, a significant difference ($p < 0.05$) was observed in the redness and yellowness; CB5 had the higher redness value, and C had the highest value in yellowness. The crumb CB5 and CB10 were significantly different from the C ($p < 0.05$) that had the highest lightness value. The crust and crumb of CB5 had a higher browning index. Net Color change value (**Table 2**). This change in the color of the crust and crumb in CB5 with respect to the control may be related to the caramelization process during baking and the Maillard reaction between amino acid and reducing sugars, the use of CrF, which is high in protein to enrich the bread may cause significant changes in the color of the products (**Figure 1**) and can be responsible for reducing the lightness in CB5 and CB10 [10].

4. Conclusions

The results of the present study established the potential for cricket flour as an ingredient high in protein for bakery products. The results indicate that substituting 5 and 10% of CrF in wheat bread influences their volume, width/height ratio, and color, and the bread hardness decreases with cricket flour compared to the control. Both percentages of cricket flour tested in bread change the properties evaluated compared to the control. These results show that using insect flours is possible, but further studies are needed including sensorial analysis, in order to understand the interactions between cereal-based matrices and insect flours to adjust insect flour percentage to obtain better-balanced products and the acceptances that can have this type of products in the consumers.

Acknowledgements

S. Cortazar-Moya thanks financial support from the National Council of Humanities, Sciences and Technologies (CONAHCYT) of Mexico to complete her doctoral studies in Food Science, and Universidad de las Americas Puebla (UDLAP) for the financial support to this research.

DOI: http://dx.doi.org/10.5772/intechopen.115472

Conflict of interest

The authors confirm that they have no conflicts of interest with respect to the work described in this manuscript.

Author details

Sheila Cortazar-Moya*, Ricardo H. Hernández-Figueroa, Victor E. Vera-Santander, Aurelio López-Malo and Jocksan Ismael Morales-Camacho
Department of Chemical, Food and Environment Engineering, Universidad de las Américas Puebla, San Andrés Cholula, Mexico

*Address all correspondence to: sheila.cortazarma@udlap.mx

References

[1] Kowalski S, Mikulec A, Mickowska B, Skotnicka M, Mazurek A. Wheat bread supplementation with various edible insect flours. Influence of chemical composition on nutritional and technological aspects. LWT. 2022;**159**:113220. DOI: 10.1016/j.lwt.2022.113220

[2] van Huis A. Edible Insects: Future Prospects for Food and Feed Security. Rome: FAO forestry paper; Food and Agriculture Organization of the United Nations; 2013

[3] González CM, Garzón R, Rosell CM. Insects as ingredients for bakery goods. A comparison study of *H. illucens*, *A. domestica* and *T. molitor* flours. Innovative Food Science & Emerging Technologies. 2019;**51**:205-210. DOI: 10.1016/j.ifset.2018.03.021

[4] Quah Y, Tong S-R, Bojarska J, Giller K, Tan S-A, Ziora ZM, et al. Bioactive peptide discovery from edible insects for potential applications in human health and agriculture. Molecules. 2023;**28**(3):1233. DOI: 10.3390/molecules28031233

[5] Rumpold BA, Schlüter OK. Potential and challenges of insects as an innovative source for food and feed production. Innovative Food Science & Emerging Technologies. 2013;**17**:1-11. DOI: 10.1016/j.ifset.2012.11.005

[6] Sogari G. Entomophagy and Italian consumers: An exploratory analysis. Progress in Nutrition. 2015;**17**(4):311-316

[7] Van Zanten HHE, Mollenhorst H, Oonincx DGAB, Bikker P, Meerburg BG, De Boer IJM. From environmental nuisance to environmental opportunity: Housefly larvae convert waste to livestock feed. Journal of Cleaner Production. 2015;**102**:362-369. DOI: 10.1016/j.jclepro.2015.04.106

[8] Al-Attabi ZH, Merghani TM, Ali A, Rahman MS. Effect of barley flour addition on the physico-chemical properties of dough and structure of bread. Journal of Cereal Science. 2017;**75**:61-68. DOI: 10.1016/j.jcs.2017.03.021

[9] Roncolini A, Milanović V, Cardinali F, Osimani A, Garofalo C, Sabbatini R, et al. Protein fortification with mealworm (*Tenebrio Molitor* L.) powder: Effect on textural, microbiological, nutritional and sensory features of bread. PLoS One. 2019;**14**(2):e0211747. DOI: 10.1371/journal.pone.0211747

[10] Institute of Food Technology of Plant Origin, Poznań University of Life Sciences, Poland, Pauter P, Różańska M, Wiza P, Dworczak S, Grobelna N, et al. Effects of the replacement of wheat flour with cricket powder on the characteristics of muffins. Acta Scientiarum Polonorum. Technologia Alimentaria. 2018;**17**(3):227-233. DOI: 10.17306/J.AFS.0570

[11] Mafu A, Ketnawa S, Phongthai S, Schönlechner R, Rawdkuen S. Whole wheat bread enriched with cricket powder as an alternative protein. Food. 2022;**11**(14):2142. DOI: 10.3390/foods11142142

[12] Hernández-Figueroa RH, Mani-López E, López-Malo A. Antifungal activity of wheat-flour sourdough (type II) from two different lactobacillus *in vitro* and bread. Applied Food Research. 2023;**3**(2):100319. DOI: 10.1016/j.afres.2023.100319

[13] American Association of Cereal Chemists. Approved Methods of the American Association of Cereal

Chemists. 10th ed. St. Paul, Minnesota, USA: AACC; 2000

[14] Zhao Y, Zhang J, Wei Y, Ai L, Ying D, Xiao X. Improvement of bread quality by adding wheat germ fermented with Lactobacillus plantarum dy-1. Journal of Food Quality. 2020;**2020**:1-8. DOI: 10.1155/2020/9348951

[15] Hernández-Figueroa RH, Mani-López E, López-Malo A. Antifungal capacity of poolish-type sourdough supplemented with *Lactiplantibacillus plantarum* and its aqueous extracts *in vitro* and bread. Antibiotics. 2022;**11**(12):1813. DOI: 10.3390/antibiotics11121813

[16] Osimani A, Milanović V, Cardinali F, Roncolini A, Garofalo C, Clementi F, et al. Bread enriched with cricket powder (*Acheta Domesticus*): A technological, microbiological and nutritional evaluation. Innovative Food Science & Emerging Technologies. 2018;**48**:150-163. DOI: 10.1016/j.ifset.2018.06.007

[17] Bárcenas ME, Rosell CM. Effect of HPMC addition on the microstructure, quality and aging of wheat bread. Food Hydrocolloids. 2005;**19**(6):1037-1043. DOI: 10.1016/j.foodhyd.2005.01.005

[18] Khuenpet K, Pakasap C, Vatthanakul S, Kitthawee S. Effect of larval-stage mealworm (*Tenebrio molitor*) powder on qualities of bread. International Journal of Agricultural Technology. 2020;**16**(22):283-296

[19] Sui X, Zhang Y, Zhou W. Bread fortified with anthocyanin-rich extract from black rice as nutraceutical sources: Its quality attributes and *in vitro* digestibility. Food Chemistry. 2016;**196**:910-916. DOI: 10.1016/j.foodchem.2015.09.113

[20] Haber M, Mishyna M, Martinez JJI, Benjamin O. The influence of grasshopper (*Schistocerca gregaria*) powder enrichment on bread nutritional and sensorial properties. LWT. 2019;**115**:108395. DOI: 10.1016/j.lwt.2019.108395

[21] Balestra F, Cocci E, Pinnavaia G, Romani S. Evaluation of antioxidant, rheological and sensorial properties of wheat flour dough and bread containing ginger powder. LWT - Food Science and Technology. 2011;**44**(3):700-705. DOI: 10.1016/j.lwt.2010.10.017

[22] Smith SD. Quantifying color variation: Improved formulas for calculating hue with segment classification. Applications in Plant Sciences. 2014;**2**(3):1300088. DOI: 10.3732/apps.1300088

Chapter

Use of Different Toxicological Assays in Heavy Metals

Teresa Larios, Benito Corona and Jose Luis Sánchez-Salas

Abstract

The increase in heavy metals in freshwater bodies due to anthropogenic activities causes significant alterations to the biota, including human beings. This review examines the adverse effects of various heavy metals of ecotoxicological relevance on diverse test models such as microcrustaceans, bacteria, polyps, herbaceous plants, algae, and fish, along with their distinctive characteristics and the advantages they offer as test organisms. Overall, there is a need to obtain indicators that are more sensitive and representative than traditional ones to generate new information about the feasibility of each bioassay. Furthermore, the current trend toward studying mixtures of substances and indirect effects allows for a more comprehensive understanding of the issue. The impact on populations is highly variable and depends on factors such as the nature of the metal, environmental characteristics, exposure time, cultivation conditions, and genetic aspects. However, most studies focus on a few species, leaving gaps in knowledge about representative species from each region.

Keywords: acute toxicity, chronic exposure, heavy metals, *Daphnia magna*, vibrio fisheri, *Lactuca sativa*

1. Introduction

Heavy metals are released into the environment, mainly through natural erosion of rocks and volcanic activity, as well as human activities such as agriculture, mining, and industry [1, 2]. Metals can be classified into four major groups based on their reported toxicity in humans: (1) major toxic metals, (2) minor toxic metals, (3) essential metals, and (4) metals related to medical therapy [3]. The main pathways for heavy metals through the atmosphere and water are dispersion and deposition processes that lead to accumulation in soil and water sediments, consequently becoming semi-natural secondary energy reservoirs [4]. The continuous accumulation and migration of heavy metals in sediments not only pose a serious threat to aquatic organisms and agricultural land irrigation but also endanger human health through drinking water or food chains [5]. It has been demonstrated that the consumption of groundwater polluted with heavy metals in areas where water is used for crop irrigation, direct contact with soils irrigated by this resource, and consumption of products derived from the same source represent a serious risk to humans due to the

IntechOpen

possible onset of both cancerous and non-cancerous diseases [6]. Additionally, some metals tend to accumulate in aquatic organisms through "bioaccumulation" and reach progressively higher concentrations as they ascend trophic levels, posing a health risk to people who consume them [7]. Heavy metals are the main aquatic toxins, so concern about the contamination of these compounds in water has been increasing in recent years [4]. Their toxicity depends not only on their total concentration, route, and frequency of exposure but also on their availability and reactivity with other components of the system [8].

Toxicity tests are used to assess the adverse effects of a chemical on living organisms under standardized and reproducible conditions that allow comparison with other chemicals or species evaluated. This parameter is a property relative to the chemical potential of a substance, which has a harmful effect on a living organism. These effects can be both inhibitory and magnifying, evaluated by the reaction of organisms, such as death, growth rate, proliferation, multiplication, morphological, physiological, or histological changes [9], and the effects can manifest at different levels, from subcellular structures or enzyme systems to complete organisms, populations, or communities.

The toxicity experienced by organisms in the tests can be of two types:

Acute toxicity: This refers to the adverse effect (lethal or sublethal) induced on the test organisms during a period of exposure to the test material, typically lasting a few days (48 to 96 hours). This toxicity is sufficiently high to elicit a rapid response in organisms and is usually determined by the median lethal concentration (LC50). It is more useful when parameters such as 95% confidence intervals or the effective concentration (EC50) are obtained, which indicates the toxic concentration sufficient to kill 50% of the test organisms.

Chronic toxicity: This refers to long-term toxic effects that can persist for around one-tenth of the species' median lifespan. These effects are related to changes in metabolism, growth, or survival capacity (death and reduced reproductive ability). There may be a latency period between exposure and effects, but these effects can be lethal. The duration for this test ranges from 10 days to 7 years [9]. The objective of a toxicity assay is to estimate the safe concentration or the concentration at which no effect is observed; however, this term is more of a biological concept than a statistical outcome, so the results will be the highest concentration at which there is no significant difference compared to the control treatment (NOEC); and the lowest concentration at which there is a significant difference when compared to the control treatment (LOEC).

The NOEC and LOEC are determined by examining the data and comparing treatments with the control to detect significant differences through hypothesis testing.

The effects can be mortality, immobilization, reduced cell count in the case of algae, or behavioral observations. These endpoints are generally determined using t-tests and analysis of variance (ANOVA) and are most often associated with chronic tests. Both NOEC/LOEC depend on concentration and do not have associated confidence intervals [10]. There are several techniques for toxicity assessment that use organisms from different trophic levels ranging from bacteria, protozoa, and crustaceans to larger animals such as frogs, fish, rats, rabbits, and human cells, among others [11]. Among them, *Daphnia magna* is a microcrustacean that has been widely used as a test organism in acute and chronic toxicity studies of various chemical compounds present in aquatic ecosystems [9]. It is the most used species in the world for toxicity testing due to its high sensitivity to toxic agents. This microorganism has a short life cycle and reproduces by parthenogenesis, making it easy to handle in the laboratory [12]. Another alternative is a bioluminescent marine bacterium known as *Vibrio fischeri*, which can quickly and sensitively detect a variety of toxic substrates. Toxic chemicals can interfere with the electron transfer system in the respiration of *V. fischeri* and inhibit its light production. This bioassay can be used as an early warning for toxicity because some structures associated with the bacterium are like those associated with mammals and fish [13].

On the other hand, many of the reactions and processes involved are general for most seeds, so the response of this species and the data obtained from the application of this test are largely representative of the effects on seeds or seedlings in general. The lettuce seed (*Lactuca sativa*) toxicity bioassay is a static acute toxicity test (120 h) in which the phytotoxic effects of pure compounds or complex mixtures on the seed germination process and seedling development during the first days of growth can be evaluated. As endpoints for assessing phytotoxic effects, inhibition in germination, radicle elongation, and hypocotyl elongation are determined. Although L. *sativa* is not a representative species of aquatic ecosystems, the information generated from this test provides data on the potential effect of pollutants on plant communities near the margins of contaminated water bodies, also being an interesting species to consider for its importance from a horticultural standpoint, and common enzymes with other organism or the photosynthesis enzyme system [9]. Additionally, it has easy and rapid germination, making it possible to conduct the test in a few days.

Due to the recent modification of NOM-001-SEMARNAT-2021 in Mexico, which establishes permissible limits of pollutants in wastewater discharges into national water bodies, effective from 2026 for toxicity testing, *Vibrio fischeri* is indicated as a model for biomonitoring; however, the main inconvenience is the economic cost of the test and equipment required [14]. It has become clear that our freshwater resources are limited, and efforts are needed to protect them. This review is developed to explore the tests used in the evaluation of heavy metal toxicity in water and to provide decision-making elements when selecting the most suitable techniques for assessing possible contamination of industrial wastewater in terms of toxicity. Subsequently, it is intended to apply them experimentally to obtain real results.

2. Heavy metals in the environment

Currently, it is estimated that over a million different substances are introduced into natural waters through anthropogenic discharges. Many of these substances may not be considered toxic, but they can alter the organoleptic characteristics of water, severely disrupt the ecosystem, and/or be directly harmful to humans [15].

According to Pabón et al. [16], heavy metals are chemical elements with high density (greater than 4 g/cm^3), mass, and atomic weight above 20 on the periodic table, and they are toxic even at low concentrations. Some of these elements include aluminum (Al), beryllium (Be), copper (Cu), iron (Fe), manganese (Mn), cadmium (Cd), mercury (Hg), and lead (Pb), among others [17]. Currently, one of the major environmental issues is the contamination of the world's water sources by heavy metals, as the toxicity of these compounds in river water is considered a serious problem for the inhabitants of communities that rely on such sources. Contrary to many organic pollutants, heavy metals generally are not removed from aquatic ecosystems through natural processes because they are non-biodegradable [18]. Instead, they are highly contaminating and undergo a global eco-biological cycle, with natural waters being the primary pathway. Nowadays, heavy metals are of great significance as indicators of the ecological quality of any water flow due to their toxicity and particularly their bioaccumulative behavior [19].

Similarly, contamination by heavy metals in soil and sediments is a key concern due to the growing threat to global food security [20]. By 2050, global agricultural production would need to double to meet the estimated ultimatum of a rising population with enriched living standards [21], but the industrialization of emerging nations is causing contamination by heavy metals in the environment, with millions of hectares of fertile soil being occupied for industrial uses [22]. Furthermore, the mining sector of heavy metals and their release into the environment affects the food chain [23]. The impacts of heavy metals on microorganisms and plants are listed in **Table 1**,

Heavy metal	**Classification of heavy metal carcinogenicity**	**Effect in humans**	**Effect in plants**
Copper	Carcinogenicity not classifiable	Wilson's disease (copper accumulated in organs instead of being excreted by bile): Lack of appetite, Fatigue, Jaundice, Kayser-Fleisher rings, Speech impairment, Difficulty in swallowing, Uncontrolled poisoning Brain damage, Demyelination Hepatic cirrhosis	Cu influences chlorosis, oxidative stress and impedes growth of plants. Dislocate cellular role and hinders enzyme activities in microorganism
Arsenic	Carcinogenic	Gastrointestinal system irritation, Lung irritation, Skin changes, decreased production of both red blood cells and white blood cells, Increased chances of cancer have been suggested Infertility and miscarriages.	Impairment cell membrane, reduction of growth and roots extension and production, restricts with perilous metabolic routes, fertility, yield, and fruit production is lost and oxidative stress.Arsenic causes enzyme deactivation in microorganisms.
Mercury	Carcinogenicity not classifiable	Vomiting and diarrhea, Genotoxic, it damages the DNA and chromosomes, Mongolism, also known as Down's syndrome.	Distresses anti- oxidative system and photosynthesis, increases lipid peroxidation, persuaded genotoxic impact, reduces plant growth, yield, nutrient uptake and homeostasis and oxidative stress.Diminution population size
Zinc	Probably not carcinogenic	Nausea and vomiting, Stomach cramps Anemia, Decrease in high-density lipoprotein (HDL)cholesterol, Pancreatic complications, Fatigue, Epigastric pain, Copper deficiency, Impaired immune function, Neutropenia.	Reduces photosynthesis, growth rate, chlorophyll content, germination rate and plant biomass. Demise, impeding of biomass and growth in microorganism
Chromium	Carcinogenic	Oral intake of chromium (VI) usually causes acute poisoning and various symptoms, including: gastrointestinal ulceration, nausea and vomiting, fever, diarrhea, vertigo, toxic nephritis, liver damage, coma death (usually at 1–3 g).	Chlorosis, tardy, wilting, biochemical lesions, decreased biosynthesis germination, diminutive growth, and oxidative stress.Elongation of lag phase, growth, and oxygen uptake reduction.

Heavy metal	Classification of heavy metal carcinogenicity	Effect in humans	Effect in plants
Cadmium	Carcinogenic	Kidneys primarily affected causing nephrotoxicity, Infertility caused by a reproductive system failure, Calcium metabolism alterations, Bone fracture Psychological disorders, Gastrointestinal disorders,	Chlorosis declines in plant nutrient, growth, and seed germination. Impairment nucleic acid, hinders cell division and transcription, carbon, and nitrogen mineralization, and denaturation of protein in microorganisms.
Lead	Possibly carcinogenic	Hypertension, Miscarriages, Premature and low births, Stillbirths, Renal impairment, Brain injury	Disturbs photosynthesis, growth, and chlorosis, reduces enzyme activities and seed germination, and oxidative stress in plants.
Nickel	Possibly carcinogenic	Lung embolisms, Asthma, Allergic reactions (jewelry), Respiratory failure, Heart disorders	Degeneration of proteins and nucleic acid and hinders enzyme activities and transcription.Reduces chlorophyll content, enzyme activities and growth and nutrient uptake.Dislocate cell membrane, and impedes enzyme activities in microorganism

Table 1.
Classification of heavy metals and their impacts in humans and plants (Reproduced from: Refs. [23, 24]).

along with their classification according to their carcinogenicity. Exposure to metal pollution is the main source of diseases and accounted for over 9 million premature deaths worldwide in 2015 [25]. The presence of heavy metals in the environment and food, as described, can trigger various intoxications causing irreversible damage to human and animal health, as severe as teratogenic effects, cancer, and even death. It is important to consider that elevated concentrations of these metals in living organisms disrupt biochemical and physiological processes, leading to various pathologies.

To assess the detrimental influence of heavy metals on humans, ecosystems, toxicity, and ecological vulnerability associated with metals, the Potential Ecological Risk Index (PERI) is primarily utilized. In this regard, the review conducted by Kumar et al. [24] focused on determining this index for heavy metals such as (Hg, Cr, Pb, Ni, Zn, As, Cd, Mn, Cu, and Co) in soils and sediments collected worldwide during the years 2020 and 2021. This dataset was also analyzed using a heatmap and principal component analysis.

The values of the Potential Ecological Risk Index in sediments were ranked as follows: Cd > Hg > Mn > Cu > As > Ni > Co > Cr > Zn, while the index in soils was as follows: Hg > Cd > Mn > Cr > Ni > Pb > Cu > As > Zn > Co. The maximum PERI was recorded in sediments from India, and the minimum in Australia. Based on the number of heavy metal pollution sites in sediments and soils, it can be inferred that heavy metal pollution is a potential threat in China and Ireland among the 17 countries studied. Finally, the authors recommend that both governmental and non-governmental organizations should launch massive awareness campaigns to stop activities contributing to sediment and soil pollution to restore ecological stability and economic viability.

Moreover, reservoirs, serving as direct source areas and initial collection areas of contaminants, are crucial for drinking water safety. Therefore, systematic observations of heavy metal pollution in reservoirs under hydrological regulation are urgently required. Studies conducted by Hao et al. [26] were collected 65 water samples from the Danjiangkou Reservoir (Shiyan, Hubei, China), which is the water diversion source through the longest basins in the world during storage and flooding discharge periods from 2015 to 2019. The temporal and spatial variation of eight dissolved heavy metals such as chromium (Cr), manganese (Mn), nickel (Ni), copper (Cu), zinc (Zn), arsenic (As), cadmium (Cd), and lead (Pb) were investigated, and their risks to human health were evaluated. The results showed that the concentrations of heavy metals decreased after 5 years of reservoir operation and varied significantly ($p < 0.05$) between the storage and flooding discharge periods. As, Mn, and Pb exceeded the standard water quality threshold. The assessment of risks to human health indicated that only arsenic posed a potential carcinogenic risk (1.29×10^{-4} μg/L) during the flooding discharge period. They also recommended further research efforts, such as controlled *in situ* experiments, to understand pollution pathways aimed at developing reservoir water management and maintenance strategies since it will continue to supply water for many years, benefiting millions of people with this resource.

The process of urbanization, along with the rapid growth of population and industry, is undoubtedly the most significant global phenomena in the development of human society due to their impact on environmental deterioration. In Mexico, the situation is no different; urbanization trends are dynamic and face risks that force low-income populations to concentrate in cities in sectors with high environmental impact [27]. Such is the case of the Balsas Hydrological Region, formed by 15 hydrographic basins, including the Alto Atoyac basin (Puebla-Tlaxcala) [28], where the Atoyac River is located, housing the Valsequillo reservoir, which is a Ramsar wetland due to its importance as a convergence point for migratory waterbirds. It is in central Mexico and is currently endangered due to constant discharge of municipal and industrial effluents from the cities of Puebla and Tlaxcala. In this context, the study conducted by the Mexican National Commissioned of Water [29, 30] evaluated 13 potentially toxic metals (PTMs) in water, using the water hyacinth plants (*Eichhornia crassipes*), and concentration on sediments at this site. A combined number of 31 samples were collected. The results showed that the levels of As in the Atoyac River and Hg in the Alseseca River are close to the limits, posing a potential threat to aquatic life. The metal content in the sediments resembles that of *E. crassipes*, especially in the roots, suggesting a constant deposition of plants at the bottom of the reservoir. Hg contributes 51% of the potential ecological risk in the study area where *E. crassipes* is present. The standard values of PTMs obtained from 2014 to 2019 have remained practically constant, indicating that the water quality in the basin is highly contaminated, surpassing the natural self-cleaning capacity of the ecosystem, causing the river to lose its natural balance and reach its environmental limit.

3. Toxicity bioassays and their regulation

3.1 Freshwater models for heavy metals

Most of the current environmental toxicological data come from controlled laboratory assays and generally involve individual chemicals and very small populations of test organisms. Even in a standard toxicity bioassay, an attempt to simulate what would happen in a large population typically involves observing only about

30 individuals per exposure or treatment [10]. Contributing to the development of ecotoxicological studies on heavy metals using species from different trophic levels, particularly freshwater organisms, for assessing the toxicity of pure chemicals, effluents, natural waters, surface and groundwater, and sediments is imperative due to the notable presence and persistence of these compounds in the environment and in humans and animals. The methodologies described below, as shown in **Table 2**, correspond to assays for determining lethal and sublethal acute effects with test organisms

Model	Description	Regulation	Advantages	Disadvantages
Daphnia	Newborns less than 24 hours old are exposed to the sample for a period of 48 hours, at the end of which the number of deceased organisms is quantified [9].	OECD C2. ISO 6341:2012 ASTM E1193-97 EPA.2021.0 NMX-AA- 087-SCFI-2010 ISO 10706: 2000 [11, 9]	Easy to cultivate under laboratory conditions, their body size is small (2-5 mm), their life cycle is short, they have high fecundity, exhibit parthenogenetic reproduction, and their maintenance costs are low [9]	They are less sensitive to solvents, and their sensitivity can vary among *Daphnia* species [31].
Hydra attenuata	The test has a maximum duration of 96 hours, during which the organisms are exposed to the toxic substance, and a microscopic examination is conducted daily to record the morphological changes produced [9].	Adaptations of the protocol by Trottier et al. [32] following recommendations from the WaterTox network [9].	Ease of cultivation in the laboratory, rapid reproduction, and primary structure favor intra and intercellular exchange, increasing its potential for detecting toxins, as well as presenting easily recognizable morphological changes under progressive intoxication conditions [9].	
Lactuca sativa	It is a static test (120 hours of exposure) in which the phytotoxic effects of compounds on the germination process of seeds and the development of seedlings during the first days of growth can be evaluated [9]	No. 208, OECD US EPA, 1989, US EPA, 600/3-88/029, Corvallis [11, 9].	The seed bioassay allows for the direct evaluation of the phytotoxicity of samples that are colored or have high turbidity, without the need for prior filtration, thus reducing interference due to pretreatment and simplifying the testing procedure [9]	
Selenastru m capricornut um	Cells exposed to samples containing toxic contaminants have their reproduction affected, altering the growth rate of the algal population [9].	EPA. 1003.0 Green alga, Environment Canada, 1992, EPS 1/ RM/24. ISO).1989. ISO 8692 [11, 9]	It is a simple and low-cost technique [9]	

Model	Description	Regulation	Advantages	Disadvantages
Larvae and juveniles of the fish species *Brachydani o rerio* and *Poecilia reticulata*	It consists of exposing the test organisms to different concentrations of the problematic compounds or percentage dilutions of effluents and water samples [9].	(OECD). 1992. OECD No. 203. (USEPA). 1996. OPPTS 850.1075. EPA 712–C–96–118 [11, 9].	The relevance of using fish as test organisms lies in the fact that, being vertebrate organisms, they are typically considered good representatives of more complex organisms.Therefore, they are easier to understand and interpret results from [9].	
Vibrio fisheri	The test is based on measuring the luminescence emitted by the bacterium *Vibrio fischeri*, which decreases when the bacteria are exposed, for a period of 5 to 30 minutes, to samples containing toxic compounds [33].	ISO 11348:2007 NMX-AA-112-SCFI-2017 [33] DIN 38412 Parte 34	Quick test with high sensitivity to various compounds.	Not suitable for samples with high turbidity and presence of color or darkness. And most notably, the high cost associated with the test [14].

Table 2.
Freshwater models for heavy metals.

Daphnia magna and *Hydra attenuata*, sublethal phytotoxic effects with *Lactuca sativa*, chronic effects with the algae *Selenastrum capricornutum* (*Pseudokirchneriella subcapitata*), and acute effects with larvae and juveniles of fish *Brachydanio rerio* and *Poecilia reticulata*, and the bacterium *Vibrio fisheri*.

3.2 Regulation for bioassays on heavy metals

When conducting toxicity assays, it is important to use standardized procedures that allow for the comparison of results both within and among laboratories. There are standardized methodologies recommended by certain international organizations such as the U.S. Environmental Protection Agency (US EPA), Environment Canada (EC), and the Organization for Economic Cooperation and Development (OECD), along with standardization and rationalization bodies such as the American Society for Testing and Materials (ASTM). These procedures also provide recommendations for sample collection, handling, and preservation, as well as the selection of reference toxicants to be used in toxicity assays.

The maximum allowable concentrations of metals in the environment and drinking water are strictly regulated in many countries. In Mexico, the NOM-001-SEMARNAT-2021 was updated for this purpose. Its stated purpose is to comply with the obligation established in Article 85 of the National Water Law, which mandates that individuals or entities that exploit, use, or take advantage of national waters in any activity are responsible for taking necessary measures to prevent their pollution and, if necessary, to reintegrate the waters in appropriate conditions and maintain the balance of vital ecosystems. Regarding the permissible limits of heavy metals, these values underwent few changes compared to the previous standard. However, the actual NOM does not consider the differentiated assimilation capacity of surface water bodies and the current state of each receiving water body for the determination of limit values, that is, its eco-hydrological condition.

The adverse effects of wastewater include reproductive impairment or even death, which can disrupt populations and trophic links in an aquatic ecosystem. Therefore, it is essential to develop methods to accurately identify the toxicity of wastewater before its discharge into water bodies [34].

This means that in Mexico, there is still a need to dedicate efforts and resources to the identification, standardization, and intercalibration of such bioassays. From this document, it is intended to begin a series of tests that will result in the selection of the most appropriate group of assays for Mexican environments.

3.3 Bioassays for water and sediments with different models

It is known that toxicity tests provide quick, easy-to-use, and cost-effective means to assess the impact of wastewater quality on aquatic ecosystems. However, there are not many reported cases where wastewater management standards are established based on diverse ecotoxicity values from bioassays. Therefore, Park et al. [35] tested a new multi-taxon approach to compare water quality indices with toxicity measurements obtained from ecotoxicity tests, conducted with four aquatic models representing various trophic levels, namely, *Aliivibrio* (bacteria), *Ulva australis* (sea lettuce), *Daphnia magna* (crustacean), and *Lemna* (aquatic plant), for 99 samples of industrial wastewater from South Korea [36].

From this study, it was determined that *Lemna* and *Daphnia magna* had a much wider range and higher toxicity unit values than the other two organisms, suggesting that they are more sensitive to industrial wastewater. Meanwhile, *Ulva* was the least sensitive model. Compared to assessments with a single test species, this approach allowed for seeing differences in the specific sensitivities of each species and considering considerations of the food chain. The use of multiple species can be strategic, as it has been shown that toxicity can vary depending on (1) the bioindicator or assessment criterion used in bioassays and (2) the sensitivity of each taxonomic group and species [37]. Finally, the authors suggest continuing to explore the sensitivity of models for each group of contaminants and considering the cost-effectiveness of each.

Continuing with wastewater effluents, the case study conducted by Aragão et al. [38] aimed to investigate the potential environmental risks that raw effluents and stabilization pond effluents from the pig effluent biodigestion process may cause to the environment. This was achieved by characterizing their chemical composition and evaluating their toxic potential through phytotoxicity, cytotoxicity, genotoxicity, mutagenicity, and biochemical biomarker tests using different concentrations of the two effluents in the plant models *Allium cepa* and *Lactuca sativa*. The toxicity of metals in plants is measured at different intensities, which can vary depending on the species and organ or tissue of the plant, as well as the type of metal, its concentration, and the environment in which the plant is exposed to the metal, including variations in soil pH and conditions [39]. Thus, it is necessary to understand the risks of metals to the environment, considering that plants can absorb and accumulate metals in leaves and fruits, intensifying physiological and morphological changes [40].

The observed results suggested that the inhibition of germination and root growth, in both treatments, may be related to the presence of heavy metals in the samples. The phytotoxicity of heavy metals in plants has been evidenced by other authors, such as *Allium cepa* [41]. The results concluded that the raw effluent and the stabilization lagoon effluent offer environmental risks. Both effluents showed phytotoxicity characteristics through the inhibition of seed germination and reduction of root growth at higher concentrations, in both plant models. They also presented cytotoxicity and genotoxicity, observed respectively, in alterations of the mitotic index and chromosomal alterations at higher concentrations. Therefore, these alterations in metabolism are possibly related to the heavy metals found in the samples, especially the presence of zinc and copper.

The study conducted by Paylar et al. [42] aimed to evaluate gene expression profiles related to acute toxicity and metal stress in response to different concentrations of Zn, Cu, and Cd using water samples collected from a lake receiving runoff water from a decommissioned mine. The study employed five toxicological models, namely, *Daphnia magna* (water flea), *Ceriodaphnia dubia* (water flea), *Caenorhabditis elegans* (nematode), *Danio rerio* (zebrafish), and *Heterocypris incongruens* (ostracod), as well as two cell lines, the human hepatocellular *carcinoma cell line* (HepG2) and the zebrafish liver cell line (ZFL), to assess acute toxicity and gene expression profiles related to metal stress. These organisms were chosen for their ecological importance and sensitivity to metal exposure.

One of the key molecular responses to mixed metal exposure is the positive regulation of methallothioneins genes (*mt* genes), such as *mt*-1/a and *mt*-2/b. Metallothioneins (MT) are low molecular weight proteins that play an essential role in metal homeostasis and detoxification. In this study, it was found that exposure to mixed metals induced the expression of MT genes in all analyzed species. The highest increase in expression was observed in *D. magna*, which showed a positive regulation of 23.7 and 6.5 times for *mt*-a and *mt*-b, respectively. Additionally, *D. magna* was found to be the most sensitive species to both high and low concentrations of mixed Zn, Cu, and Cd metals compared to other vertebrate and invertebrate organisms tested under the same conditions. The differences observed among the studies can be attributed to the fact that each metal exposure produces specific effects on different organisms depending on the presence and abundance of specific targets. In conclusion, the species *D. magna* is an appropriate model organism for assessing the risk of low-level exposure to mixed metals due to its high sensitivity [43].

On the other hand, regarding sediments, the study conducted by Palma et al. [44] explores the use of a set of observational, chemical, and ecotoxicological assays with *Heterocypris incongruens* (Ostracod), *Vibrio fischeri* (bacteria), *Pseudokirchneriella subcapitata* (microalgae), and *Thamnocephalus platyurus* (macrocrustacean). The study included sediment analysis from a reservoir and streams in the Guadiana Basin, over two consecutive years with different climatic conditions in 2017 (dry year) and 2018 (normal year) [45, 46].

Freshwater ostracods are considered important bioindicators for evaluating anthropogenic stressors such as heavy metals, pesticides, and hydrocarbons. In relation to heavy metals, several studies in literature have reported their high sensitivity to total Cr, Pb, and Cd. The results of sublethal toxicity coincided with the determined degree of contamination, with the Álamos stream site inducing the greatest inhibition of ostracod growth, mainly during 2018, followed by Mn sediments. The temporal analysis allowed the conclusion that the reservoir sediments collected in 2018 induced a higher toxic response (higher growth inhibition rate) in the organisms compared to those from 2017 [47].

The toxicity of pore water was evaluated using species from different trophic levels and various toxicological criteria. The results showed that although the *V. fischeri* bioassay is known to be highly sensitive for detecting heavy metal toxicity, generally being a good option for assessing this type of risk, that was not the case in this study. *T. platyurus* was more sensitive to samples from the wet period of 2017. In fact, the mortality of *T. platyurus* showed a significant positive correlation with the concentrations of Zn in the sediments. Regarding the growth rate of the green microalga *P. subcapitata*, the results revealed overall high sensitivity to sediments from the Guadiana Basin.

In conclusion, the results highlighted that climatic differences occurring in different years did not induce pronounced variability in the organization of the data. The data analysis showed that, to facilitate managing authorities to recognize existing problems in the basin area and implement solutions, the tools must include three fields of research: observational, analytical, and biological.

4. Conclusion

It should be considered that the toxic effect on biological systems is exerted by the combined action of all harmful substances present in the environment, including those that are not toxic themselves but affect the chemical or physical properties of the system and, consequently, the living conditions of organisms. Also, effects not directly related to substances should be considered, such as deterioration or damage caused by changes in temperature or radiation.

The results of bioassays refer, first, to the organisms used in the assay and the conditions stipulated in the testing procedure. A harmful effect evaluated through standardized biological assays can indicate levels of hazard transferable and assimilable to organisms that are part of natural systems and biocenosis. If the use of ecotoxicity methods is to be extended to the management of industrial wastewater, the ease of implementation and cost-effectiveness of different methodologies must be considered.

In general, this study highlights the importance of evaluating the effects of heavy metals, as well as metal mixtures, in different species and cell lines and the importance of examining conserved gene expression responses to metal exposure in aquatic environments. The findings of this study can contribute to the development of more effective environmental risk assessments and management strategies for metal mixtures in aquatic ecosystems.

Acknowledgements

The authors thank CONAHCYT for the postgraduate scholarship to carry out this research and the University of the Americas Puebla for providing support.

Conflict of interest

The authors declare that they have no competing financial interests or known personal relationships that could have influenced the work presented.

Author details

Teresa Larios, Benito Corona and Jose Luis Sánchez-Salas*
UDLAP, Puebla, Mexico

*Address all correspondence to: jluis.sanchez@udlap.mx

References

[1] Bradl H. Handbook of Heavy Metals in the Environment: Sources and Origins, Interaction and Remediation. Vol. 6. Germany: Elsevier; 2005. pp. 1-269

[2] Romanova TE, Shuvaeva OV, Belchenko LA. Phytoextraction of trace elements by water hyacinth in contaminated area of gold mine tailing. International Journal of Phytoremediation. 2016;**18**(2):190-194. DOI: 10.1080/15226514.2015.1073674

[3] Curtis KD. Handbook of Toxicology the Basic Science of Poisons. Seven ed. Kansas: McGraw-Hill; 2008

[4] Elzwayie A, Afan HA, Allawi MF, El-Shafie A. Heavy metal monitoring, analysis and prediction in lakes and rivers: State of the art. Environmental Science and Pollution Research. 2017;**24**(13):12104-12117. DOI: 10.1007/s11356-017-8715-0

[5] Zhang Z et al. Relationships between soil properties and the accumulation of heavy metals in different *Brassica campestris* L. growth stages in a karst mountainous area. Ecotoxicology and Environmental Safety. 2020;**206**(August):111150. DOI: 10.1016/j.ecoenv.2020.111150

[6] Mora A, García-Gamboa M, Sánchez-Luna MS, Gloria-García L, Cervantes-Avilés P, Mahlknecht J. A review of the current environmental status and human health implications of one of the most polluted rivers of Mexico: The Atoyac River, Puebla. Science of the Total Environment. 2021;**782**:146788. DOI: 10.1016/j.scitotenv.2021.146788

[7] Pérez Cruz YG, Rangel Ruiz LJ, Gamboa Aguilar J. Metales en almejas y sedimentos en la Reserva de la Biósfera 'Pantanos de Centla' Tabasco, México. Hidrobiológica. 2013;**23**(1):1-8

[8] Cabezas JG, Alonso J, Pastor J, Sastre-Conde I, Lobo M. Absorción y acumulación de metales pesados en tres especies vegetales en suelos enmendados con lodos de depuradora. In: Environmental Biotechnology and Engineering [Online]. Madrid, España: CSIC Publisher; 2004. pp. 6-8. Available from: https://digital.csic.es/handle/10261/52285 [Accessed: April 25, 2022]

[9] Castillo G, Pica Y, Antonio R, Ramírez L. Estandarización Ensayos Toxicológicos y Métodos de Evaluación de Calidad de Aguas, Intercalibración, Resultados y Aplicaciones. Mexico: IMTA; 2004

[10] Peluso ML. Bioensayos de toxicidad. La Plata, Buenos Aires, Argentina: National University of La Plata Publishing House; 2021. pp. 268-308. Available from: http://sedici.unlp.edu.ar/handle/10915/131173

[11] Agudelo EA, Gaviria-Restrepo LF, Barrios-Ziolo LF, Cardona-Gallo SA. Techniques to determine toxicity in industrial wastewater contaminated with dyes and pigments. DYNA (Colombia). 2018;**85**(207):316-327. DOI: 10.15446/dyna. v85n207.71915

[12] U.S. Environmental Protection Agency. Methods for Measuring the Acute Toxicity of Effluents and Receiving Waters to Freshwater and Marine Organisms. 5th ed. Washington DC: U.S. Environmental Protection Agency; 2002

[13] Backhaus T, Froehner K, Altenburger R, Grimme LH. Toxicity

testing with *Vibrio fisheri*: A comparison between the long term (24 H) and the short term (30 min) bioassay. Chemosphere. 1997;**35**:2925-2938

[14] Reinado Lansac E. Aplicación de la bacteria *Vibrio fischeri* en la medida de la toxicidad en agua [thesis]. Spain: Universidad de Zaragoza; 2011

[15] Rosas H. Estudio de la contaminación por metales pesados en la cuenca del Llobregat. Catalunya: Universidad Politécnica de Catalunya. Departamento de Ingeniería, Minería y Recursos Naturales; 2001

[16] Pabón SE, Benítez R, Sarria A, Gallo JA. Contaminación del agua por metales pesados, métodos de análisis y tecnologías de remoción. Una revisión., Entre Ciencia e Ingeniería. 2020;**14**(27):9-18

[17] Londoño Franco LF, Londoño Muñoz PT, Muñoz García FG. Los riesgos de los metales pesados en la salud humana y animal. Biotecnología en el Sector Agropecuario y Agroindustrial. 2016;**14**(2):145. DOI: 10.18684/bsaa(14)145-153

[18] Förstner U, Prosi F. Heavy metal pollution in freshwater ecosystems. In: Biological Aspects of Freshwater Pollution. Luxemburg: Elsevier Publisher; 1979. pp. 129-161. DOI: 10.1016/b978-0-08-023442-7.50011-6

[19] Moalla SMN, Awadallah RM, Rashed MN, Soltan ME. Distribution and chemical fractionation of some heavy metals in bottom sediments of Lake Nasser. Hydrobiologia. 1998;**364**:31-40

[20] Hou D, Al-Tabbaa A. Sustainability: A new imperative in contaminated land remediation. Environmental Science & Policy. 2014;**39**:25-34. DOI: 10.1016/j.envsci.2014.02.003

[21] Tilman D, Balzer C, Hill J, Befort BL. Global food demand and the sustainable intensification of agriculture. Proceedings of the National Academy of Sciences of the United States of America. 2011;**108**(50):20260-20264. DOI: 10.1073/pnas.1116437108

[22] Hou D, Li F. Complexities surrounding China's soil action plan. Land Degradation and Development. 2017;**28**(7):2315-2320. DOI: 10.1002/ldr.2741

[23] Briffa J, Sinagra E, Blundell R. Heavy metal pollution in the environment and their toxicological effects on humans. Heliyon. 2020;**6**(9):e04691. DOI: 10.1016/j.heliyon. 2020.e04691

[24] Kumar V, Pandita S, Setia R. A meta-analysis of potential ecological risk evaluation of heavy metals in sediments and soils. Gondwana Research. 2022;**103**:487-501. DOI: 10.1016/j.gr.2021.10.028

[25] Somma R et al. The first application of compositional data analysis (CoDA) in a multivariate perspective for detection of pollution source in sea sediments: The Pozzuoli Bay (Italy) case study. Chemosphere. 2021;**274**:129955. DOI: 10.1016/j.chemosphere.2021.129955

[26] Hao R et al. Dynamics of dissolved heavy metals in reservoir bays under different hydrological regulation. Journal of Hydrology. 2021;**595**:126042, 1-12. DOI: 10.1016/j.jhydrol.2021.126042

[27] McNeill JR, Population, and the natural environment: Trends and challenges, Population and the Natural Environmental: Trends and Challenges, Dec 2011;**32**(Suppl. 1):183-201. DOI: 10.1111/j.1728-4457.2006.tb00007.x

[28] Comisión Nacional del Agua. Estadística del Agua en la cuenca del Río

Balsas, 2010. 2010. [Online]. Available from: www.conagua.gob.mx

[29] Tabla-Hernandez J, Rodriguez-Espinosa PF, Mendoza-Pérez JA, Sánchez-Ortíz E, Martinez-Tavera E, Hernandez-Ramirez AG. Assessment of potential toxic metals in a Ramsar wetland, Central Mexico, and its self-depuration through *Eichhornia crassipes*. Water (Switzerland). 2019;**11**(6):1248-1269. DOI: 10.3390/w11061248

[30] Roces MEA-B, Moreno AET et al, 1 er Informe estratégico cuenca del Alto Atoyac (Tlaxcala y Puebla): Región de emergencia sanitaria y ambiental; Problemática socioambiental y recomendaciones para su atención integral, México. Mexico D.F.: CONAHCYT; 2023. Available from: https://cdn.conahcyt.mx/enis/toxicologia/resa-atoyac/inicio/descargables/informe-caa.pdf

[31] Martins J, Oliva Teles L, Vasconcelos V. Assays with *Daphnia magna* and *Danio rerio* as alert systems in aquatic toxicology. Environment International. 2007;**33**(3):414-425. DOI: 10.1016/j.envint.2006.12.006

[32] Trottier S, Blaise C, Kusui T, Johnson EM. Acute toxicity assessment of aqueous samples using a microplate-based Hydra attenuata assay. Environmental Toxicology & Water Quality. 1997;**12**:265-271. DOI: 10.1002/(SICI)1098-2256(1997)12:3<265::AID-TOX10>3.0.CO;2-9

[33] Comite Técnico de Normalización Nacional de Medio Ambiente y Recursos Naturales. NMX-AA-112-SCFI-2017 Análisis de Agua y Sedimentos - Evaluación de Toxicidad Aguda con *Vibrio fischeri* - Método de Prueba. Comite Técnico de Normalización Nacional de Medio Ambiente y Recursos Naturales; 2017

[34] Ward AJ, Arola K, Thompson Brewster E, Mehta CM, Batstone DJ. Nutrient recovery from wastewater through pilot scale electrodialysis. Water Research. 2018;**135**:57-65. DOI: 10.1016/j.watres.2018.02.021

[35] Park J et al. Evaluating ecotoxicological assays for comprehensive risk assessment of toxic metals present in industrial wastewaters in the Republic of Korea. Science of the Total Environment. 2023;**867**:161536. DOI: 10.1016/j.scitotenv.2023.161536

[36] Schreiber B, Fischer J, Schiwy S, Hollert H, Schulz R. Towards more ecological relevance in sediment toxicity testing with fish: Evaluation of multiple bioassays with embryos of the benthic water fish (*Misgurnus fossilis*). Science of the Total Environment. 2018;**619-620**:391-400. DOI: 10.1016/j.scitotenv.2017.11.122

[37] Di Salvatore M, Carafa AM, Carratù G. Assessment of heavy metals phytotoxicity using seed germination and root elongation tests: A comparison of two growth substrates. Chemosphere. 2008;**73**(9):1461-1464. DOI: 10.1016/j.chemosphere.2008.07.061

[38] Aragão FB et al. Toxic risk evaluation of effluents from a swine biodigester in the plant models *Lactuca sativa* and *Allium cepa*. Environmental Monitoring and Assessment. 2024;**196**(1):64-78. DOI: 10.1007/s10661-023-12173-x

[39] Nagajyoti PC, Lee KD, Sreekanth TVM. Heavy metals, occurrence, and toxicity for plants: A review. Environmental Chemistry Letters. 2010;**8**(3):199-216. DOI: 10.1007/s10311-010-0297-8

[40] Freitas AS, Fontes Cunha IM, Andrade-Vieira LF, Techio VH. Effect of SPL (spent pot liner) and its main

components on root growth, mitotic activity, and phosphorylation of histone H3 in *Lactuca sativa* L. Ecotoxicology Environmental Safety. 2016;**124**:426-434. DOI: 10.1016/j.ecoenv.2015.11.017

[41] da Cunha Neto AR, da Silva IG, Calvelli JVB, Martins GEC, Carvalho M, Barbosa S. Toxicity of heavy metals that affect germination, development and cell cycle of *Allium cepa* L. Bulletin of Environmental Contamination and Toxicology. 2023;**111**(2):22-29. DOI: 10.1007/s00128-023-03775-9

[42] Paylar B et al. Assessing organism differences in mixed metal sensitivity. Science of the Total Environment. 2023;**905**:167340, 1-11. DOI: 10.1016/j.scitotenv.2023.167340

[43] P.-E. Olsson, ke Larsson, and C. Haux, Metallothionein and heavy metal levels in rainbow trout (*Salmo gairdneri*) during exposure to cadmium in the water, Marine Environmental Research, vol. 24, pp. 151-153, 1988

[44] Palma P et al. Integrative toolbox to assess the quality of freshwater sediments contaminated with potentially toxic metals. Environmental Research. 2023;**217**:114798, 1-13. DOI: 10.1016/j.envres.2022.114798

[45] Ruiz F et al. Freshwater ostracods as environmental tracers. International Journal of Environmental Science and Technology. 2013;**10**(5):1115-1128. DOI: 10.1007/s13762-013-0249-5

[46] Baran A, Tarnawski M. Assessment of heavy metals mobility and toxicity in contaminated sediments by sequential extraction and a battery of bioassays. Ecotoxicology. 2015;**24**(6):1279-1293. DOI: 10.1007/s10646-015-1499-4

[47] Codina JC, Cazorla FM, Pérez-García A, García G, De Vicente A. Heavy metal toxicity and genotoxicity in water and sewage determined by microbiological methods. Environmental Toxicology and Chemistry. 2000;**19**(6):1552-1558